Cloud Computing for Engineering Applications

Benito A. Stradi-Granados

Cloud Computing
for Engineering Applications

 Springer

Benito A. Stradi-Granados
School of Materials Science
Office of the Vice President for Research
Institute of Technology of Costa Rica
Cartago, Costa Rica

Departments of Chemistry and Chemical Engineering
University of Costa Rica
San José, Costa Rica

ISBN 978-3-030-40447-5   ISBN 978-3-030-40445-1   (eBook)
https://doi.org/10.1007/978-3-030-40445-1

This Springer imprint is published by the registered company Springer Nature Switzerland AG
The registered company address is: Gewerbestrasse 11, 6330 Cham, Switzerland

Acknowledgements

The author would like to thank the current and former Vice Presidents for Research for giving support to this initiative in chronological order; they are Dagoberto Arias Aguilar, Milton Villareal Castro, Paola Vega Castillo, Alexander Berrocal Jiménez, and Jorge Alfredo Chaves Arce. Similarly, coordination with the IT department (DATIC) has been most important. A number of people from DATIC have collaborated over the years: Gustavo Bolaños Solano, Alexander Valerín Castro, William Hernández Gómez, Alfredo Villareal Rodríguez, Victor Corrales Thames, and Andrea Cavero Quesada. The administrative support from Mabel Valerín Gómez, Dunia Zúniga Valverde, Jairo Ramírez Sojo, Silvia Campos Brenes y Kenya Román Marchena is greatfully acknowledged. The example on recursive programming was provided by Jose A. Stradi-Granados from the Computer Engineering Department. Similarly, the collaboration of the Helen Kellogg Institute for International Studies at the University of Notre Dame that has spanned for many years is greatfully acknowledged.

Introduction

Modern engineering education requires significant computational capacities. Schools of Engineering in their different areas of specialization teach topics in resistance of materials, solid mechanics, fluid flow, product design, and many more. These subjects have in common that they have become increasingly more demanding of computational capacities as industrial products become more complex and/or improvements are needed. These new computational capacities allow for designs and simulations better suited for immediate application in a commercial environment. Fluid flow analysis is an example of a computationally demanding subject particularly when the expectation is to implement full Navier–Stokes equations and arrive to a representation of the fluid flow with features such as turbulence, phase changes, and non-Newtonian behavior. There are numerous other examples where there is a need to predict or model as close as possible the phenomenon of interest as it is seen in reality; those examples include design of aircrafts, heart valves, prosthetic devices, among many others. Finite element discretization is a technique that has found multiple applications in the modeling of engineering phenomena. Finite element analysis is a process by which a partial differential equation is transformed into a set of algebraic equations. A series of points are detailed in the body of the subject of study. These points are called nodes that contain the variable of interest to solve for. Their number and position of the nodes depend on the particulars of the problem, the level of accuracy required for the solution, and nonlinearity of the phenomena under study. An example would be the computation of the temperature profile of a solid that is heated. The unknown in this case is the temperature, and the nodes would be the locations at which the temperature is to be computed on the surface and inside the solid. Each node is an unknown; consequently, there are as many equations as nodes or degrees of freedom. This is the result of changing the complexity from solving a set of partial differential equations to a set of algebraic equations. The problem is not trivial at all because the problem has to be well posed with adequate boundary conditions, initial conditions, and sufficiently small-size discretization to achieve the desired precision in the solution. The nodes are interconnected by funtionalities, the ensemble of which is referred to as a mesh. The finer the mesh implies the more nodes, at the same time, the closer the values that would have been found with an analytical solution. The appeal is then to use as many nodes as possible to have a higher precision, and this implies a larger problem with more variables and in turn higher computational requirements. This is a complex matter, and there are books dedicated to finite element analysis. The finite element strategy to solve partial differential equations sometimes is confused with finite differences; the latter coincides with the former to the first degree and with well-defined geometries, generally squares or rectangles.

The development of finite element equations for a particular problem is lengthy. This fact results in the inability to efficiently use class or training time to solve even small-size problems. The learning process instead becomes an exercise on equation rewriting. An exercise that is particularly tedious, prone to errors, and time consuming even for problems of about ten nodes. For larger problems, the finite element analysis can only be done with the assistance of computers. Over the last 30 years, there are programs that have evolved into software packages to aid in the solution of modeling problems that involve multiple nodes and complex shapes. There is also a larger emphasis in the last 10 years to be able to solve problems that involve varied phenomena such as fluid flow and electromagnetic interactions simultaneously, just to cite an example. These problems are given the name Multiphysics problems, making emphasis on the capacity to solve problems of systems simultaneously undergoing multiple physical phenomena.

The Multiphysics capacity is not always tied to finite element methods because there are also finite volume elements. Nonetheless, the subject matter boils down to efficiently solve very large sets of equations, where efficiency implies rapidly and with limited resources. The evolution of software packages that can handle Multiphysics is an ongoing project and is no small contribution of many specialists to solve complicated problems.

Simple problems are solved by hand to learn the ideas and reasoning that come with the science. This expertise is later used on larger problems. More complex problems require computational capacities that are generally only partially solved by average desktop or laptop computers with eight or a few more gigabytes of memory and similarly limited processing power. The result is that training and class time sometimes are limited to examples that reasonably use the limited computing resources available. More complex tasks require specialized resources with more random access memory (RAM) and more computer processing unit (CPU) power. This is generally a substantial dollar proposition. The cost increases if specialized software is needed; these packages usually start with a free trial version. Once the customer is sold on that particular application then a higher bill ensues for licenses with all kinds of features, such as having limited use only for teaching and sometimes not formally permitting use for academic research. To equip a laboratory with computers capable of handling mid-size problems is no small matter, and to keep the resources available indefinitely is very demanding on resources and funding. The payback is a trained workforce with the knowledge of the science and the expertise to apply computational tools to important problems like the deformation of an automobile during a collision.

Naturally, there are instances where the funds to keep laboratories and facilities to meet the highest standards of availability and performance are not a certainty for every budgeting cycle. Similarly, it is possible that the benefit to cost ratio to own significant computational facilities is not sufficiently high compared to other needs, for example, providing student housing. In either case, the possibility to offer computational services on demand would be an improvement provided that the associated costs can be contained to a minimum and far away from a full purchase of the actual computer system at an early stage.

The subject of this book is the use of *Cloud Computing for Engineering Applications*. Cloud computing is a term used to refer to the use of computers that do not belong to the infrastructure of the user. Computers in a cloud are simply computers that are accessed remotely and the use of which is possible by paying a fee. This means that the examples discussed later in this paper were developed at a remote site; thus paying only for the use of those facilities without incurring in local infrastructure expenditures. The monetary cost seems low at the beginning and results very attractive. Longer term use should be assessed on a cost-benefit economic analysis. There are good practices that empower the user to extract a maximum value of every dollar invested in cloud computing. Under that premise, appropriate training is needed prior to any cloud computing use in order to maximize the results from the resources that are available. There is a very real possibility of incurring very high costs that would defeat the objective of using cloud services at any stage. Common errors include the use of very large systems without proper prior testing in lower capacity systems, use of solid-state storage where regular hard drives suffice, incorrectly planned networks that require redesign of deployment, installation and commissioning of equipment, and, of course, leaving your problem running on a remote facility while on vacation while generating only rubbish. All of these come generally with massive amounts of money and effort lost.

Two examples are discussed to summarize our experience using cloud computing. The results do not constitute an endorsement of any particular platform, software product, or brand. Similarly, the cloud platform, software products, and any other proprietary entity are the entire property of its owner or licensor. These results are generated based on our experience alone. The reader should endeavor to determine those benchmarks for cloud computing usage and software products that may deem necessary.

The first example is the simulation of the fluid flow around a turbine. The two fluids are water and air that cool a turbine that suffers from heating due to friction and driving motor. Heat flows from the turbine metal to the water. The final solution is shown in Figs. 1 and 2. The problem has 3,616,800 degrees of freedom; this means these are unknowns in a linear system for which a solution is needed. The problem is not of the largest complexity and illustrates the sheer size of the matrices that are generated internally even for very modest applications. The problem can be solved in steady state (final condition, Figs. 1 and 2) or as a time-dependent problem. A time-dependent problem consumes much more computational time solving the problem. The user can address these needs with machines of higher computing capacity and more memory (RAM). In this case, the 56 gigabytes of memory are never fully used, and the processor E5 is used more fully and documented at 89% usage. There is a feature that stands above the two features just mentioned. This is a significant use of hard drive space. In this case where Multiphysics is used, there are temporary files in the main drive (C drive) that reach and go over 100 gigabytes. These are large files and require a suitable hard drive for the principal drive (C drive in most systems). It is unimportant to the system whether the software was installed in another drive (i.e., D) in our experience. The software works where the operating system was installed and will most likely fill it until it runs out of space what is a certainly possible for small hard drives. A system drive in the terabyte range is usually the right size to implement to avoid problems. Generally, there is no need to use solid-state drives (SSD) as part of the system; traditional hard drive disks (HDD) would suffice at a fraction of the cost for most training purposes. More importantly, the user has to check whether the charge for the hard drive (traditional or solid state) will be for the entire drive or only for the used portion, and of course, the charges linked to that particular drive. In our case, a hard disk drive properly formatted (fast format only) would generate charges only for the space that is used at a fraction of the cost of a solid-state drive.

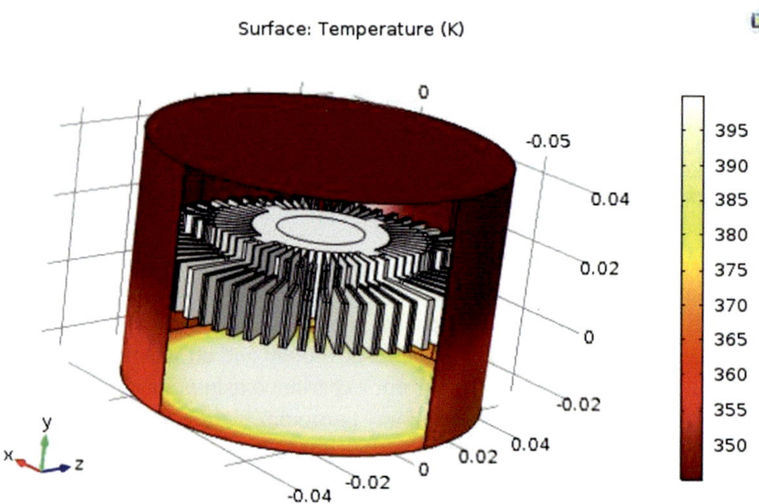

Fig. 1 Cooling of turbine with air

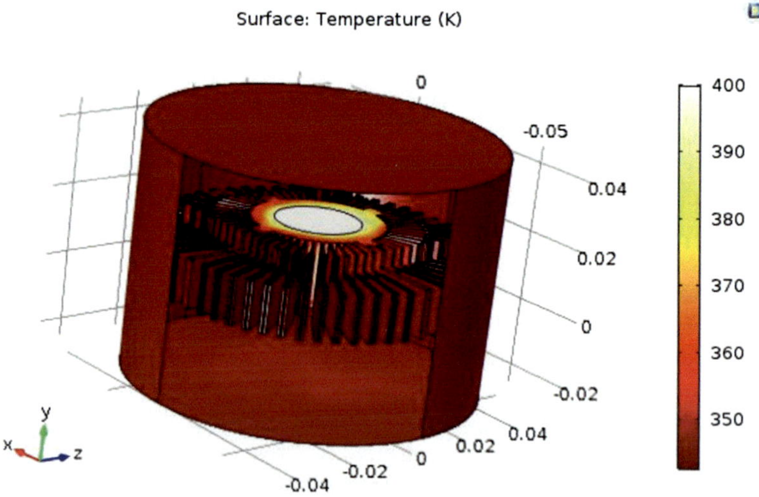

Fig. 2 Cooling of turbine with water

The time-dependent solution is a more demanding computational task. The traditional belief is that more computing power understood as more processing capacity and more random-access memory do equate necessarily and linearly with a reduction in computing time. There is a bottleneck where many software packages cannot fully utilize all computing processing power (CPU) and memory available (RAM). There are limitations that leave any surplus capacity as an unused resource. This is where resources are idled and savings should be pursued in order to keep costs low. There are two features that the user has to keep in mind to become an efficient user. The burn rate has to be monitored and the maximum spending limit has to be known. Figure 3 illustrates how a production run consumes resources and reaches a maximum spending limit for the month. This is a very important feature to control costs and also to distribute computing time among researchers with varied views related to the use of institutional funds.

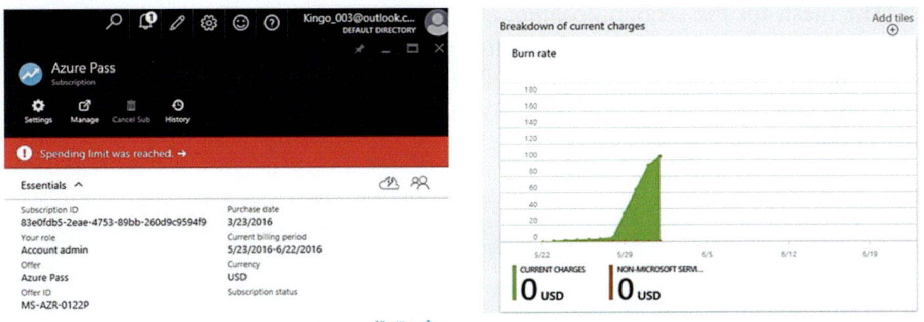

Fig. 3 Spending limit and burn rate

A point made earlier is clearly illustrated in Fig. 3 which is that if a cloud system is left to burn resources indefinitely then it would result more expensive than purchasing a particular system. As a consequence, a culture of accounting for benefit versus cost is needed to invest in those areas where more benefits are generated and reduce those areas that are unproductive. In this particular case to solve the entire problem (steady state and unsteady state) consumed about 300 dollars. This is a fraction of the cost of any computer system of similar capacities without taking into account the waiting periods generated by bureaucratic purchasing requirements and other pecuniary and non-pecuniary costs.

The second example is the cooling process of a set of metal blocks in a water bath in the absence of phase change. There are four hot blocks that are placed in a cold water bath. This is a transient temperature process where the blocks cool off and the water heats up. The problem has been solved in steady and unsteady conditions. The solution discussed is the unsteady-state solution because the previous example is concentrated on the steady-state solution of that particular problem. Five computing systems were used to solve the unsteady-state problem and compare computing times. These systems are named G1, G2, G3, G4, and G5 and are available in cloud services. The characteristics of these systems are given in Table 1.

Table 1 Description of large virtual machines

Computing system	Number of computational cores (CPUs)	Random memory access size (RAM)
G1	2	28
G2	4	56
G3	8	112
G4	16	224
G5	32	448

In Institutions with limited local computation resources, the first impulse is to jump to the largest computational system from the start. This is the G5. That choice is understandable as the expectation is that there would not be any future unseen difficulties solving the problem with such a large computing system, nonetheless with a higher financial investment.

The mesh for the second problem is illustrated in Fig. 4. The mesh is dense in order to render an accurate approximation of the temperature variations and profiles on the blocks. The number of degrees of freedom is 509,543, and thus it is a modest-size problem where there is a need for determining the variation of temperature with time.

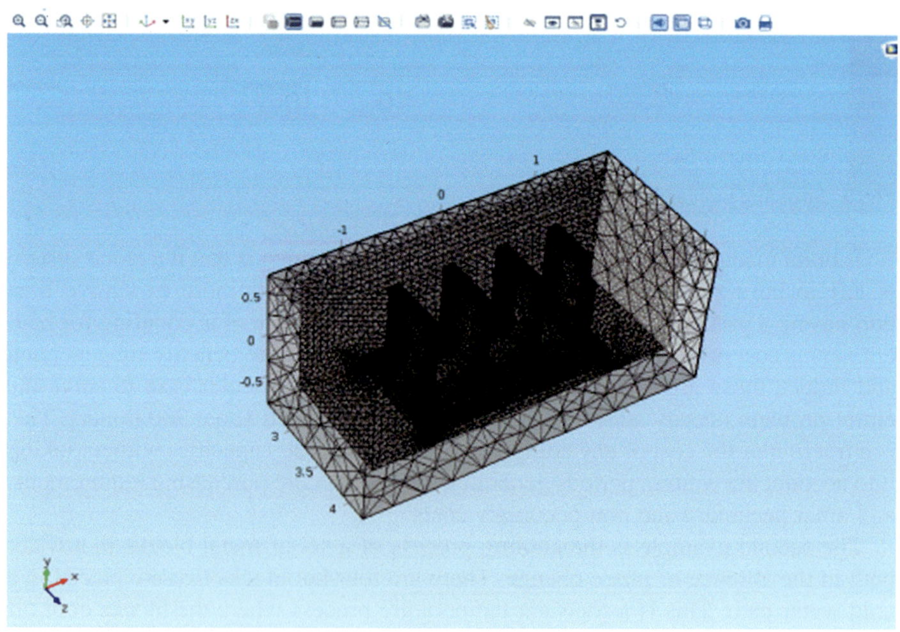

Fig. 4 Mesh density for the cooling process of metal blocks

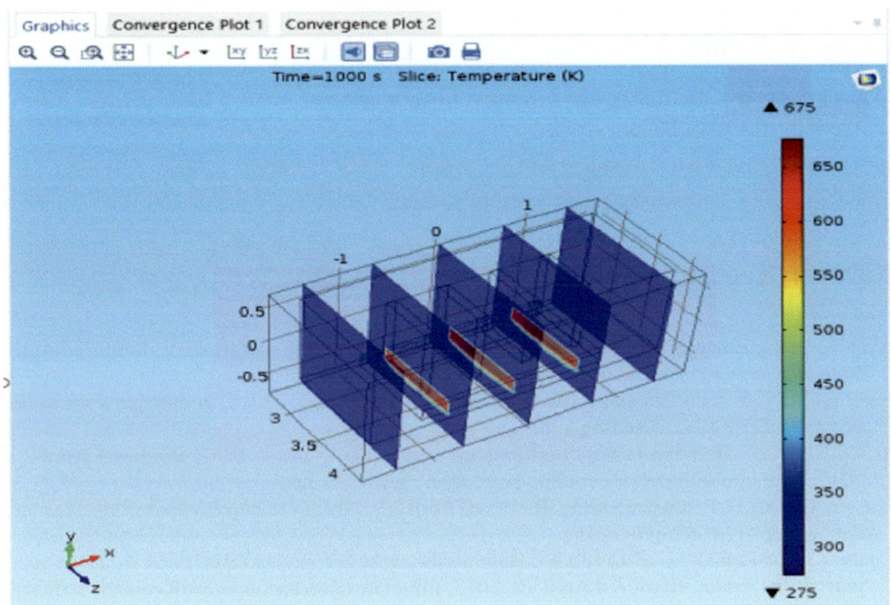

Fig. 5 Temperature profile at 1000 s

The problem that has been run on all five systems is given in Table 1. The important results are that in the largest systems the random access memory is used very minimally most of the time, at 5%; this would be about 22 gigabytes which is the size of the memory of the smallest system. The temperature profile at 1000 s is given in Fig. 5.

The computational times are given in Table 2. The message is that the largest system is not necessarily the best for a specific problem and that more affordable smaller systems can perform in a similar fashion. At this point, it would be important to determine the maximum number of cores a particular application is able to manage during computations. This is an example that serves to accumulate experience and as a result be efficient in the use of cloud computing for engineering applications.

Table 2 Computational times for large virtual machines

Computing system	Computational time (s)
G1	14,074
G2	7922
G3	5465
G4	4820
G5	9277

School of Materials Science   Benito A. Stradi-Granados
Office of the Vice President for Research

Institute of Technology of Costa Rica
Cartago, Costa Rica

Departments of Chemistry and Chemical Engineering
University of Costa Rica
San José, Costa Rica

Further Reading

Aida, K. (n.d.). Effect of job size characteristics on job scheduling performance. Retrieved April, from http://www.cs.huji.ac.il/~feit/parsched/jsspp00/p-00-1.pdf

Beal, V. (2015). What is Clustering? Webopedia. Retrieved April 18, 2015, from http://www.webopedia.com/TERM/C/clustering.html

Inside HPC (2015). What is high performance computing?—Inside HPC. Retrieved April 18, 2015, from http://insidehpc.com/hpc-basic-training/what-is-hpc/

Janssen, C. (2015). Batch processing. Retrieved April 22, 2015, from http://www.techopedia.com/definition/5417/batch-processing

Lindsey, D. & Kelbley, J. (Ed.) (2008). High performance computing (HPC) and windows compute cluster server. Retrieved April 18, 2015, from https://technet.microsoft.com/en-us/magazine/2008.02.ccs.aspx

Rouse, M. (Ed.) (2006). What is cluster?—Definition from WhatIs.com. Retrieved April 18, 2015, from http://searchexchange.techtarget.com/definition/cluster

Rouse, M. (2007). What is high-performance computing (HPC)?—Definition from WhatIs.com. Retrieved April 18, 2015, from http://searchenterpriselinux.techtarget.com/definition/high-performance-computing

The University of Sydney (2015). How does a high performance computing (HPC) system work?—Ask Sydney (ICT)—The University of Sydney. Retrieved April 18, 2015, from http://staff.ask.sydney.edu.au/app/answers/detail/a_id/548/~/how-does-a-high-performance-computing-%28hpc%29-system-work%3F

Contents

List of Figures

List of Tables

Chapter 1
From the Office to the Cloud, Why Should You Care?

The Short Answer: Because It Can Benefit You

1.1 What Is the Cloud?

The cloud is a remote site with thousands of computer servers that makes its hardware, software, and technical support available to you *for a fee*.

> Cloud services are provided for a fee.
>
> It is your job to plan your work and to make the most out of every dollar spent on those services.

The growth (achieved in the past or expected in the future) for cloud-services market value from the period 2008 to 2020 increases in excess of 100%. This means that the demand and in turn the number of users are expected to increase significantly over this period of time.

There are four main cloud services offerings depending on the specific needs of a business: IaaS, infrastructure as a service; PaaS, platform as a service; SaaS, software as a service; and BPaaS, business process as a service. Engineering applications can make use of any and all of them depending on the particular needs of the user. For the individual researcher, IaaS over the past years has worked to satisfy the need for additional computational power. As it develops, SaaS is expected to come and play a significant role in savings, particularly for applications with expensive licenses. Data storage illustrates the high demand for cloud services with millions of users per cloud-storage service.

© Springer Nature Switzerland AG 2020
B. A. Stradi-Granados, *Cloud Computing for Engineering Applications*,
https://doi.org/10.1007/978-3-030-40445-1_1

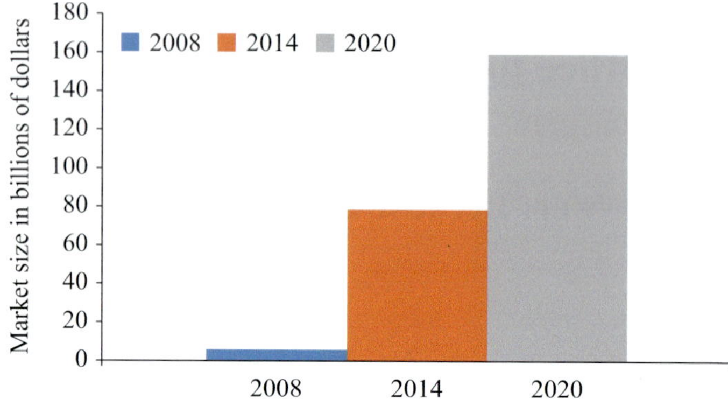

https://www.statista.com/statistics/510350/worldwide-public-cloud-computing/

Fig. 1.1 Market cap size for cloud services

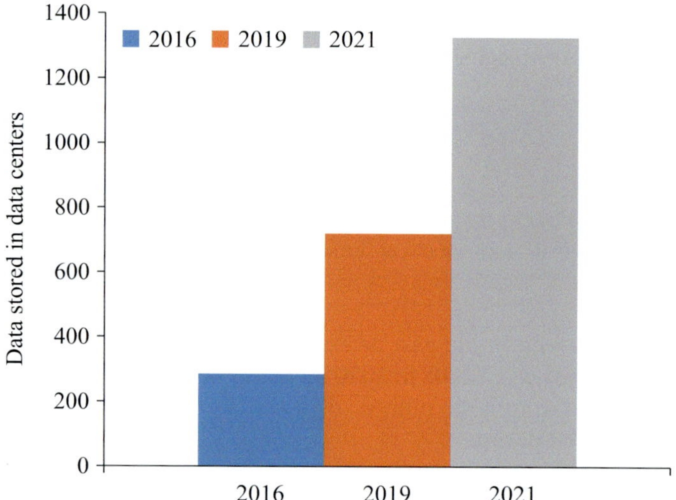

https://www.cisco.com/c/en/us/solutions/collateral/serviceprovider/global-cloud-index-gci/
white-paper-c11-738085.html

Fig. 1.2 Data stored in data centers

1.2   Cloud Facilities

Cloud facilities provide

- An environment of reasonable cost to test new ideas
- A flexible size platform that can change size according to growth:

  - More users
  - More virtual machines

- More processors
- More software
- Faster connectivity
- Technical support: basic support is of regular/low quality, premium support is more costly

- Multiple technologies in one site with a single menu
- The possibility to add particular applications as needed
- Developer environment to make and erase as needed

1.3   Cloud Computing

Cloud computing requires

- A reasonable local infrastructure

  - Fast internet connection (4 Mb/s or higher per user)
  - 8–16 GB workstation
  - 2–3 GHz CPU i5 or superior
  - Large HD display
  - 1 TB HDD (SSD only well justified)
  - Ergonomic mouse
  - Software to test the small-scale models

- A mid-level training in cloud utilization for users

  - Accounts creation
  - Money management
  - Creating new machines
  - Configuring adding features
  - Modifying machine sizes and their capabilities
  - User interface use
  - Basic script commands

- A set of soft skills from the users to include

  - Integrity (use only your resources, do not take those of others)
  - Good judgment (use only what is necessary, you may not need a 400 GB RAM to get started)
  - High level of frugality (avoid overspending)
  - Perform strict quality control in the initial stages of research (make sure your applications work in a small size computer before jumping to the full size in the cloud)
  - Be results oriented (achieve milestones within the allotted time and budget)

1.4 Role of Higher Management

The role of higher management (Office of the Provost, Office of the Vice President for Research) is to provide the resources (financial and otherwise) for the researchers to proceed with approved projects.

A few desirable features of higher management are:

- Takes into account cost/benefit according to the pertinent institutional policies.
- Takes a long-term approach and reassigns resources toward most successful projects after careful evaluation.
- Performs monitoring at a local stage: makes sure a smaller size version of the problem can be solved (homogeneous, parameter variation (homotopy)).
- Follows up at the large scale: makes possible to access the necessary cloud services.
- Mandates required training during working hours.
- Disciplines severely repeated wasteful use of resources.

1.5 Experience Developing Services

- Demonstration of a specialized cluster locally.
- Demonstration of implementation in a cloud environment.
- Training on campus for new tools in cloud environment.

1.6 Cloud Services: IaaS, PaaS, SaaS, BPaaS

- IaaS: infrastructure as a service

  - There is control of everything: operating system, software, RAM, storage
  - This is the service utilized in this text

- PaaS, SaaS, and BPaaS are not developed in this text, but as a group they serve an increasingly important number of users

  We are going to use Microsoft Azure in this text.

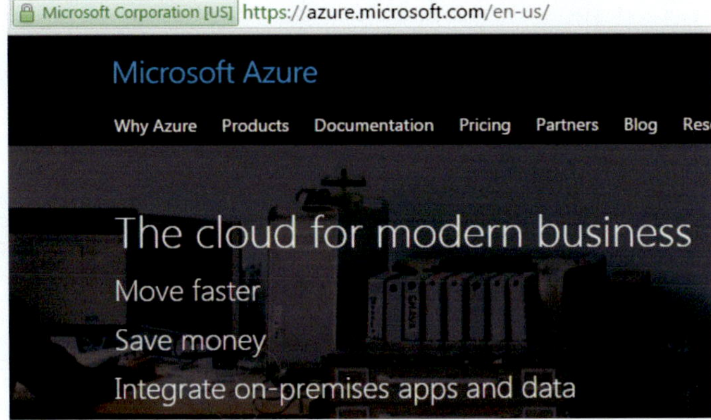

https://azure.microsoft.com/en-us/

Fig. 1.3 Microsoft Azure website

1.7 Starting Cloud Computing

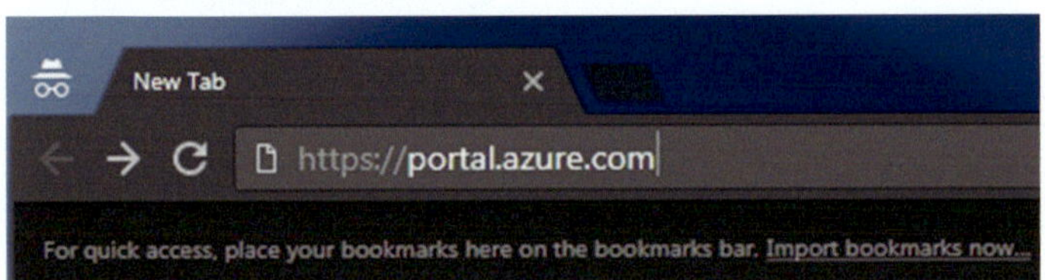

https://portal.azure.com

Fig. 1.4 Azure portal

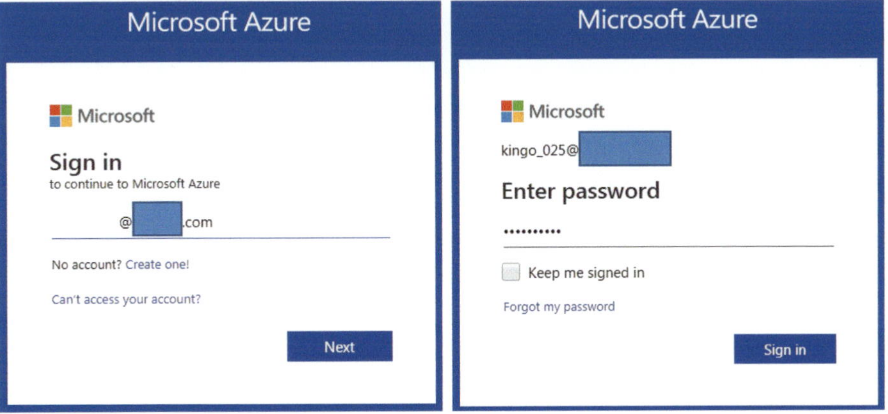

Fig. 1.5 Azure portal log-in information

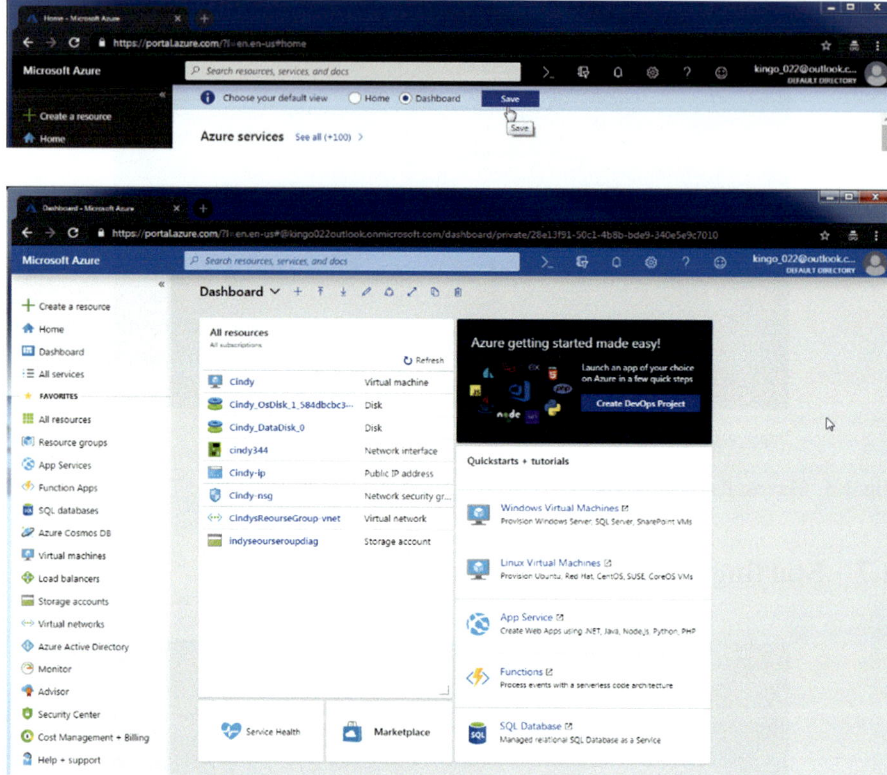

Fig. 1.6 Azure initial page: Dashboard

Now you arrived to the main page of Microsoft Azure.

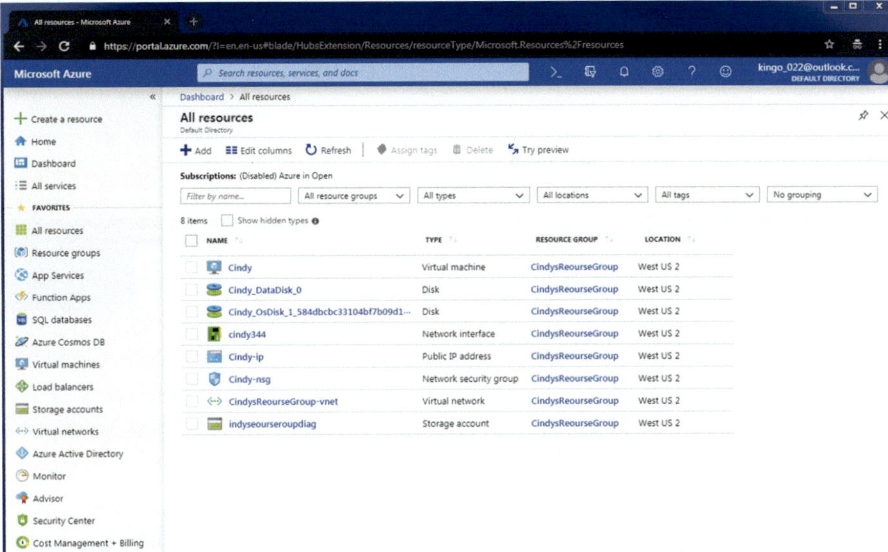

Fig. 1.7 Microsoft Azure summary of resources

On a new tab, go to Cost Management + Billing. This is the classic azure page; our interest is that it keeps track of the account.

Fig. 1.8 Account status in Microsoft Azure

We go back to the Microsoft Azure portal. This is called the Management Portal or new portal. The Management Portal is the new (Microsoft Windows) Azure interface that comes to substitute the Classic Portal (the one where the account status appears).

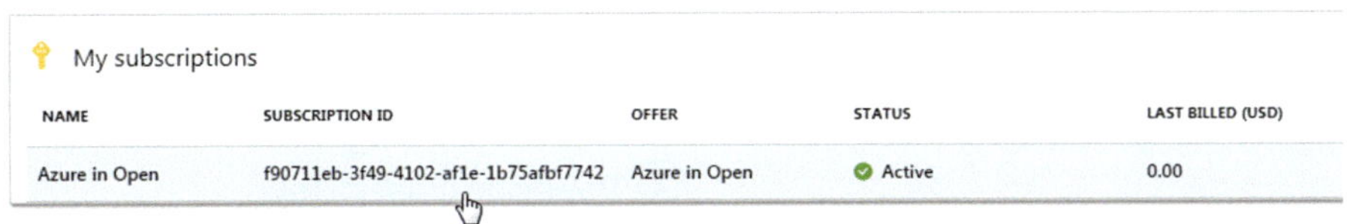

Fig. 1.9 Subscription selection

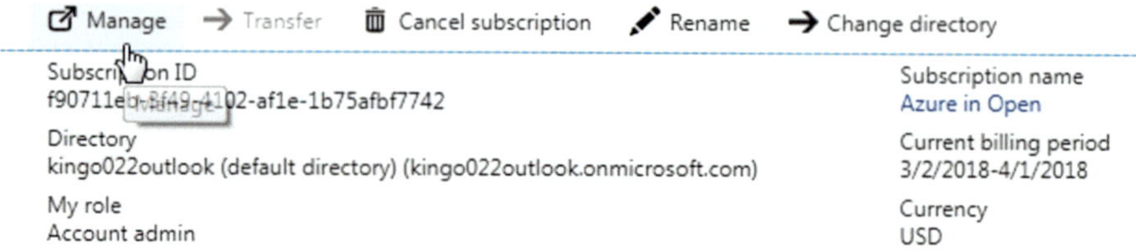

Fig. 1.10 Redirecting to the management site

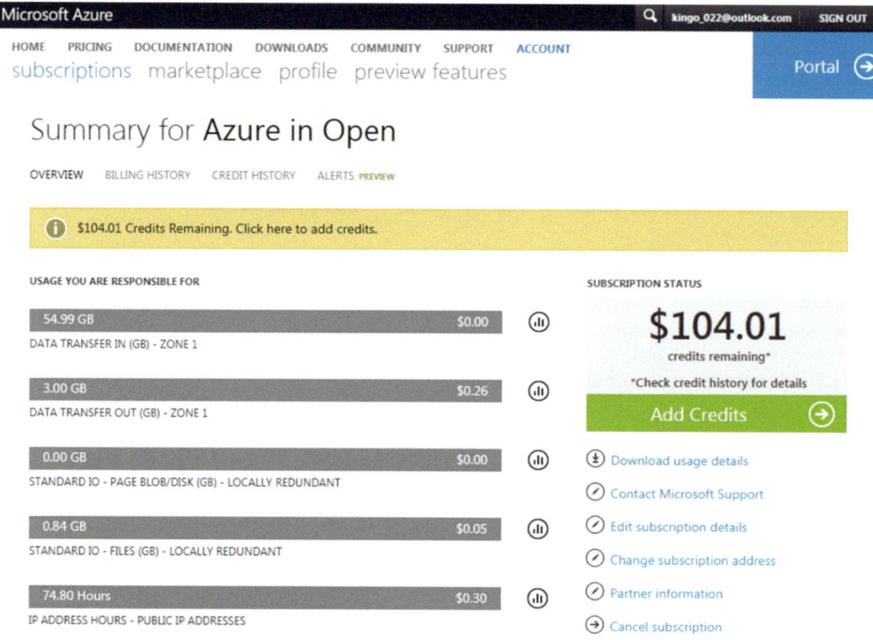

Fig. 1.11 Summary of expenses and budget

Fig. 1.12 Addition of funds to current subscription

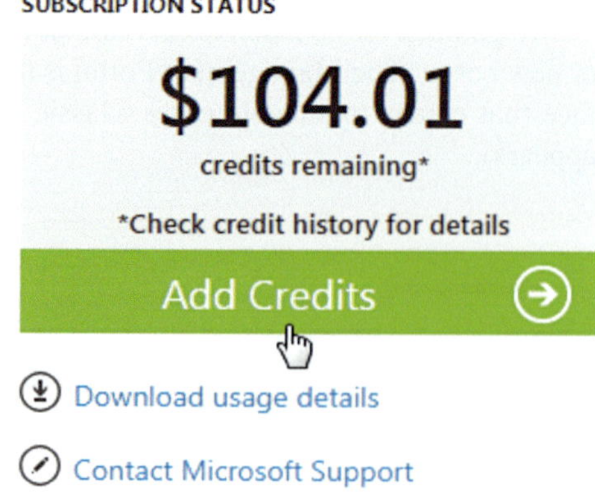

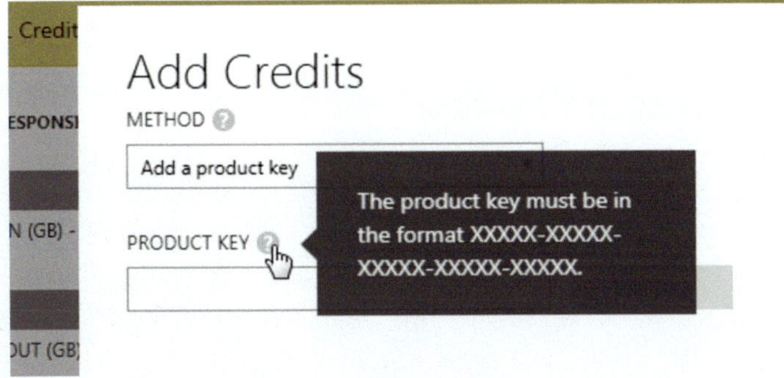

Fig. 1.13 Assignment of funds with sufficient granularity

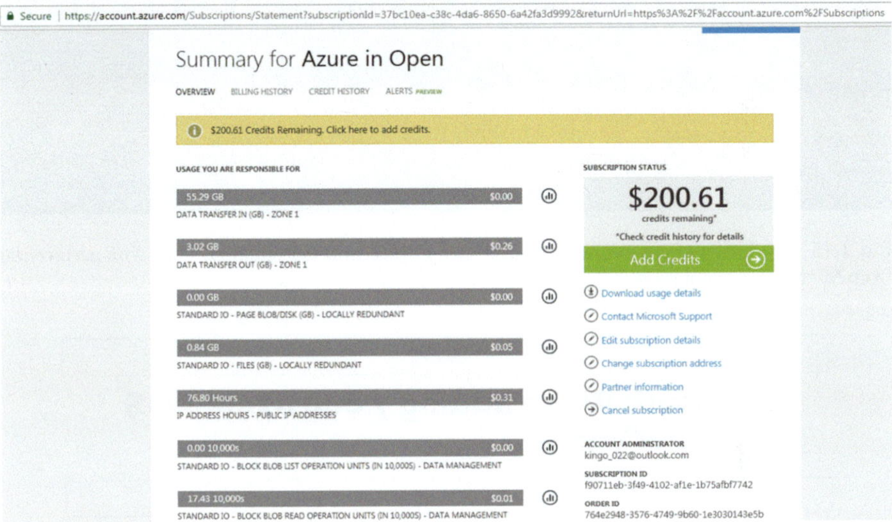

Fig. 1.14 Funding utilization and description of expenditures

All quests start with a purpose, a question, a goal that is pursued. This quest has to render benefits that are significant, worthy of investing your time and effort. Generally the fruits of your labor realize in the *form of rewards*.

Today we would like to break the chains created by low computing capacity. We would like to get the power we need instantly (or close to it).

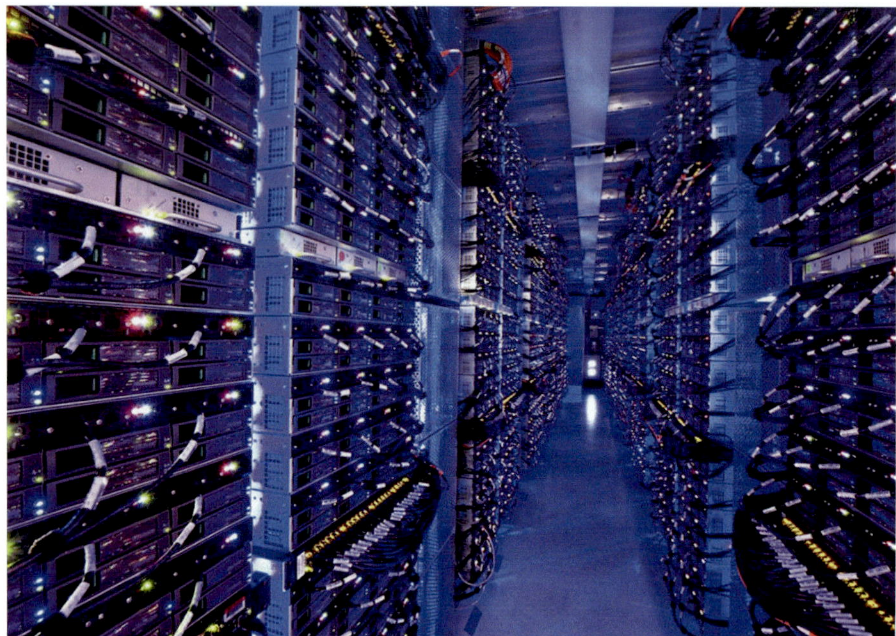

Fig. 1.15 Inside of a Microsoft Azure data center. ©Microsoft Corporation. Used with permission from Microsoft Corporation

2013¶

Microsoft: We're adding 7,000 Azure IaaS users per week

2018¶

TECH

Microsoft's Revenue Gets Lift From Cloud

Software giant exceeds $100 billion in annual revenue for first time

https://www.zdnet.com/article/microsoft-were-adding-7000-azure-iaas-users-per-week/
https://www.wsj.com/articles/microsofts-cloudcontinues-to-fuel-growth-1532032095

Fig. 1.16 Microsoft Azure utilization across the world

Microsoft Downgrades Windows' Role in Cloud-Focused Reorganization

In a sweeping reorganization, Windows will be part of a Core OS group; Terry Myerson leaves the company

https://www.wsj.com/articles/microsoft-diminishes-windows-role-in-cloud-focused-reorganization-1522335763

Fig. 1.17 Microsoft refocusing business on cloud infrastructure

Microsoft Cloud Hits Superscale As Huge Customers Migrate Mission–Critical SAP Workloads To Azure

https://www.forbes.com/sites/bobevans1/2018/02/28/microsoft-cloud-hits-superscale-as-huge-customers-migrate-mission-criticalsap-workloads-to-azure/#85026aa5993c

Fig. 1.18 Microsoft Azure utilization by large companies

1.8   Azure Global Distribution

Azure global distribution is one of the largest in the world.

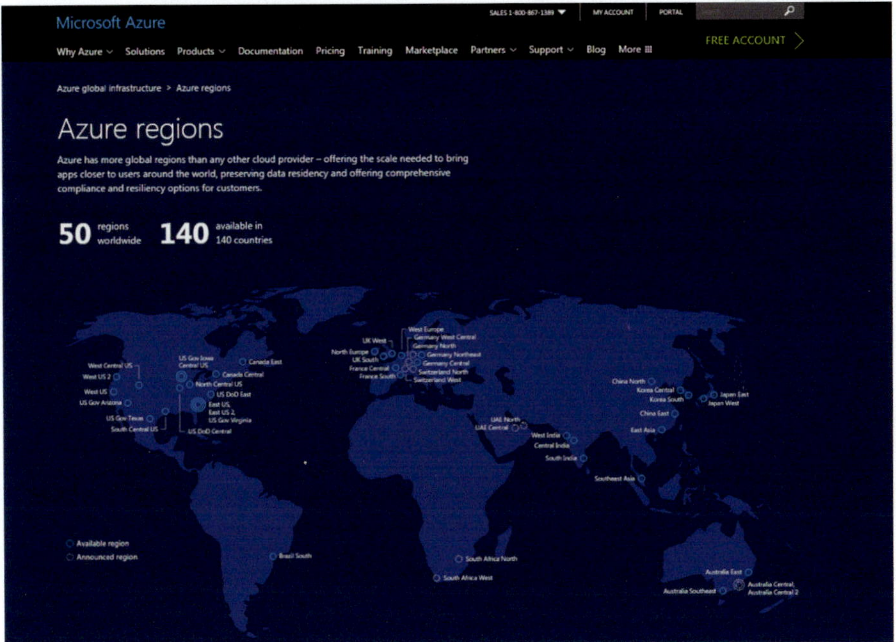

Fig. 1.19 Microsoft Azure global infrastructure. ©Microsoft Corporation. Used with permission from Microsoft Corporation

1.9   Current Successes

These are actual cases that need more computational capacity than your 8 Gb, i5, desktop.

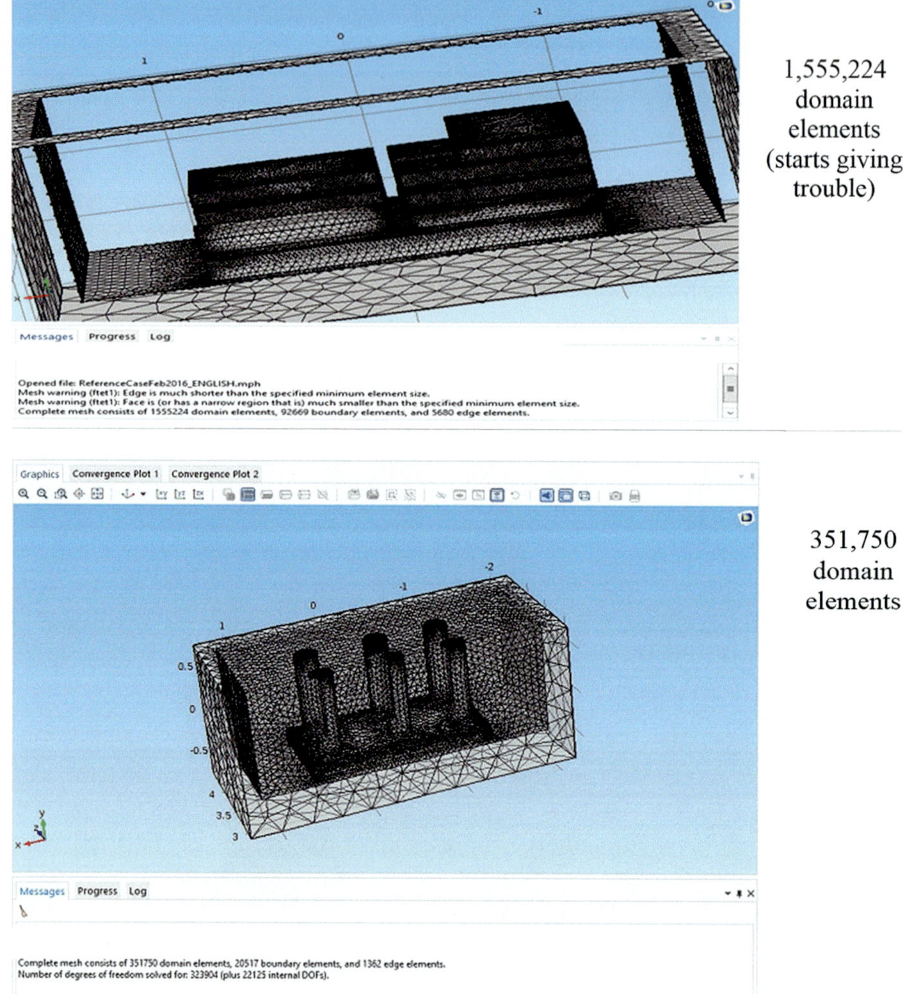

1,555,224 domain elements (starts giving trouble)

351,750 domain elements

Fig. 1.20 Increasing system complexity with number of nodes

1.10   Extract of Azure Virtual Machines

There are a series of machines which are available in Microsoft Azure; they differ in configuration and cost.

Sizes for Windows virtual machines in Azure

03/01/2018 • 2 minutes to read • Contributors all

This article describes the available sizes and options for the Azure virtual machines you can use to run your Windows apps and workloads. It also provides deployment considerations to be aware of when you're planning to use these resources. This article is also available for Linux virtual machines.

Type	Sizes	Description
General purpose	B, Dsv3, Dv3, DSv2, Dv2, DS, D, Av2, A0-7	Balanced CPU-to-memory ratio. Ideal for testing and development, small to medium databases, and low to medium traffic web servers.
Compute optimized	Fsv2, Fs, F	High CPU-to-memory ratio. Good for medium traffic web servers, network appliances, batch processes, and application servers.
Memory optimized	Esv3, Ev3, M, GS, G, DSv2, DS, Dv2, D	High memory-to-CPU ratio. Great for relational database servers, medium to large caches, and in-memory analytics.
Storage optimized	Ls	High disk throughput and IO. Ideal for Big Data, SQL, and NoSQL databases.
GPU	NV, NC, NCv2, NCv3, ND	Specialized virtual machines targeted for heavy graphic rendering and video editing, as well as model training and inferencing (ND) with deep learning. Available with single or multiple GPUs.
High performance compute	H, A8-11	Our fastest and most powerful CPU virtual machines with optional high-throughput network interfaces (RDMA).

https://docs.microsoft.com/en-us/azure/virtual-machines/windows/sizes

Fig. 1.21 Extract of the virtual machines available in Microsoft Azure

Choose your language, workload, operating system

With support for Linux, Windows Server, SQL Server, Oracle, IBM, and SAP, Azure Virtual Machines gives you the flexibility of virtualization for a wide range of computing solutions—development and testing, running applications, and extending your datacenter. It's the freedom of open-source software configured the way you need it. It's as if it was another rack in your datacenter, giving you the power to deploy an application in seconds instead of weeks.

https://azure.microsoft.com/en-us/services/virtual-machines/

Fig. 1.22 Extract of operating systems in Microsoft Azure

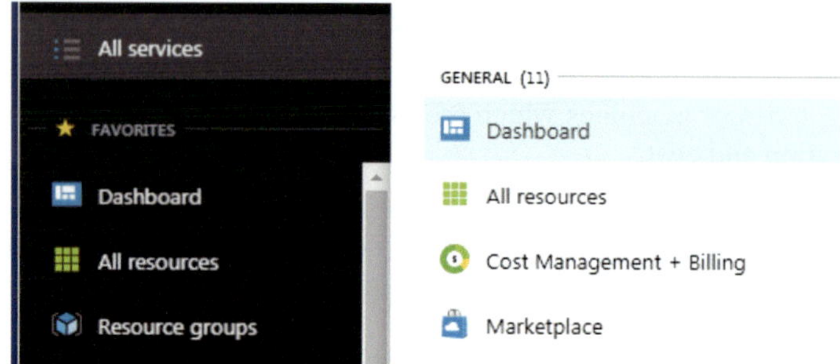

Fig. 1.23 A large selection of applications by selecting marketplace

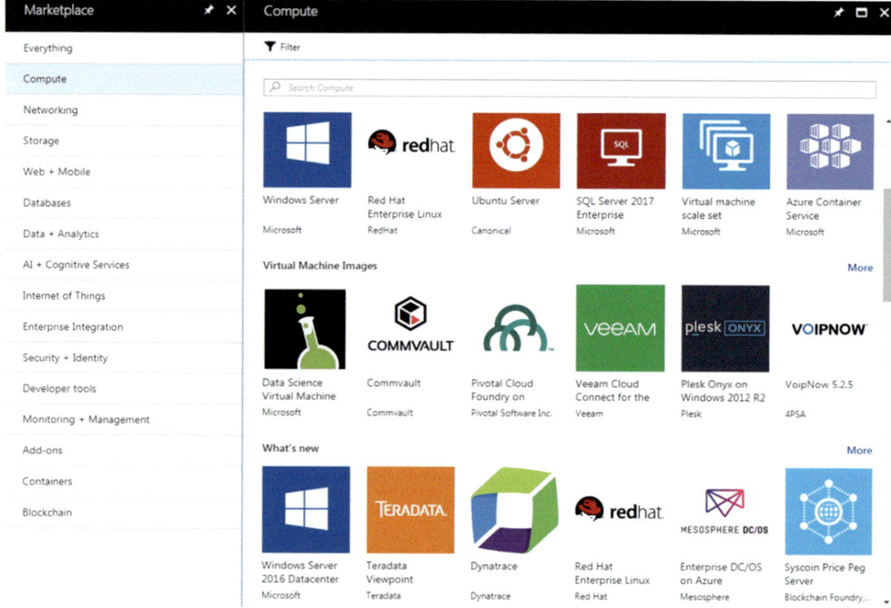

Fig. 1.24 Sample of virtual machines available from the marketplace

Fig. 1.25 Kingo

We will develop in detail examples in the text.

1.11 Use of Microsoft Azure Computational Capabilities in the Doctoral Curriculum

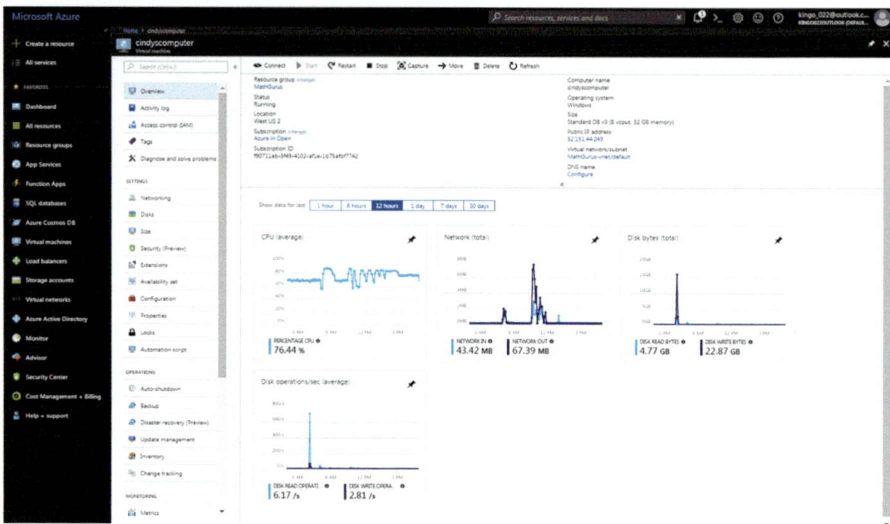

Fig. 1.26 Mathematics doctoral student jamming her simulations in Windows Azure

1.12 Commissioning a Virtual Machine in Microsoft Azure (Azure VM)

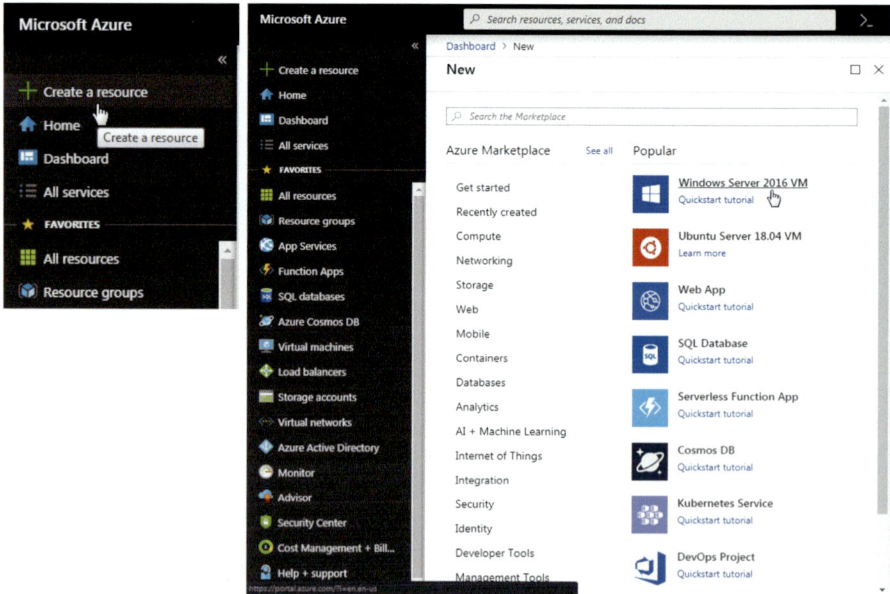

Fig. 1.27 Adding a virtual machine to our Microsoft Azure account

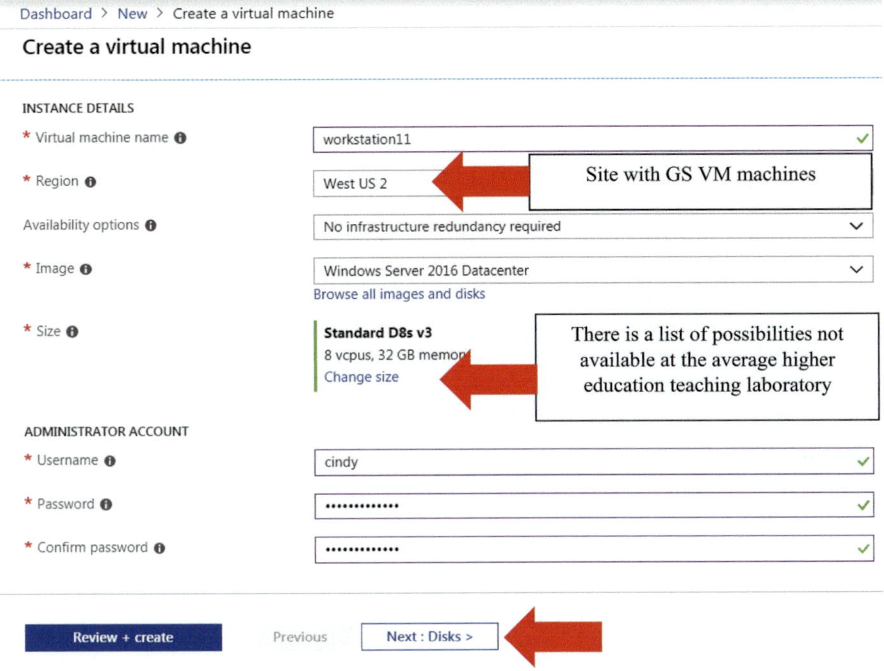

Fig. 1.28 Properties of the virtual machine (VM)

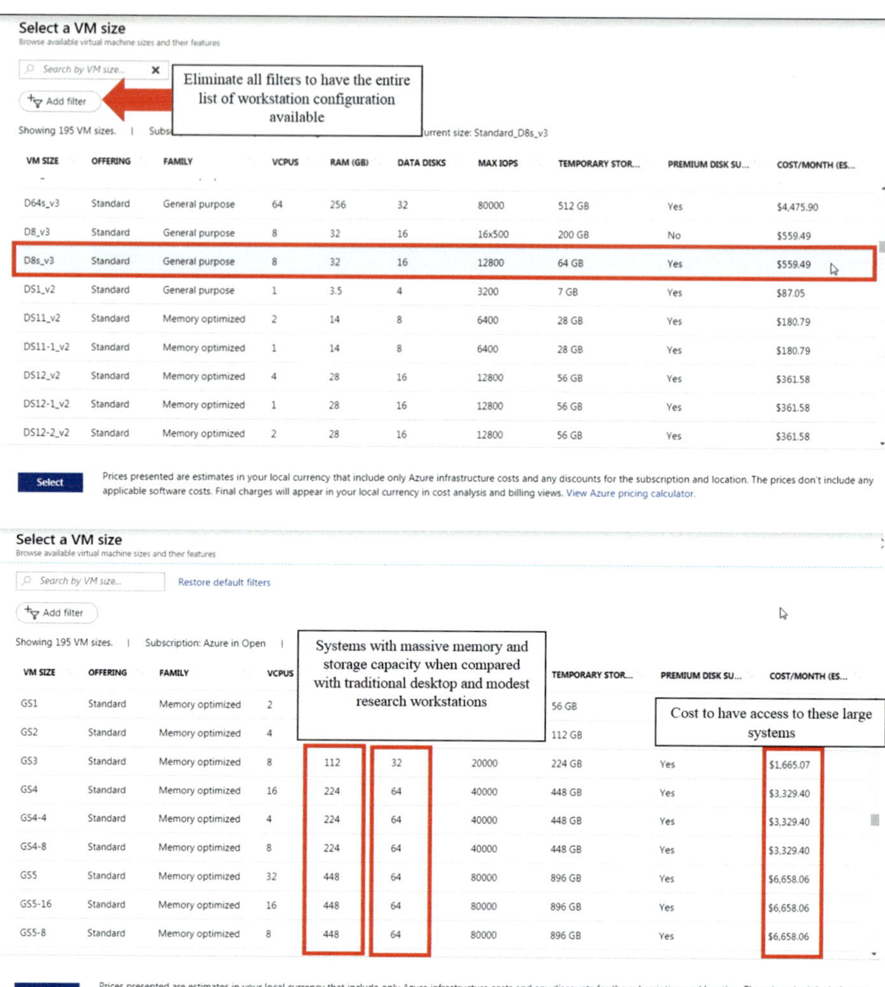

VM SIZE	OFFERING	FAMILY	VCPUS	RAM (GB)	DATA DISKS	MAX IOPS	TEMPORARY STOR...	PREMIUM DISK SU...	COST/MONTH (ES...
D64s_v3	Standard	General purpose	64	256	32	80000	512 GB	Yes	$4,475.90
D8_v3	Standard	General purpose	8	32	16	16x500	200 GB	No	$559.49
D8s_v3	Standard	General purpose	8	32	16	12800	64 GB	Yes	$559.49
DS1_v2	Standard	General purpose	1	3.5	4	3200	7 GB	Yes	$87.05
DS11_v2	Standard	Memory optimized	2	14	8	6400	28 GB	Yes	$180.79
DS11-1_v2	Standard	Memory optimized	1	14	8	6400	28 GB	Yes	$180.79
DS12_v2	Standard	Memory optimized	4	28	16	12800	56 GB	Yes	$361.58
DS12-1_v2	Standard	Memory optimized	1	28	16	12800	56 GB	Yes	$361.58
DS12-2_v2	Standard	Memory optimized	2	28	16	12800	56 GB	Yes	$361.58

VM SIZE	OFFERING	FAMILY	VCPUS	RAM (GB)	DATA DISKS	MAX IOPS	TEMPORARY STOR...	PREMIUM DISK SU...	COST/MONTH (ES...
GS1	Standard	Memory optimized	2				56 GB		
GS2	Standard	Memory optimized	4				112 GB		
GS3	Standard	Memory optimized	8	112	32	20000	224 GB	Yes	$1,665.07
GS4	Standard	Memory optimized	16	224	64	40000	448 GB	Yes	$3,329.40
GS4-4	Standard	Memory optimized	4	224	64	40000	448 GB	Yes	$3,329.40
GS4-8	Standard	Memory optimized	8	224	64	40000	448 GB	Yes	$3,329.40
GS5	Standard	Memory optimized	32	448	64	80000	896 GB	Yes	$6,658.06
GS5-16	Standard	Memory optimized	16	448	64	80000	896 GB	Yes	$6,658.06
GS5-8	Standard	Memory optimized	8	448	64	80000	896 GB	Yes	$6,658.06

Fig. 1.29 Selection of the configuration of the virtual machine

View all and search for a DS3 which is a descent size machine and you can do all your testing at a lower cost than that of the largest machines.

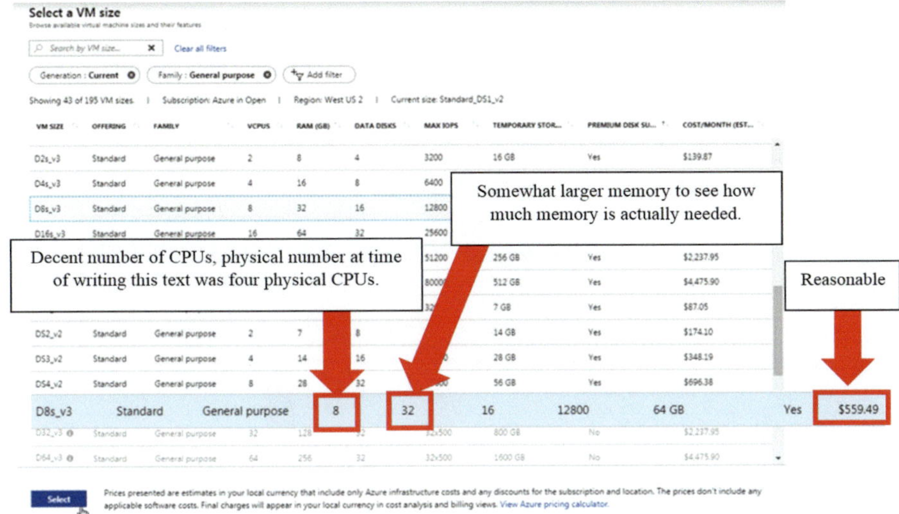

Fig. 1.30 Selecting a D8S_V3 machine for our initial tests

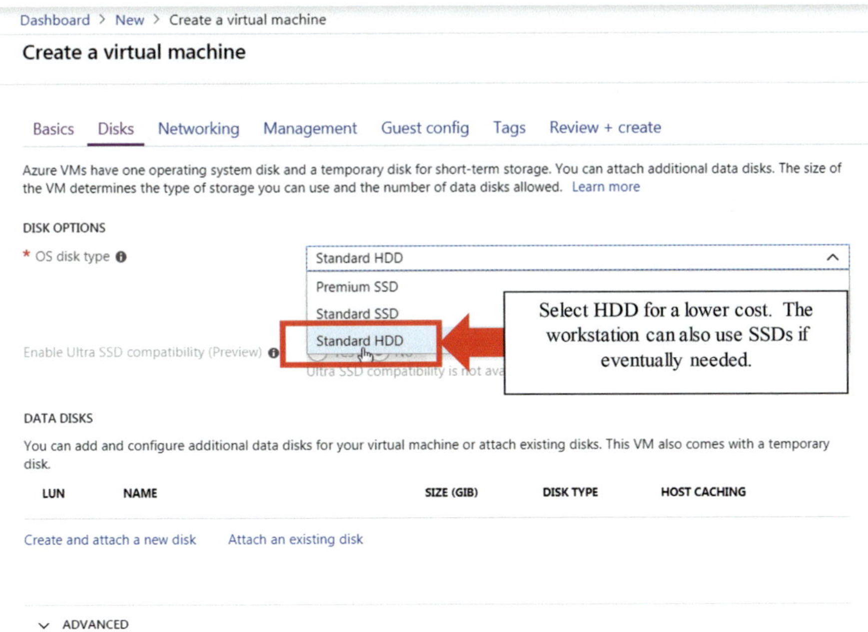

Fig. 1.31 Specifying storage characteristics

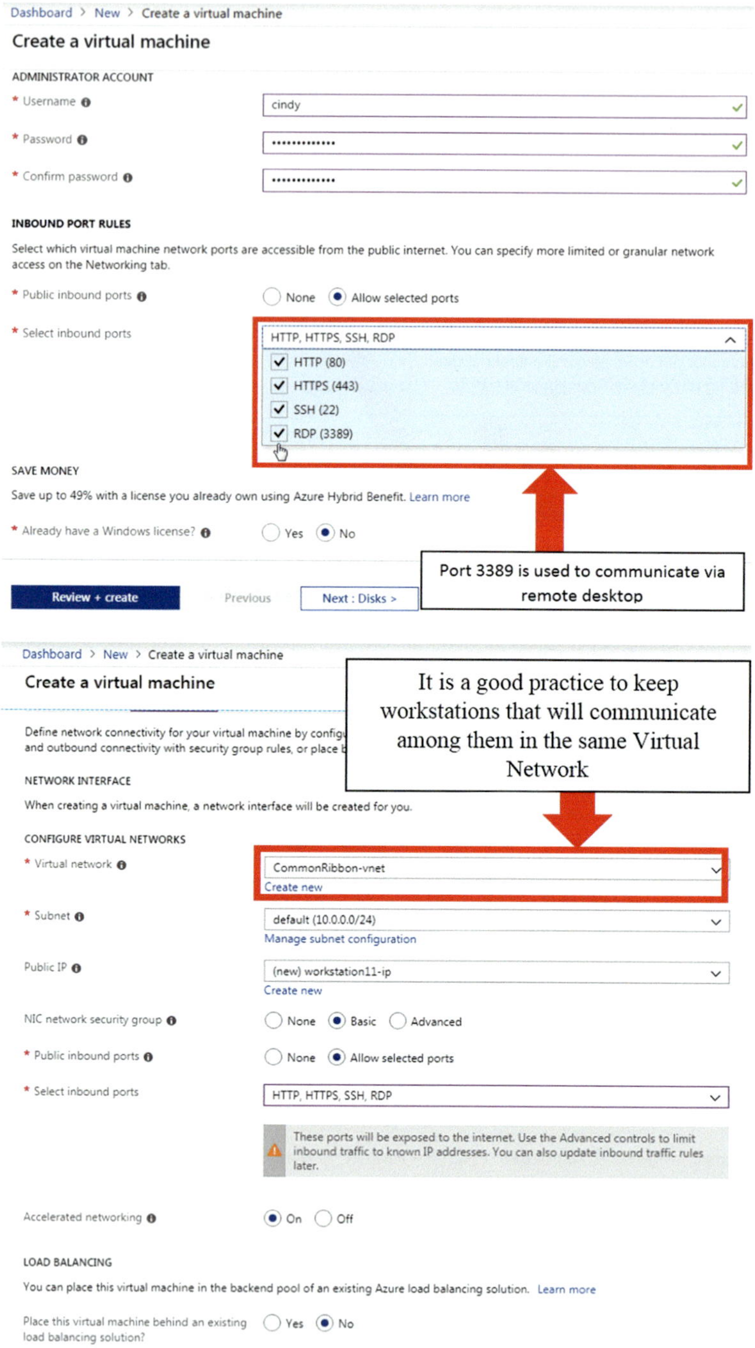

Fig. 1.32 Specifying network characteristics

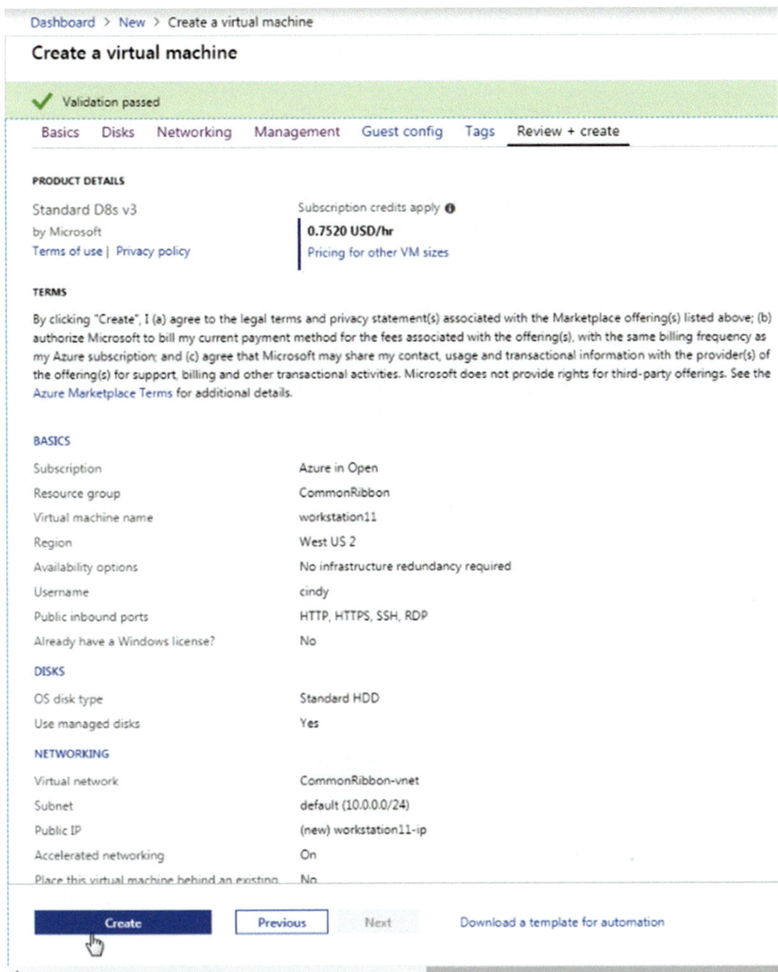

Fig. 1.33 Characteristics and cost of the new Azure VM

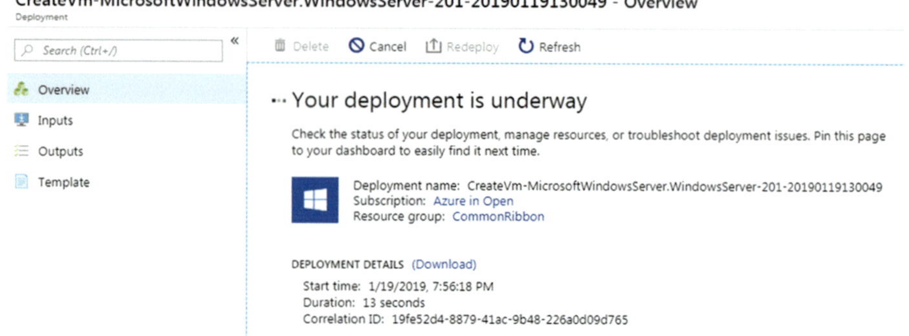

Fig. 1.34 Messaging of Azure VM submitting deployment

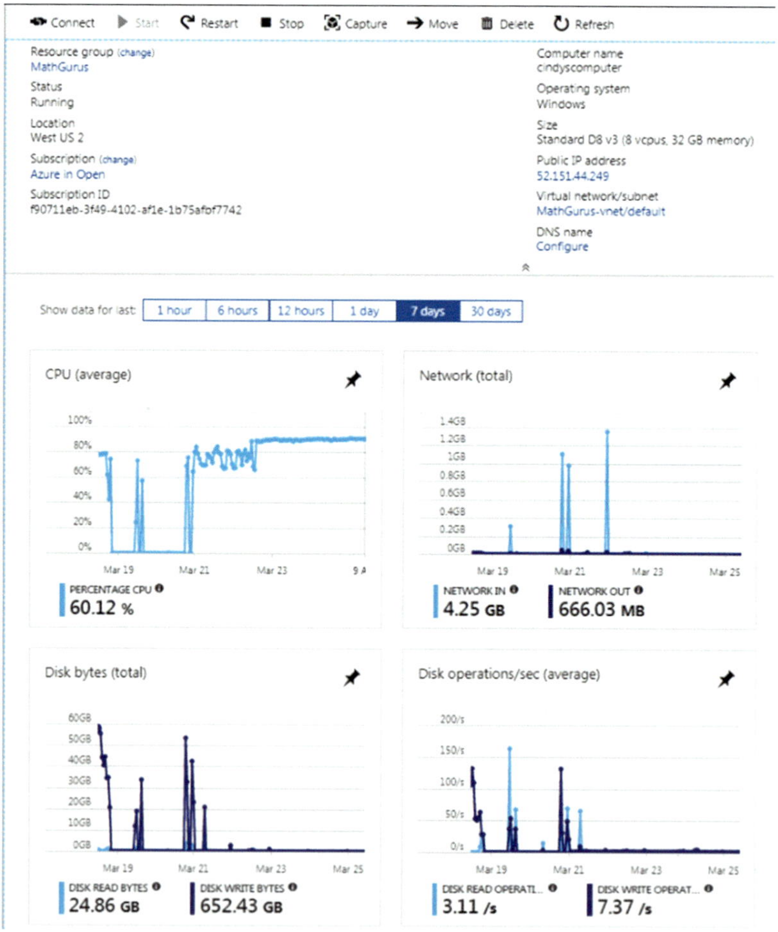

Fig. 1.35 Deployment with full CPU use

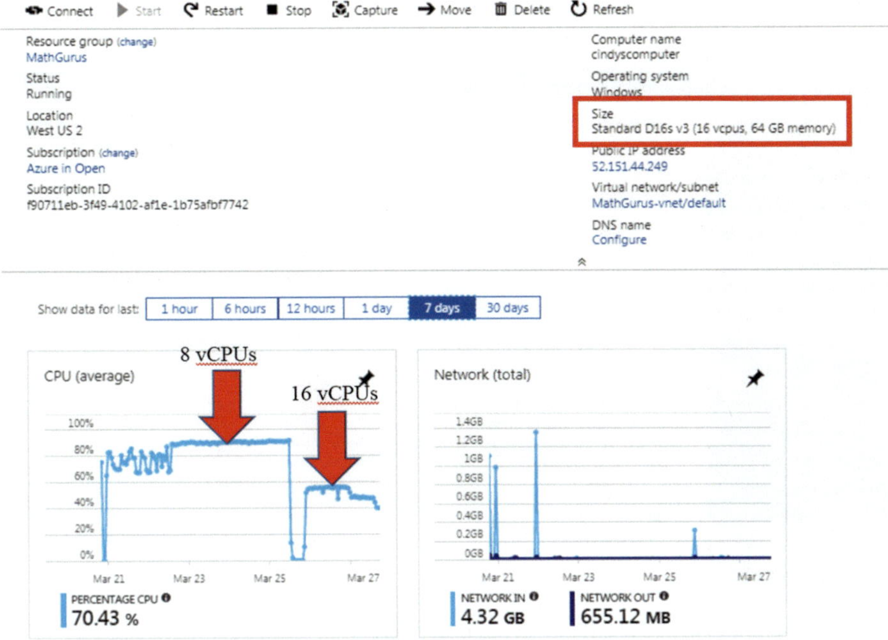

Fig. 1.36   Increasing VM size to lower CPU demand

A faculty member with no initial equipment now can work full time in her project in less than 3 days with minimal funding. Move from a smaller machine to a larger machine as needed and now is no longer CPU bottlenecked with minimal resources.

1.13   Configuration Summary

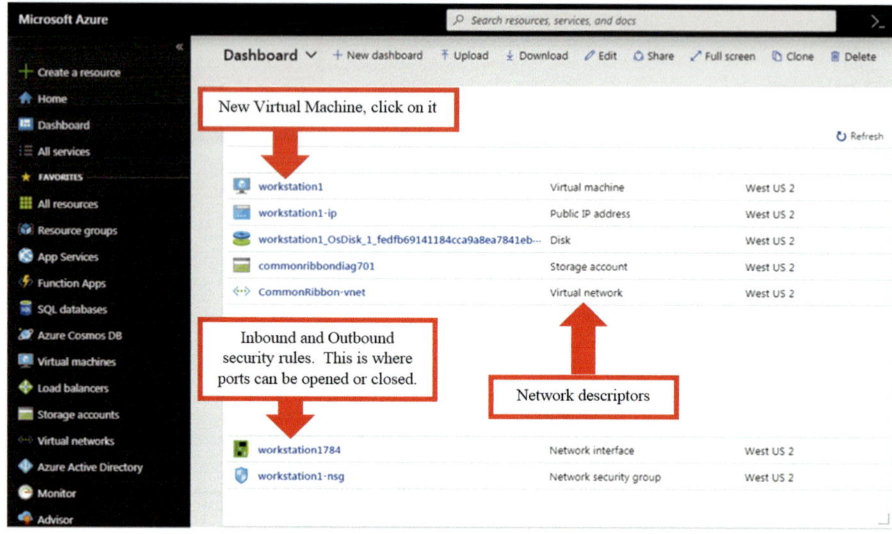

Fig. 1.37 Deployment of the new virtual machine (VM)

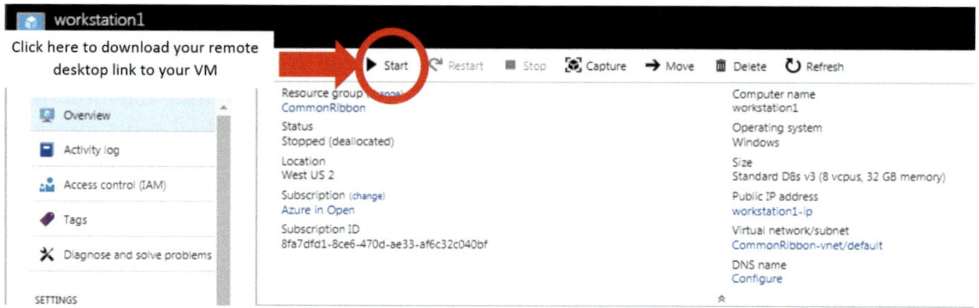

Fig. 1.38 Start the Azure VM

Fig. 1.39 Virtual internet address (VIP) and connection port

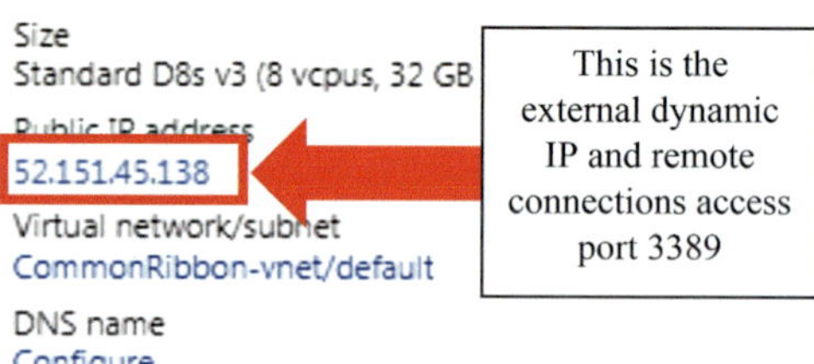

1.14 Accessing Your Virtual Machine

Fig. 1.40 Download
Remote Desktop Console
using Connect

Fig. 1.41 Downloaded
Remote Desktop Console
link

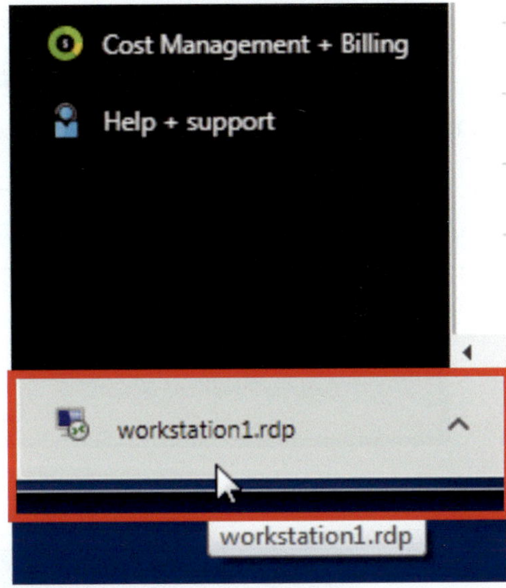

Fig. 1.42 Remote Desktop
Console link

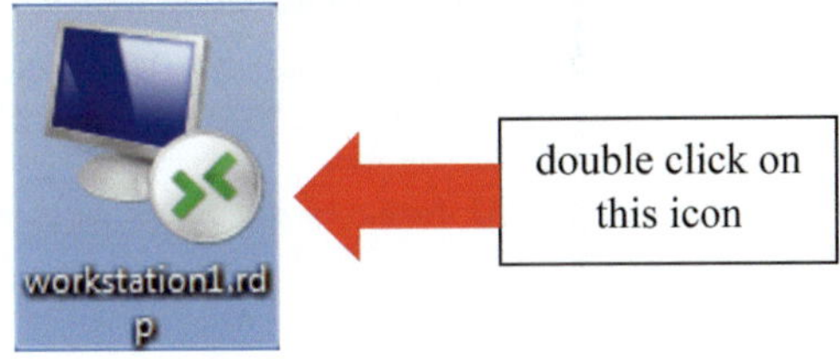

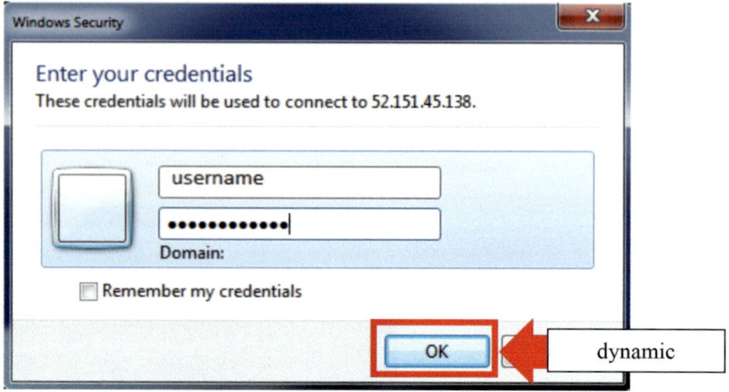

Fig. 1.43 Remote Desktop Console User and Password

Fig. 1.44 Accessing your virtual machine remotely

Fig. 1.45 Accessing your VM remotely disregard warnings

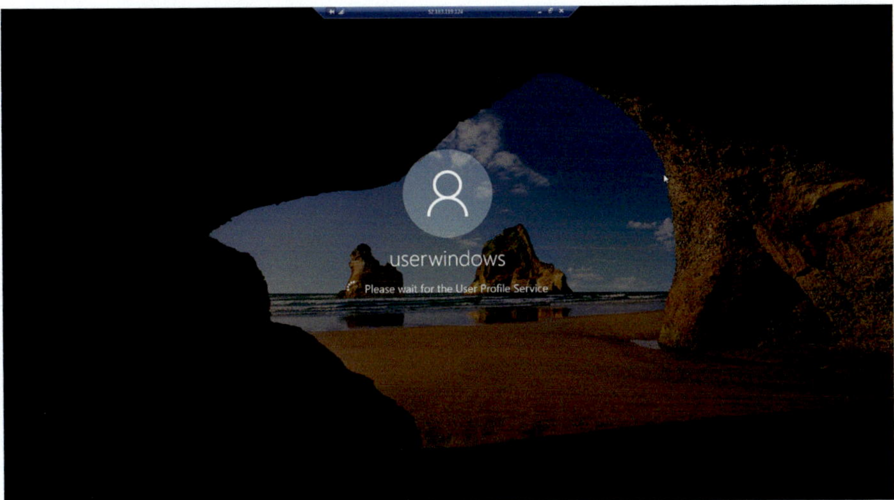

Fig. 1.46 Connection established to remote VM

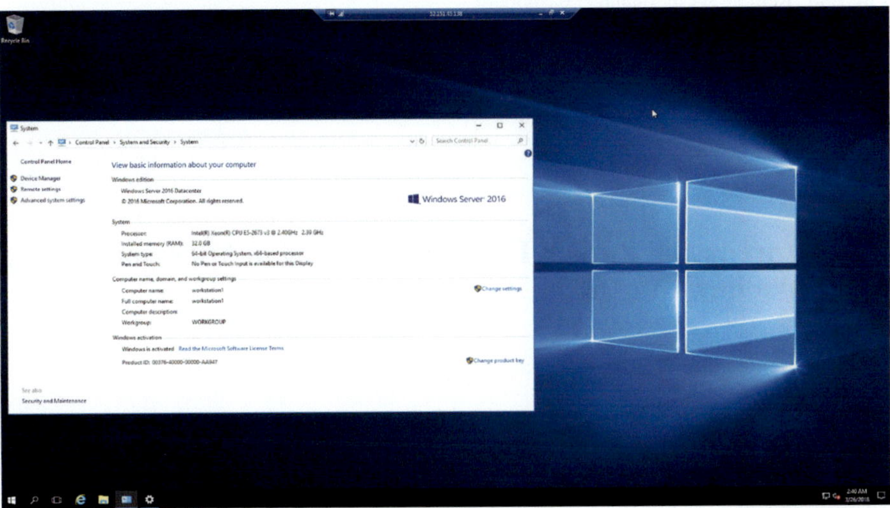

Fig. 1.47 Remote VM system configuration

1.15   Configuring Your Virtual Machine

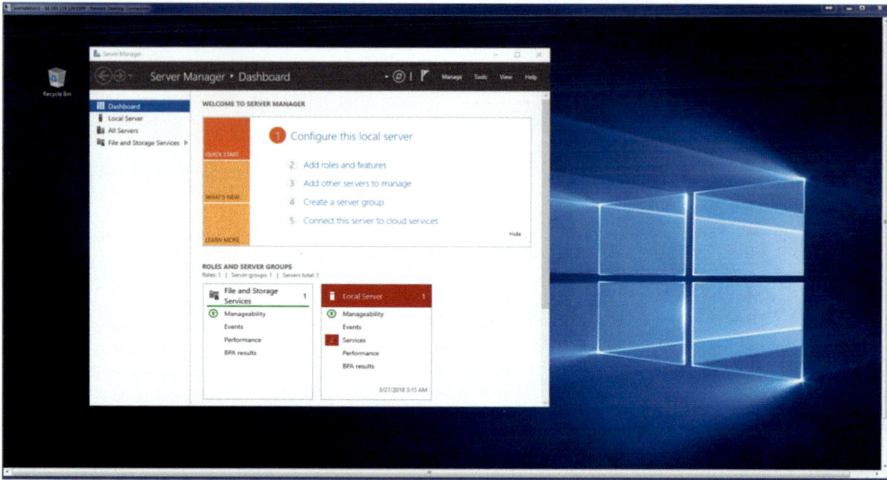

Fig. 1.48 Remote server manager dashboard

Think of this, now you have a virtual machine to work with in just a few minutes. In addition, this is a fully operational computer and with a legal operating system and you are on your way to do your job.

1.16   Purchasing Rather Than Provisioning in Microsoft Windows Azure

Let us consider that you would like to buy your own.

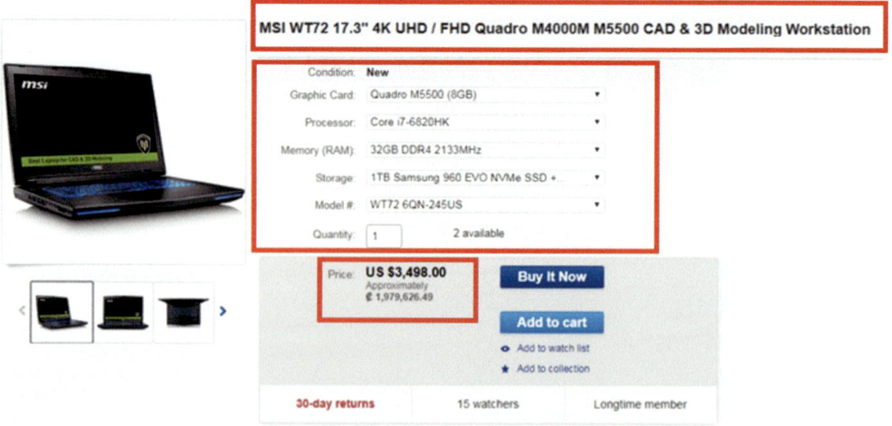

Fig. 1.49 Purchasing a laptop for office work

Fig. 1.50 Cost of current operating system license

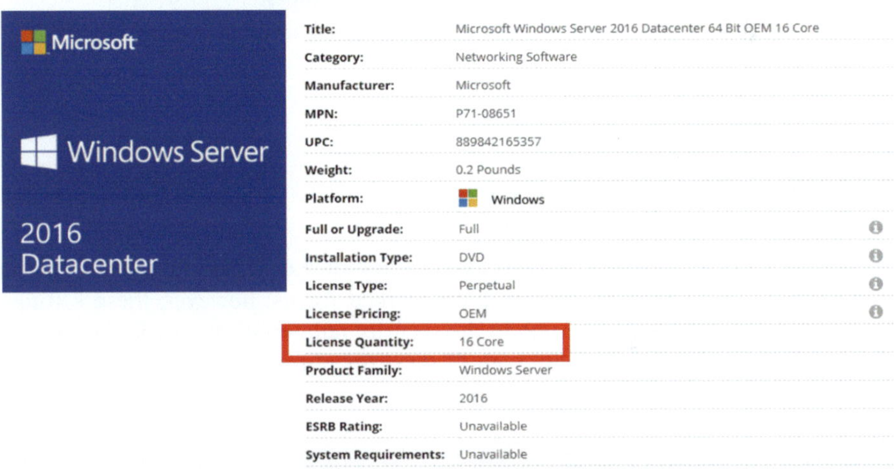

Fig. 1.51 Limitations in number of cores for a traditional installation

Other sellers offer only keys or strip down versions, fewer cores, and other shady alternatives. In good theory, one should be able to obtain a better price from a local authorized vendor for educational purposes.

Of course, you can find offers of license numbers that do not send the installation disks along; these are not offers that can be considered in scholarly work.

1.17   Virtual Machine Dynamic IP Address

I logged off and logged on next morning (you do not have to do it)

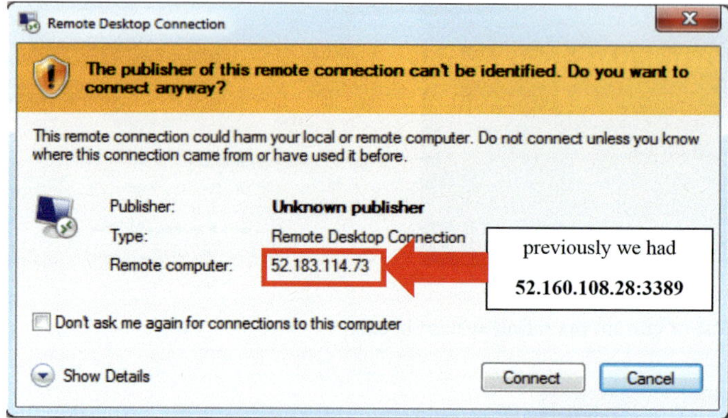

Fig. 1.52 Virtual IP address changes dynamically

1.18   Security and Initial Settings

We go back to the Virtual Machines (VM), click on Local Server. The variations implemented on security as follows have worked for us; however, these settings may need to be different according to the security policies of the particular institution or organization. Consult with your IT department for their suggestions.

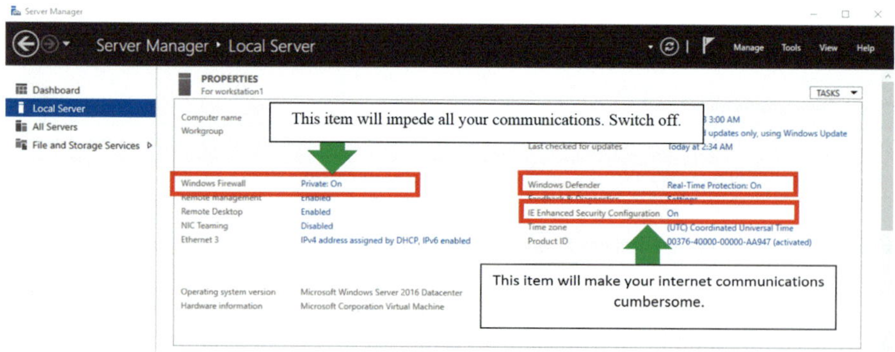

Fig. 1.53 Initial settings in Windows Server 2016 Datacenter

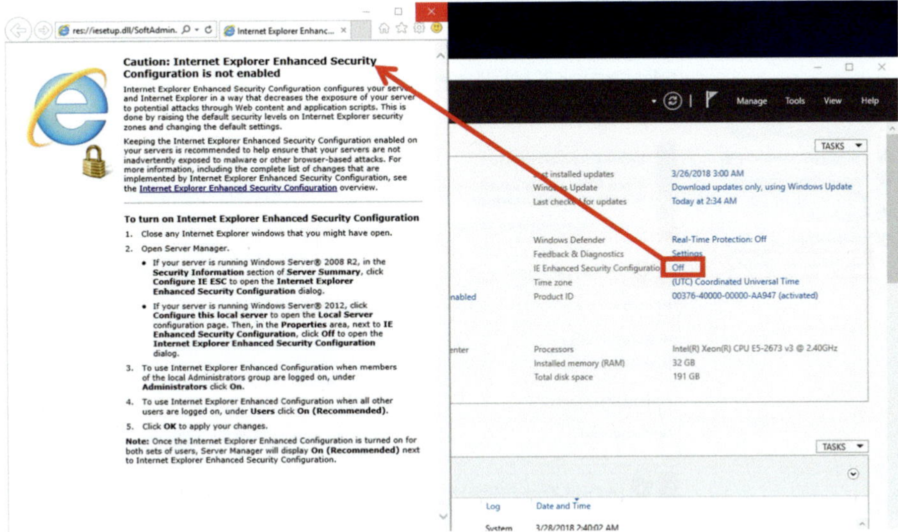

Fig. 1.54 Impeded internet connection by IE Enhanced Security

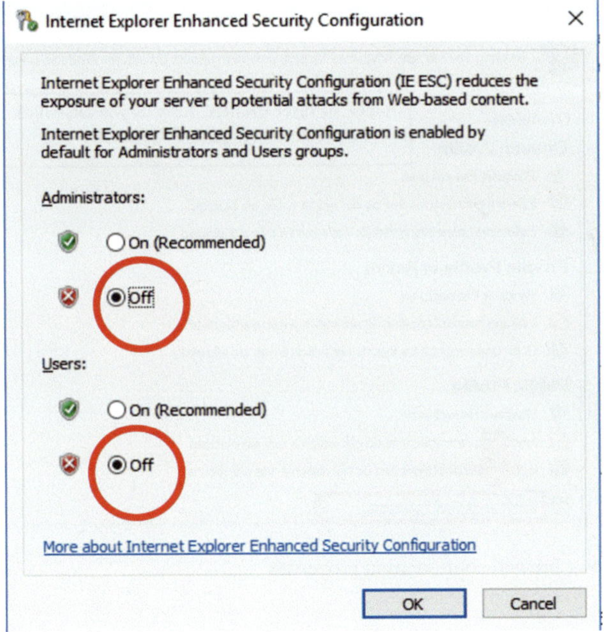

Fig. 1.55 Disabling IE Enhanced Security configuration

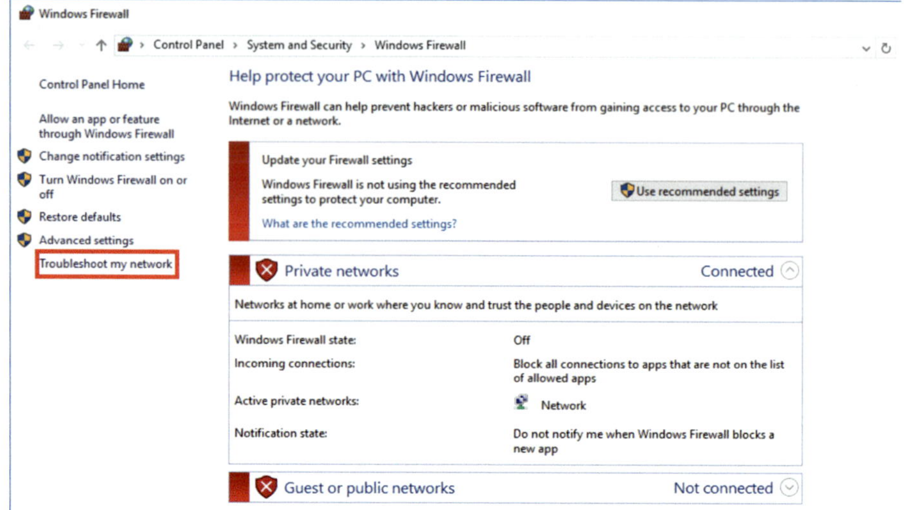

Fig. 1.56 Firewall configuration

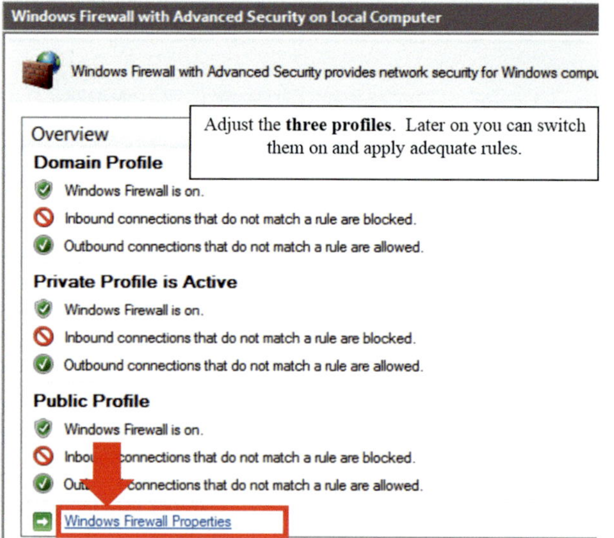

Fig. 1.57 Windows firewall configuration properties

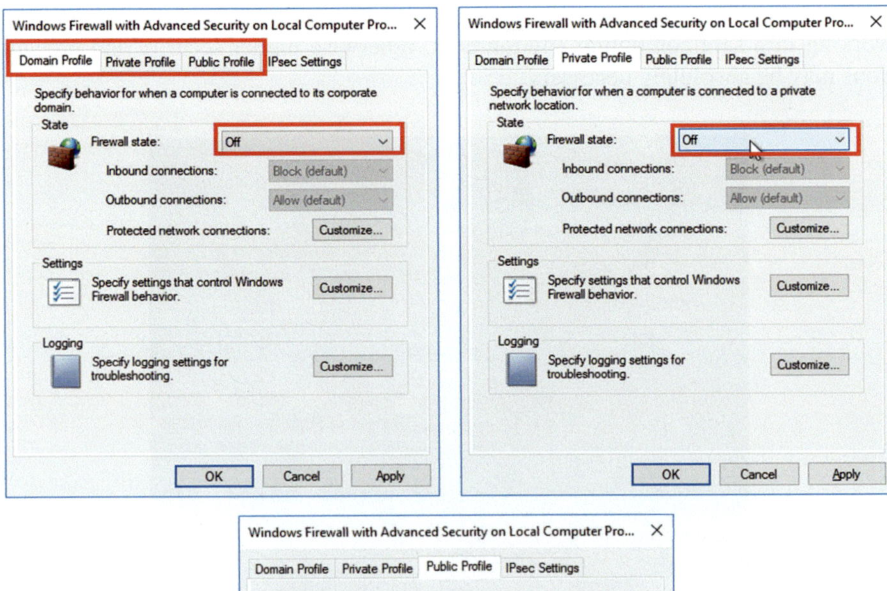

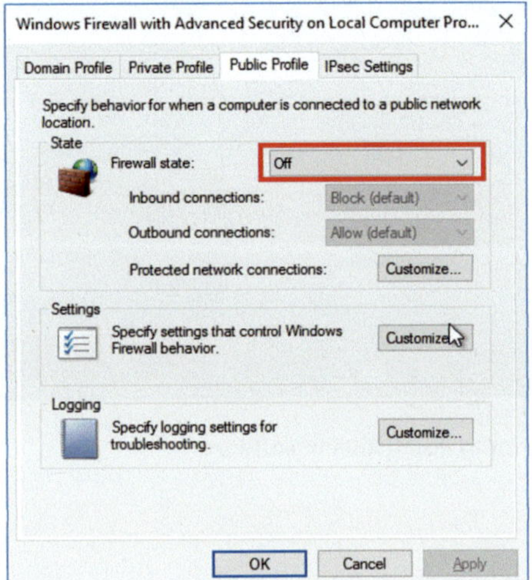

Fig. 1.58 Disabling firewall protections as a temporary measure

Firewall settings are complete at this point. Yet again make sure that you are working in a safe computing environment; otherwise higher security and precautions may be absolutely necessary.

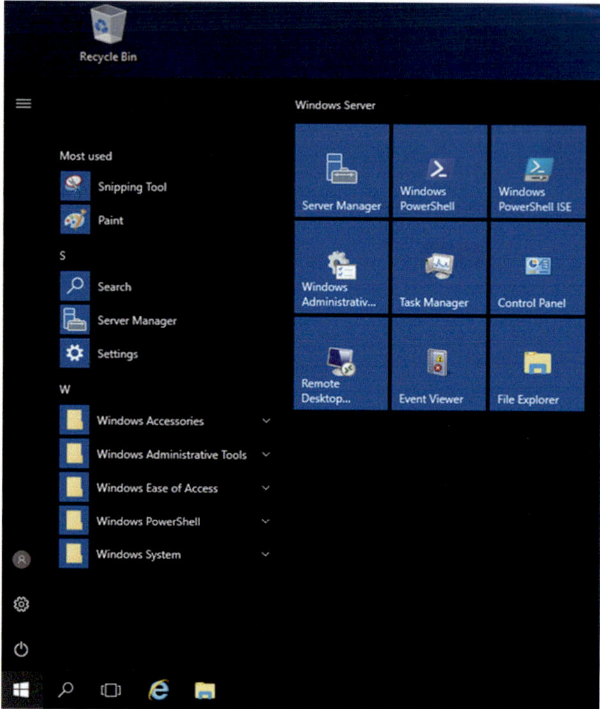

Fig. 1.59 Server state prior to installation of software

Go and get Google Chrome.

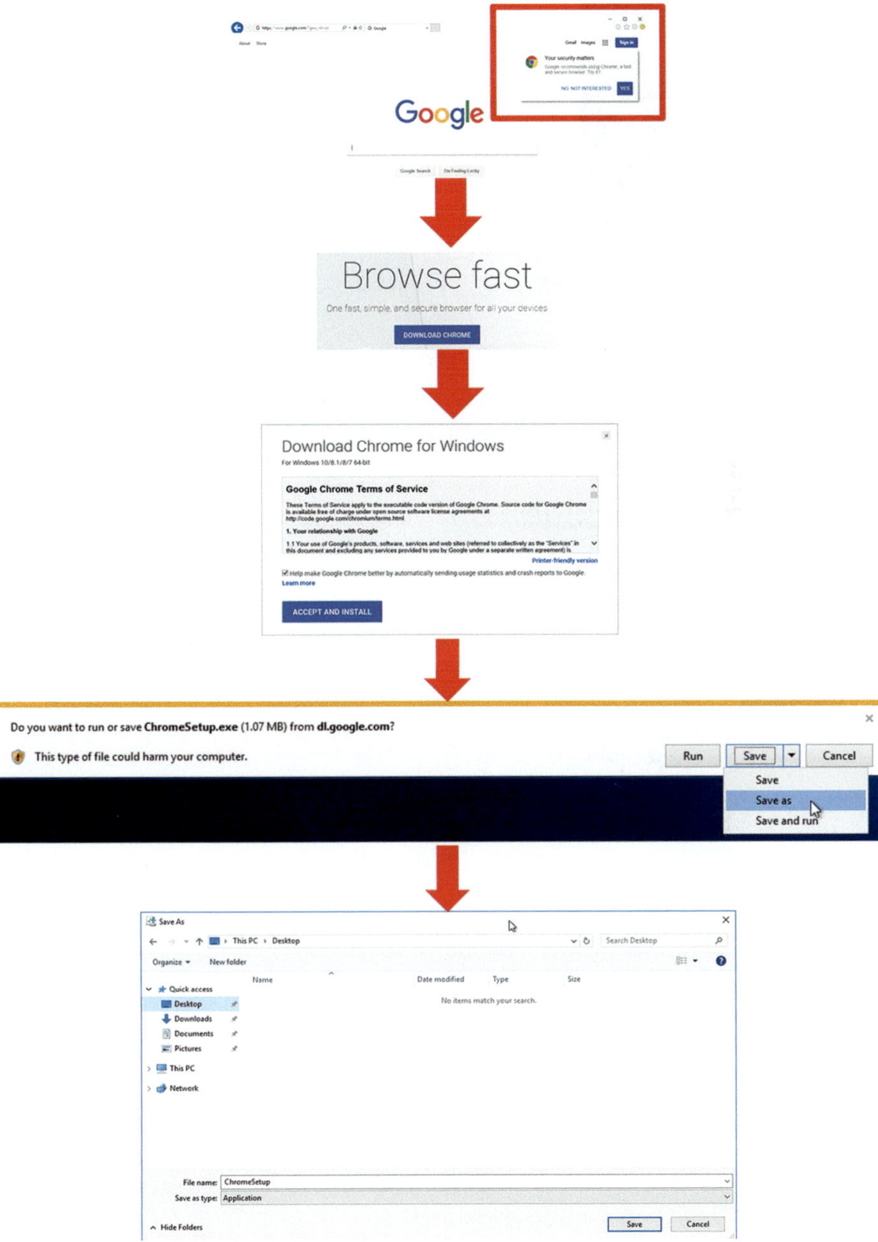

Fig. 1.60 Install Google Chrome

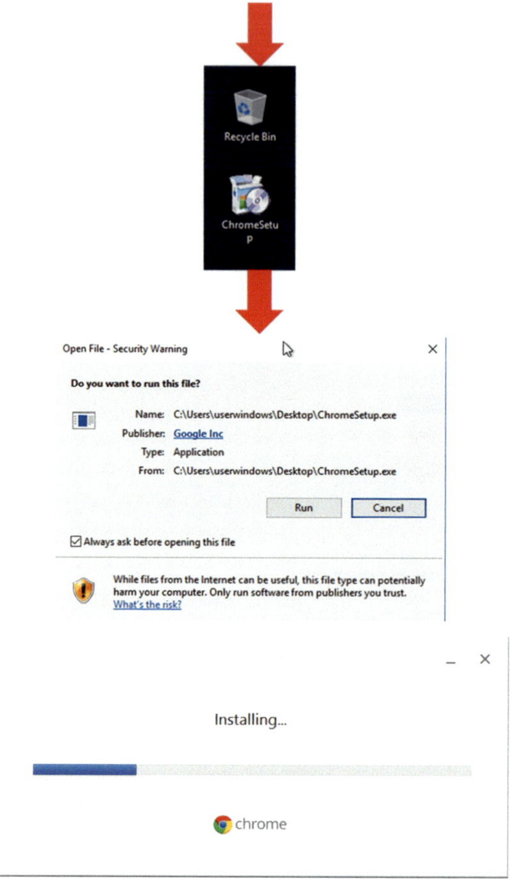

Fig. 1.60   (continued)

1.19   Computer Configuration for COMSOL Multiphysics©

We want to download the COMSOL Multiphysics©. This gives an idea of the time require to upload an image in a fast network.

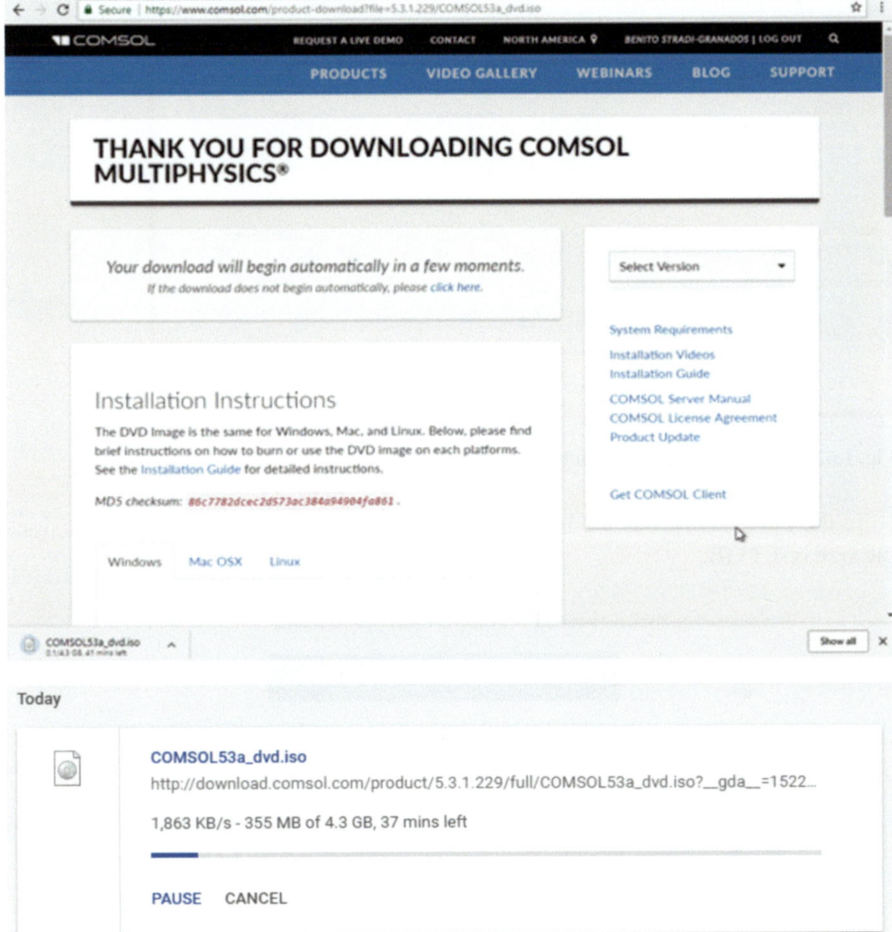

Fig. 1.61 Downloading COMSOL Multiphysics©

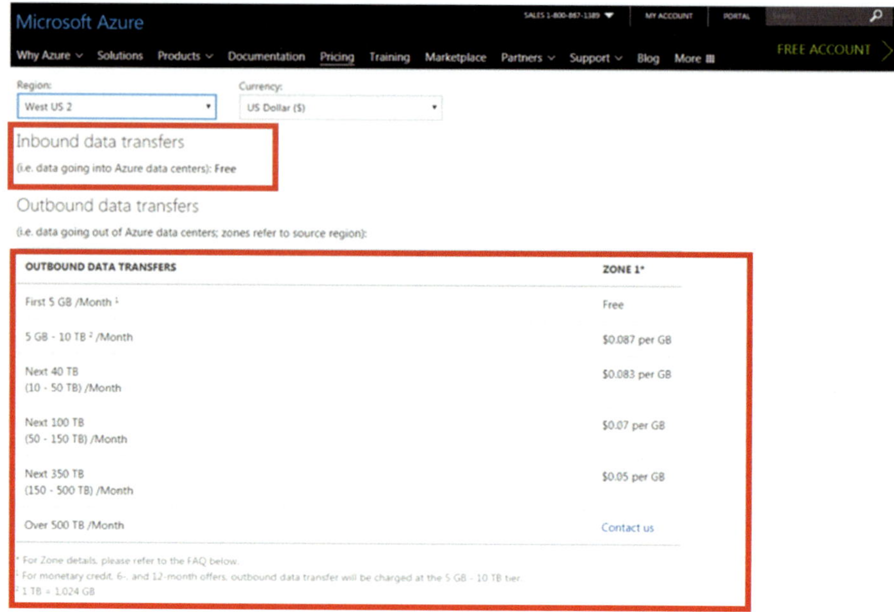

Fig. 1.62 Costs of inbound and outbound data transfers

In the past, COMSOL 5.2 image had 3.6 GB; later for the 5.3 and 5.4 versions the size is 4.3 GB.

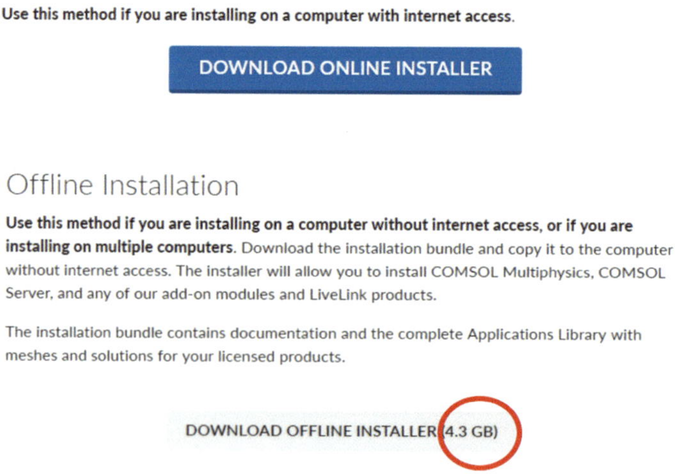

Fig. 1.63 Downloading COMSOL Multiphysics© for off-line installation

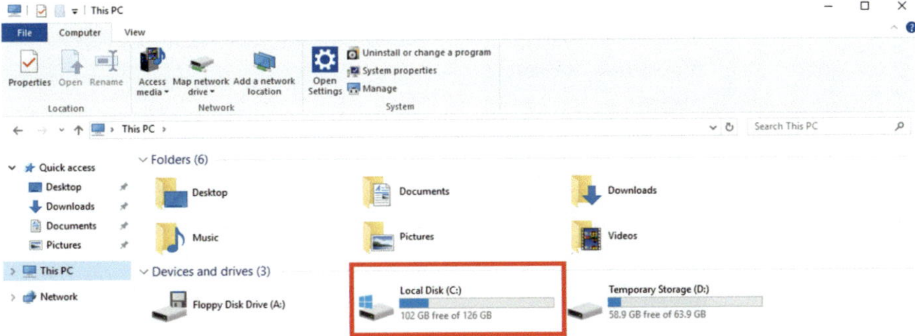

Fig. 1.64 Standard disk size in COMSOL Multiphysics©

This is a small C drive disk. In our experience, applications are designed to create temporary files in the root disk; this is the C drive. It makes no difference whether you install under other drive letter like D. In this case, in particular, the D drive will disappear with all its contents when the VM (Virtual Machine) is stopped and deallocated.

A C-drive with a size of 2 TB is prepared. This is a HDD drive not a SSD drive. The HDD drive is fast formatted. This process guarantees that the 2 TB of disk space will be available but that payment is made only for the portion that is used. If a SSD drive is used or if the HDD is surface formatted, payment for the entire disk is applied from installation in our experience.

The VM has to be in the state stopped and deallocated in order to resize the operating system drive.

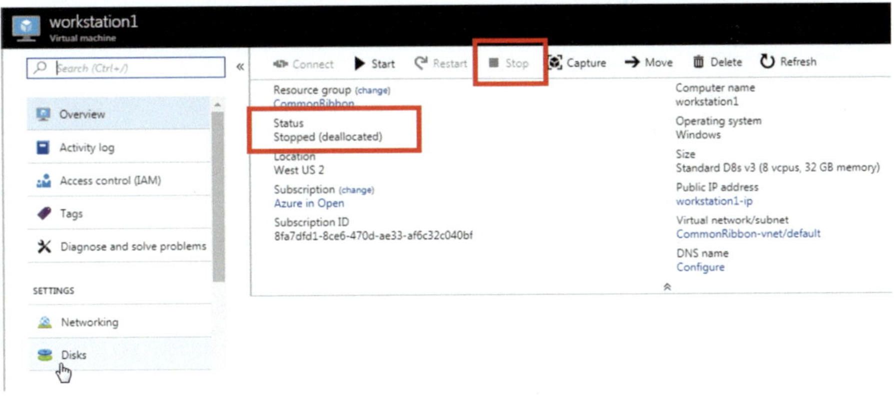

Fig. 1.65 Stop and deallocate VM prior to OS disk resizing

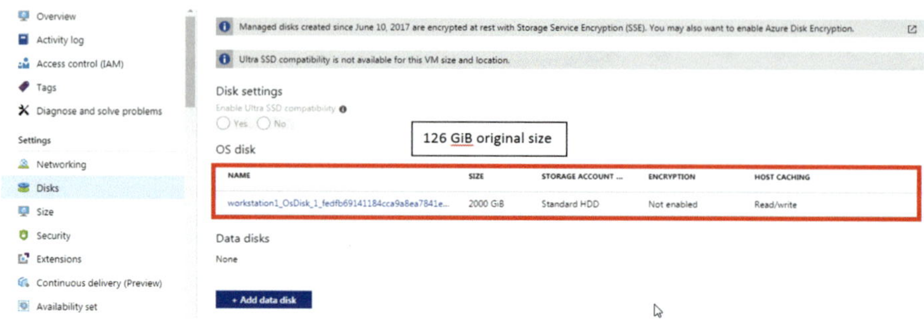

Fig. 1.66 Select OS disk

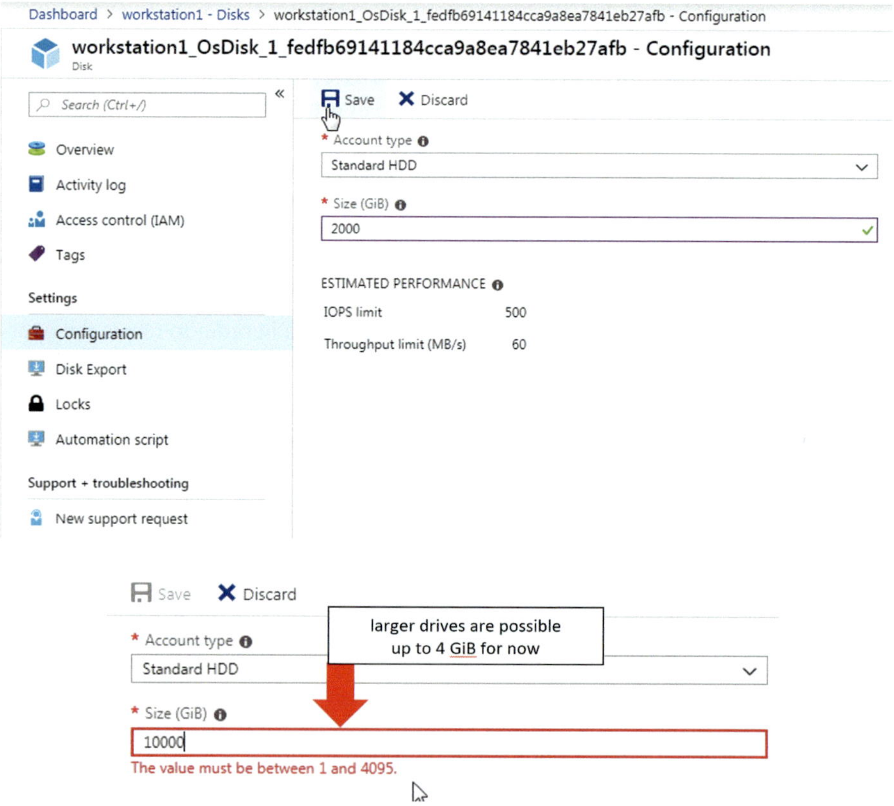

Fig. 1.67 Increase size of the OS disk

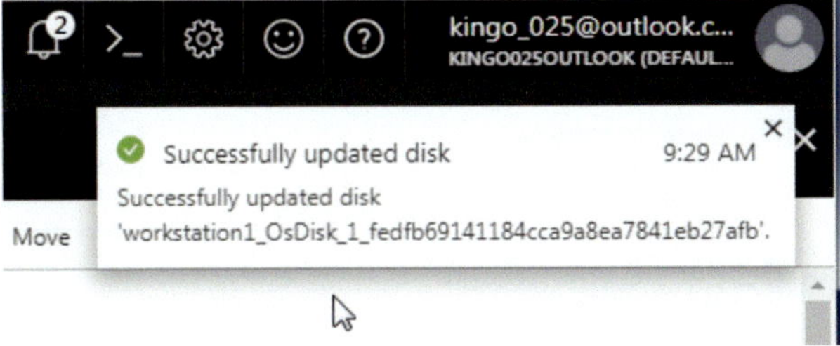

Fig. 1.68   Commissioning confirmation of a larger OS disk

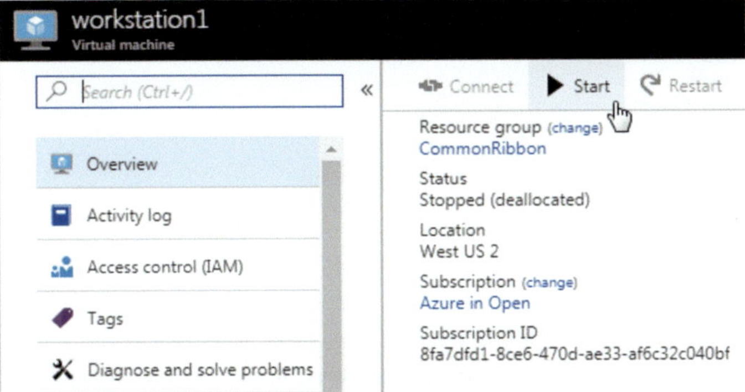

Fig. 1.69   Restart VM with larger OS disk

The VM File Explorer for This PC reveals that the disk size for the C drive remains the same. Some additional steps are needed.

Fig. 1.70   Windows administrative services

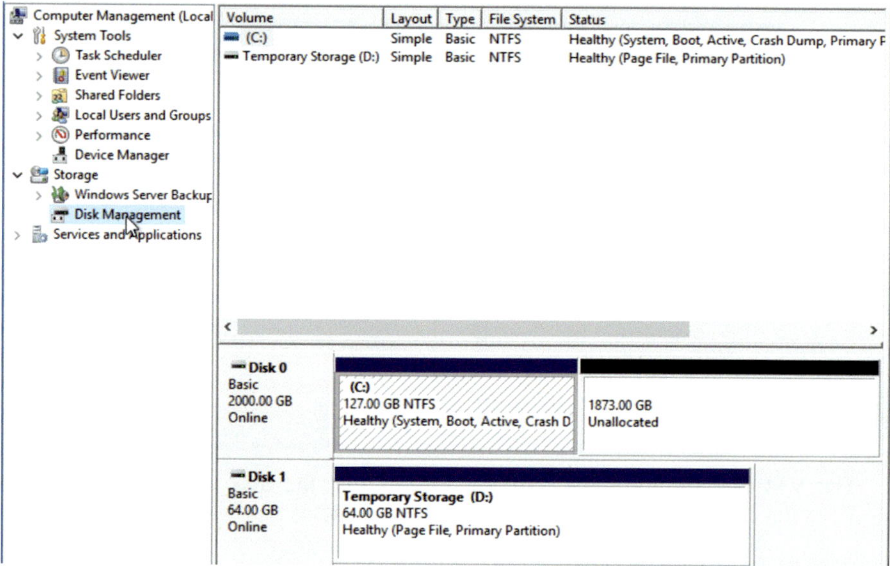

Fig. 1.71 Computer management

Fig. 1.72 Current partition of OS drive

The issue is that the additional space has been reserved but not yet allocated. With the cursor over the (C:) drive proceed to right click and select Extend Volume.

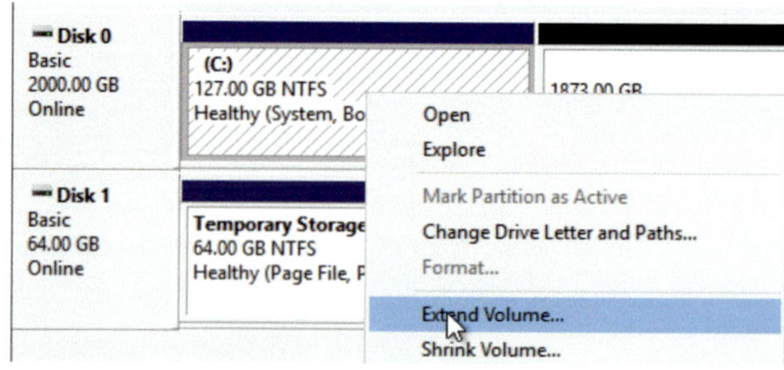

Fig. 1.73 Extending OS drive volume

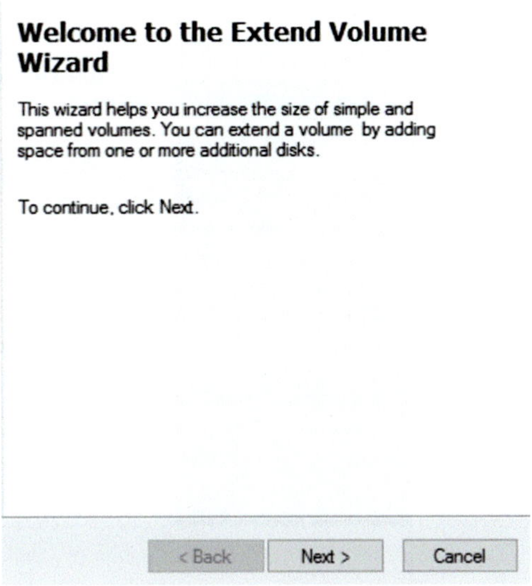

Fig. 1.74 Extend volume
wizard (screen 1)

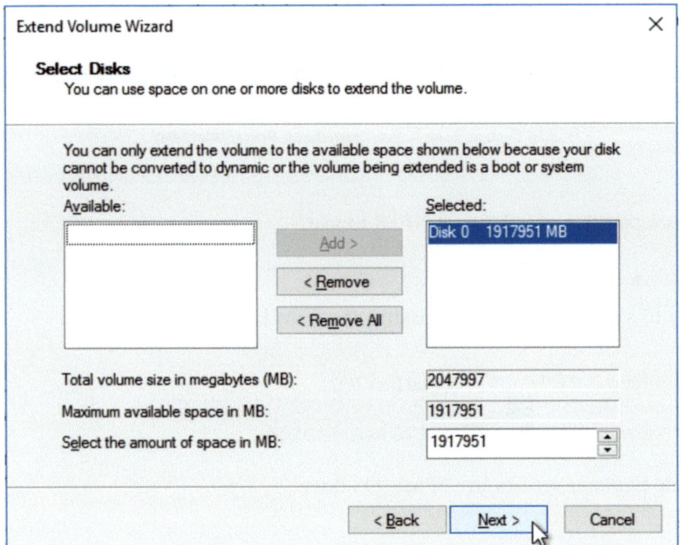

Fig. 1.75 Extend volume wizard (screen 2)

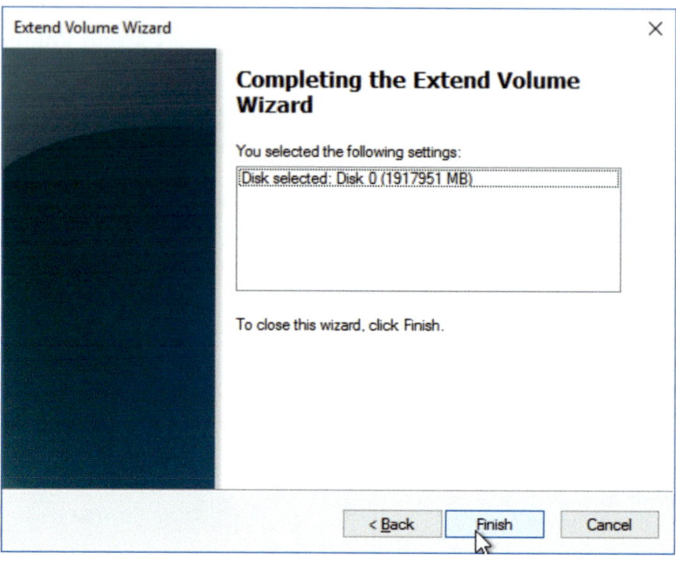

Fig. 1.76 Extend volume wizard (screen 3)

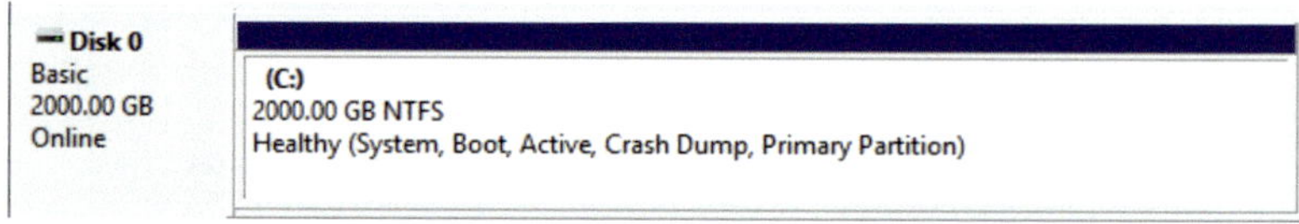

Fig. 1.77 New partition of OS disk in NFTS format

Proceed to verify the new size using the File Explorer ( )

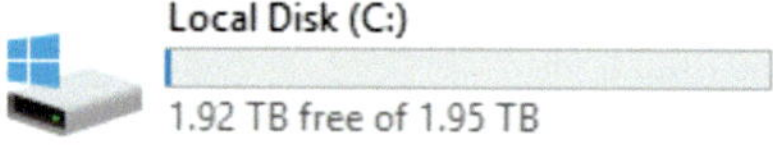

Fig. 1.78 File Explorer verifies size of new OS drive

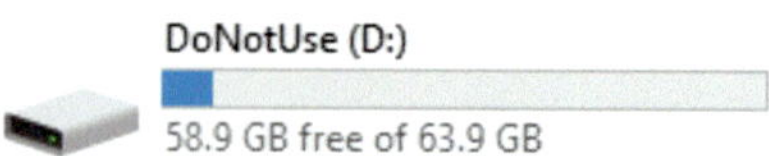

Fig. 1.79 Marked DoNotUse to prevent information loss

The image of COMSOL is in the VM desktop. In previous versions of the Windows, a program like MagicIsoCD/DVD Manager would be needed. In the server version of Windows, the mounting of images is a service that is available for immediate use.

Proceed to click on the COMSOL image on the desktop and right click on it. Select the Mount label from the menu.

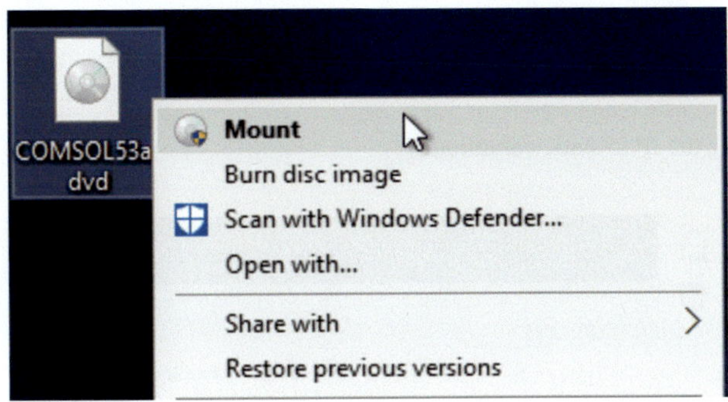

Fig. 1.80 Mounting software disk image

The File Explorer for the This PC will show the COMSOL image mounted on drive E.

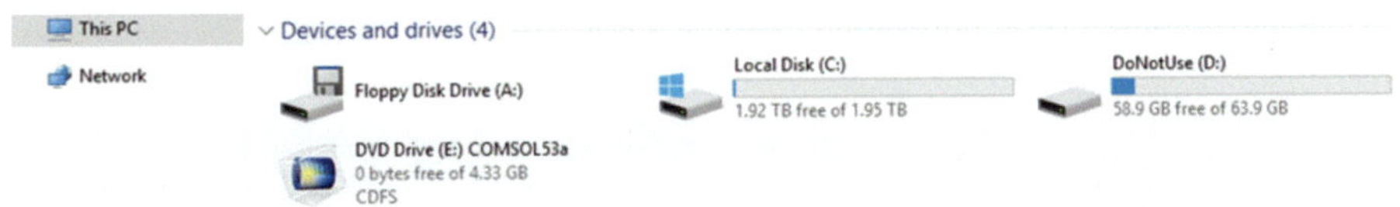

Fig. 1.81 COMSOL Multiphysics© image mounted in VM

Depending on how licensing is handled, a fixed IP address (static IP address) may be needed by some programs. This means that in addition to a valid virtual private network (VPN) connection, other restrictions are typically applied.

In your current machine, click on the current Public IP address and make select to make the address static.

Fig. 1.82 Dynamic address on Azure Dashboard

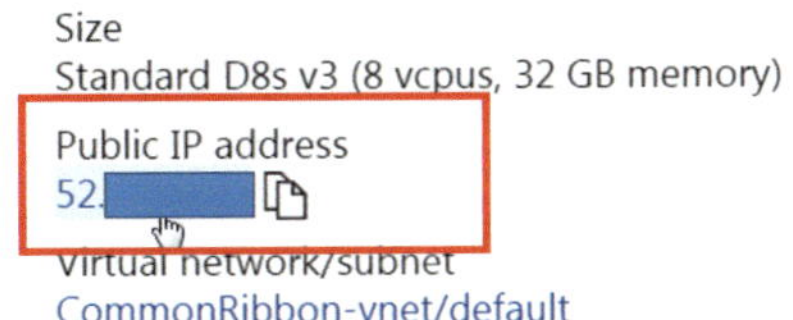

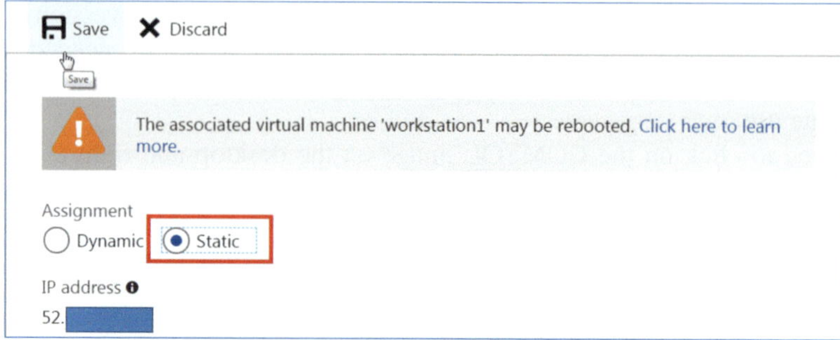

Fig. 1.83 Setting the IP to static condition

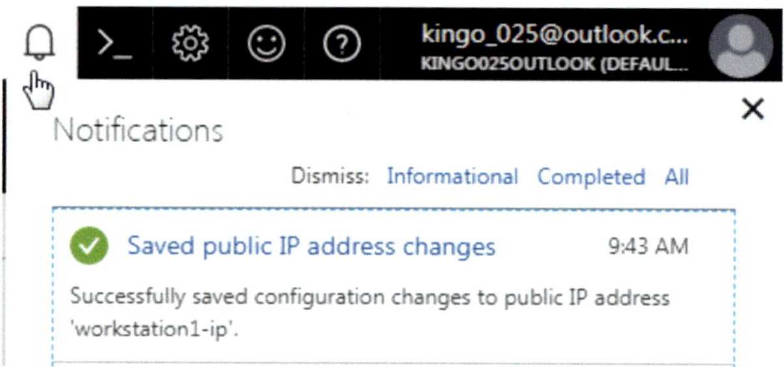

Fig. 1.84 Confirmation of static IP address

These would incur in additional charges a few dollars per month.

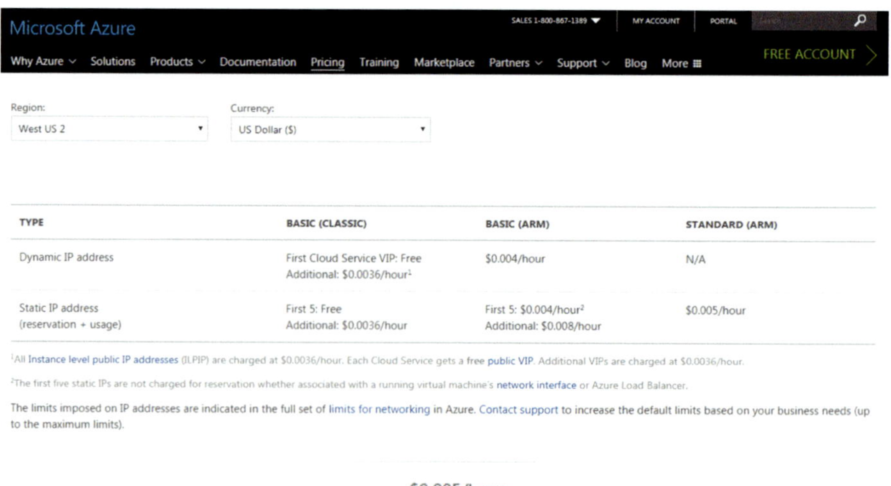

TYPE	BASIC (CLASSIC)	BASIC (ARM)	STANDARD (ARM)
Dynamic IP address	First Cloud Service VIP: Free Additional: $0.0036/hour[1]	$0.004/hour	N/A
Static IP address (reservation + usage)	First 5: Free Additional: $0.0036/hour	First 5: $0.004/hour[2] Additional: $0.008/hour	$0.005/hour

[1] All Instance level public IP addresses (ILPIP) are charged at $0.0036/hour. Each Cloud Service gets a free public VIP. Additional VIPs are charged at $0.0036/hour.

[2] The first five static IPs are not charged for reservation whether associated with a running virtual machine's network interface or Azure Load Balancer.

The limits imposed on IP addresses are indicated in the full set of limits for networking in Azure. Contact support to increase the default limits based on your business needs (up to the maximum limits).

Fig. 1.85 Cost of static IP address

1.20   Installation of COMSOL Multiphysics® 5.4

There are a number of licensing options for COMSOL Multiphysics®, the best one is probably the one that the user can afford and gets the desired results. The most recent specific options can be located in the COMSOL Multiphysics® website.

Fig. 1.86 COMSOL Multiphysics® licensing options

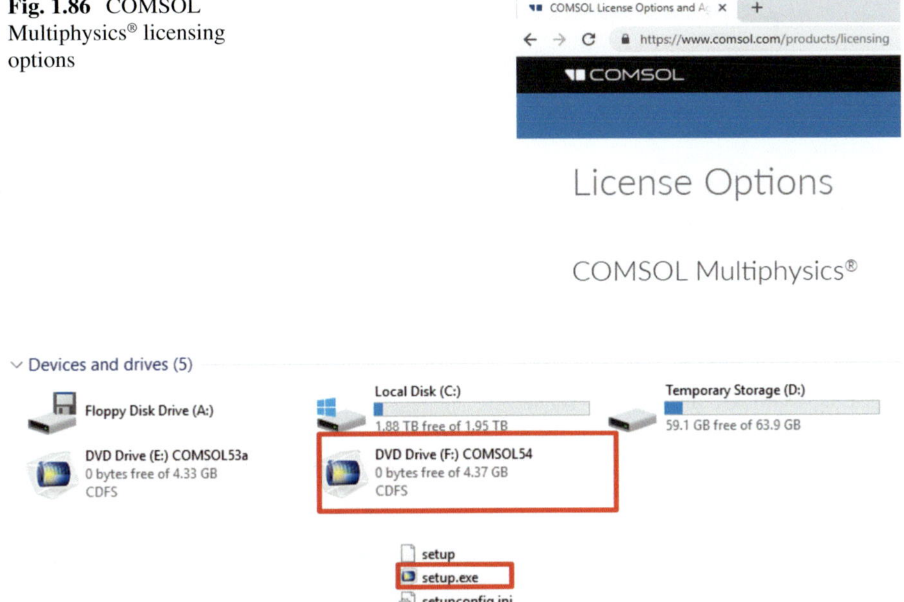

Fig. 1.87 Mounted DVD image and setup

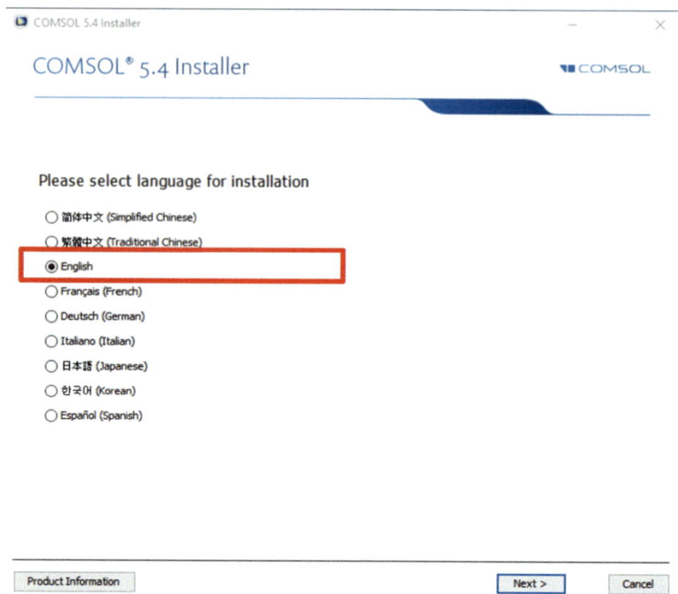

Fig. 1.88 COMSOL installation—language

Fig. 1.89 COMSOL
installation—version

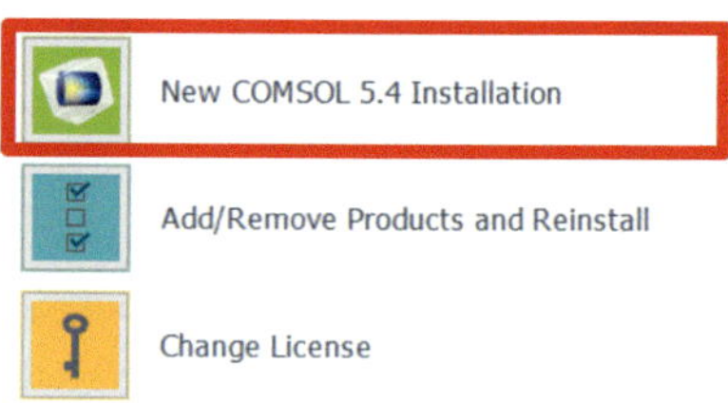

This is the most important page where the license number is entered; without that information nothing else works.

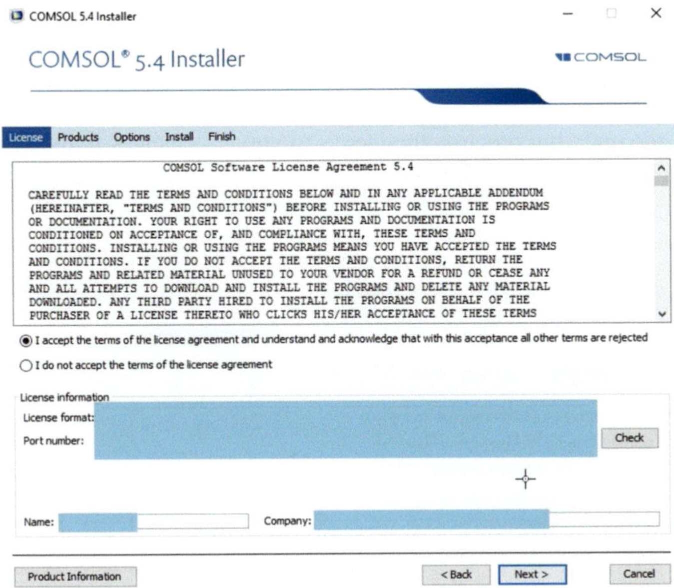

Fig. 1.90 COMSOL installation—license credentials

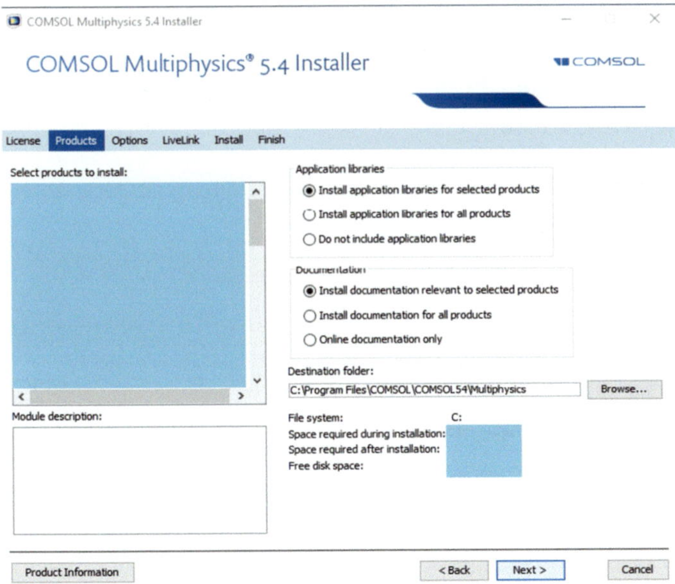

Fig. 1.91 COMSOL installation—module selection

Even if the Destination folder were created in another drive, the temporary files will be created on the C drive in our experience and without execution of other changes. To change those addresses requires an additional process without guarantee that eventually will not create performance problems.

Control the updating process; the idea is to prevent changes in the system in the middle of a production period.

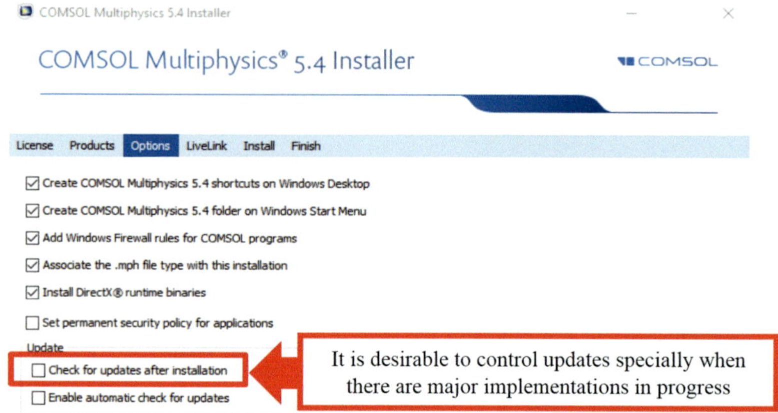

Fig. 1.92 COMSOL installation—update installation

Fig. 1.93 COMSOL installation—Livelink

There are no other programs installed at this point, the LiveLink are not used in this implementation.

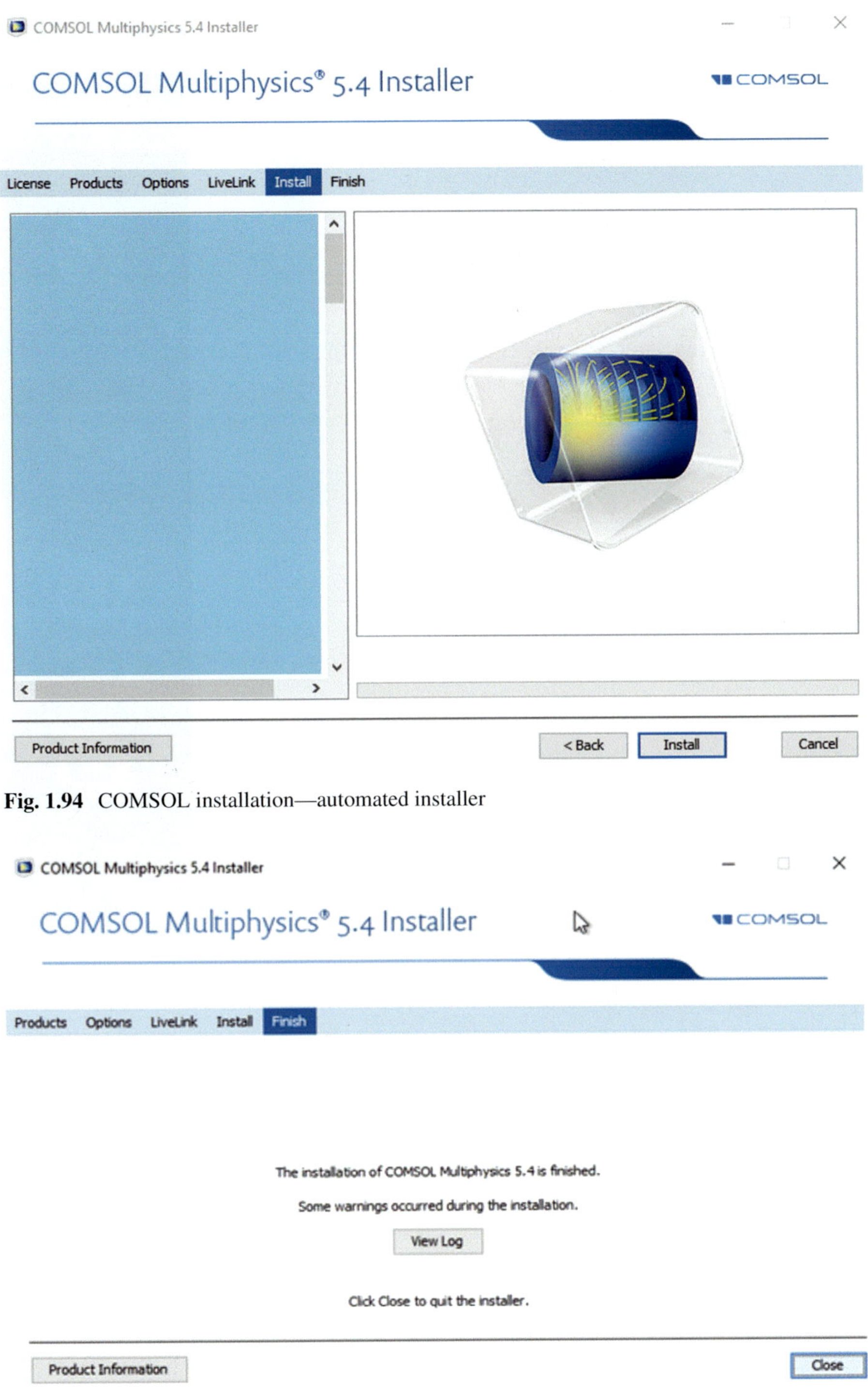

Fig. 1.94 COMSOL installation—automated installer

Fig. 1.95 COMSOL installation—installation log

Fig. 1.96 COMSOL
desktop icon

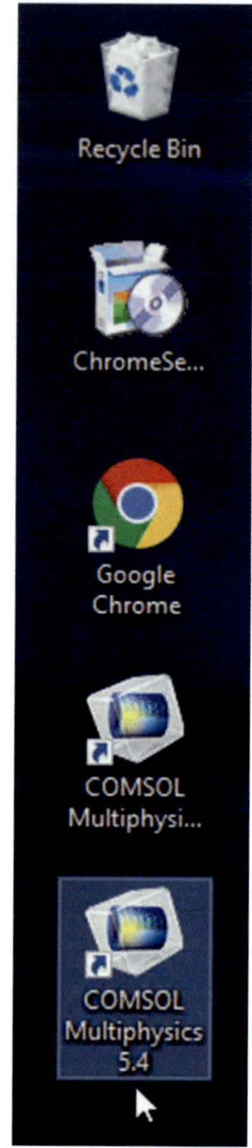

Fig. 1.97 Three-dimensional graphics warning (this warning has appeared in version 5.3a, not for 5.4 so far)

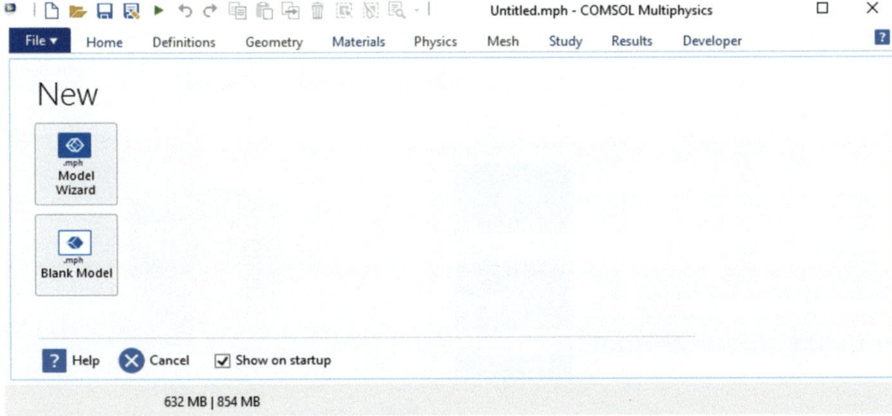

Fig. 1.98 Run model wizard to verify installation

Start the COMSOL Multiphysics© 5.4 and verify it is working.

Fig. 1.99 Restart
workstation

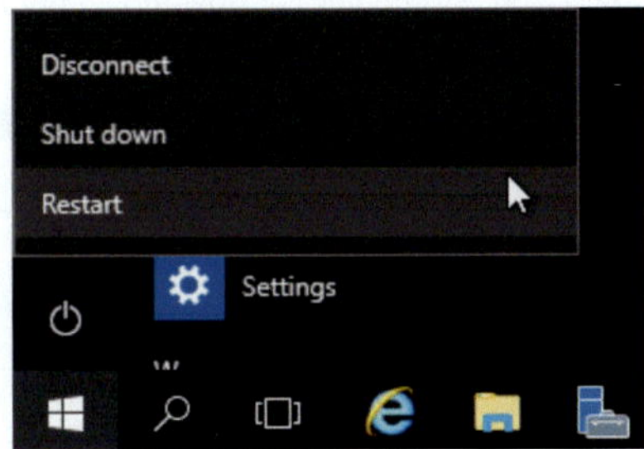

The first installation is completed. Restart the VM and review if all information and programs run in this COMSOL Multiphysics©.

The program should start normally.

1.21   COMSOL Example: Computer Fan Modeling

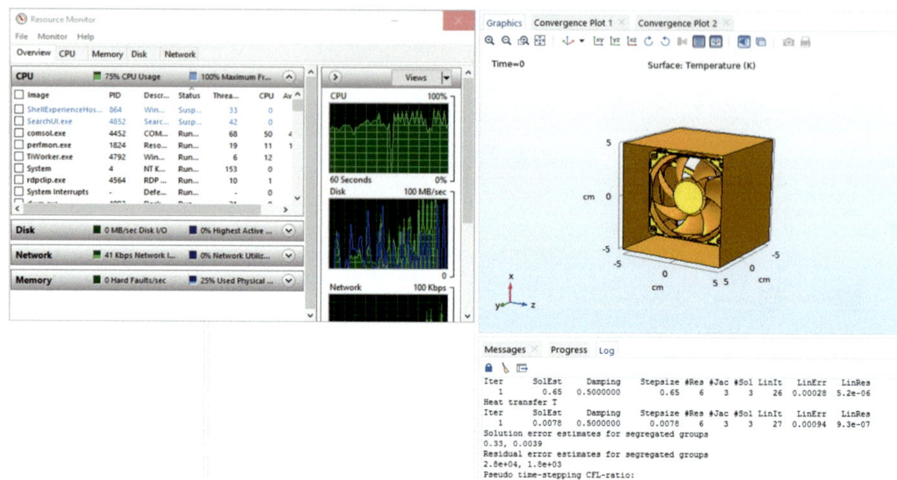

Fig. 1.100 Computer fan flow model not rotating—high CPU usage

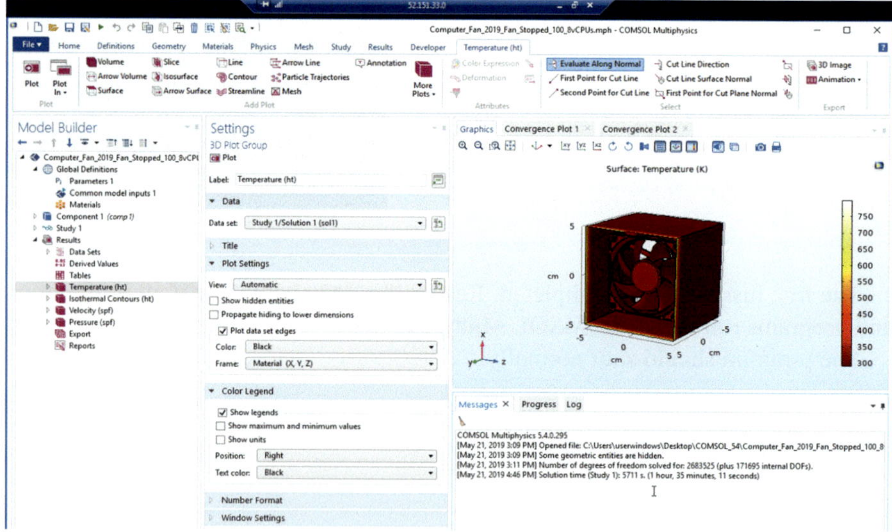

Fig. 1.101 Computer flow model not rotating—simplest laminar model

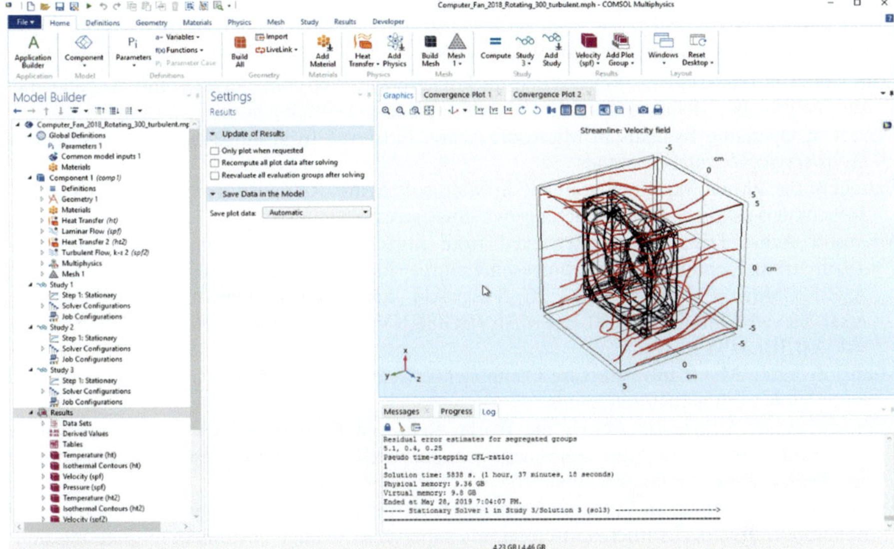

Fig. 1.102 Processing time not rotating—simplest laminar model

Fig. 1.103 Computer flow model rotating—simplest laminar model

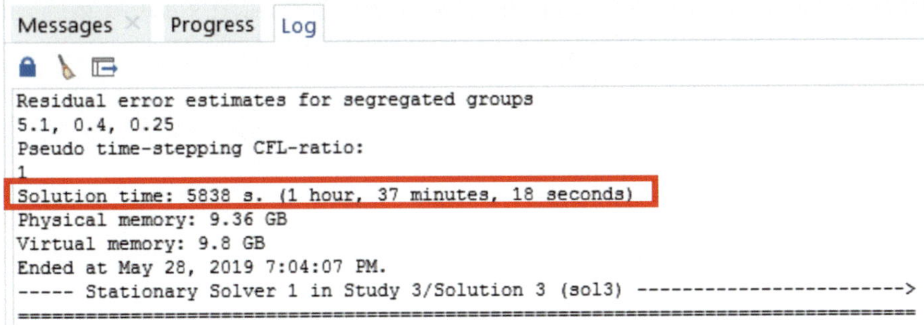

Fig. 1.104 Processing time rotating—simplest laminar model

This example worked in a Windows Azure VM (D8s v3 VM, 8 vcpus, 32 GB memory). Other smaller systems may render processing times that are much larger. Consequently, designs with increasing detailed geometric and physical features call for appropriate (though not necessarily expensive) computational facilities to generate results in suitable amounts of time.

Further Readings

Azure portal. Retrieved from https://portal.azure.com

COMSOL. Multiphysics© v.5.3a. www.comsol.com. COMSOL AB, Stockholm, Sweden.

Cloud Ranger (2018). Retrieved from https://www.youtube.com/watch?v=wlIKSNbS-Gg&list=PLPie5drlGW253H0BX_MCUnfLYE-FsxnW9

Data stored in data centers. Retrieved from https://www.cisco.com/c/en/us/solutions/collateral/service-provider/global-cloud-index-gci/white-paper-c11-738085.html

Extract of Operating Systems in Microsoft Azure. Retrieved from https://azure.microsoft.com/en-us/services/virtual-machines

Extract of the Virtual Machines available in Microsoft Azure. Retrieved from https://docs.microsoft.com/en-us/azure/virtual-machines/windows/sizes

Microsoft Azure Data Center. Retrieved from https://www.google.com/search?q=microsoft-+azure+data+center+picture&tbm=isch&source=iu&ictx=1&fir=sU-ksLJHLJeKLM%253A%252CbMne3PbPxfKAKM%252C_&usg=AI4_-kRfmAEZcyMDpwoFJxjHWgb34Aj9wA&sa=X&ved=2ahUKEwiA-8LEgPnfAhVEGt8KHVdSATUQ9QEwAHoECAUQBA#imgrc=sU-ksLJHLJeKLM:

Microsoft Azure Major Infrastructure Components. Retrieved from https://azure.microsoft.com/en-gb/global-infrastructure/regions

Microsoft Azure Utilization Across the World. Retrieved from https://www.zdnet.com/article/microsoft-were-adding-7000-azure-iaas-users-per-week/   https://www.wsj.com/articles/microsofts-cloud-continues-to-fuel-growth-1532032095

Microsoft Azure Utilization by Large Companies. Retrieved from https://www.forbes.com/sites/bobevans1/2018/02/28/microsoft-cloud-hits-superscale-as-huge-customers-migrate-mission-critical-sap-workloads-to-azure/#85026aa5993c

Microsoft Azure Website. Retrieved from https://azure.microsoft.com/en-us/

Microsoft Refocusing Business on Cloud Infrastructure. Retrieved from https://www.wsj.com/articles/microsoft-diminishes-windows-role-in-cloud-focused-reorganization-1522335763

Total size of public cloud computing market from 2008 to 2020 (in billions U. S. dollars). Retrieved from https://www.statista.com/statistics/510350/worldwide-public-cloud-computing/

Chapter 2
Generating a Model

2.1 Building a Model in COMSOL Multiphysics©

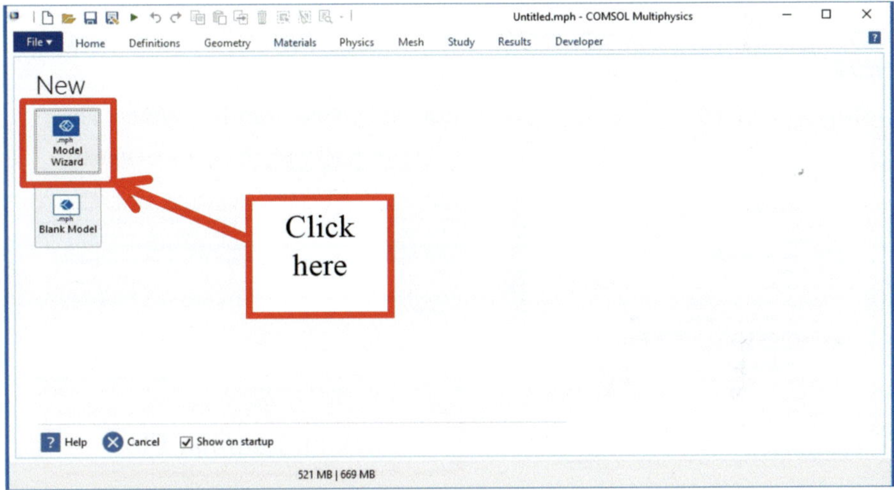

Fig. 2.1 Model wizard

© Springer Nature Switzerland AG 2020

B. A. Stradi-Granados, *Cloud Computing for Engineering Applications*,

https://doi.org/10.1007/978-3-030-40445-1_2

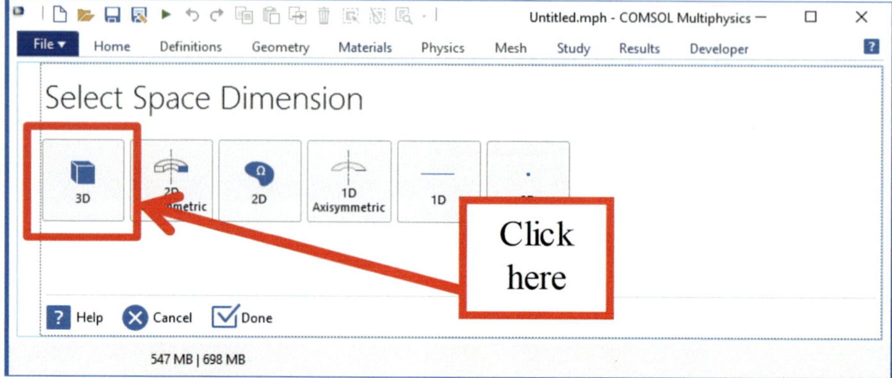

Fig. 2.2 Three-dimensional model setup

2.2   Defining the Physics Involved

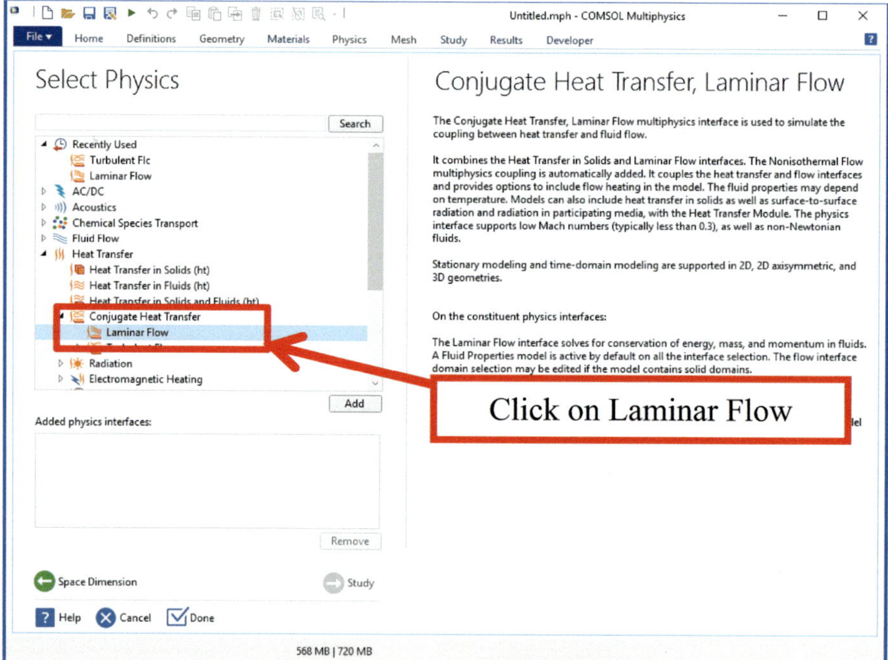

Fig. 2.3 Heat transfer model selection

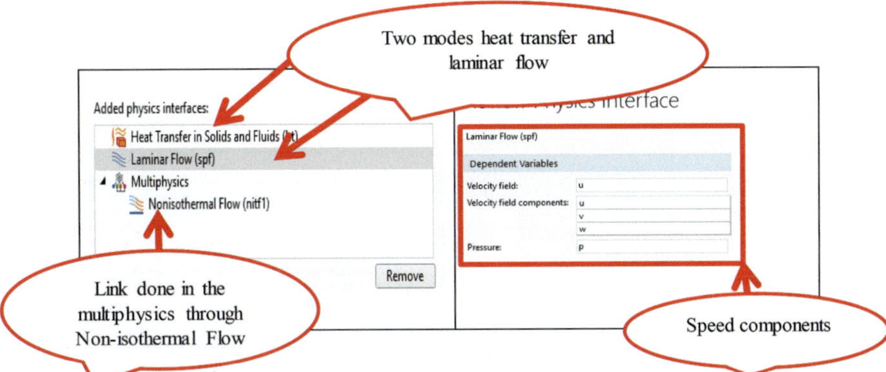

Fig. 2.4 Add physics

Fig. 2.5 Heat transfer in solids—Variable T—Create study

2.3   Defining the Study

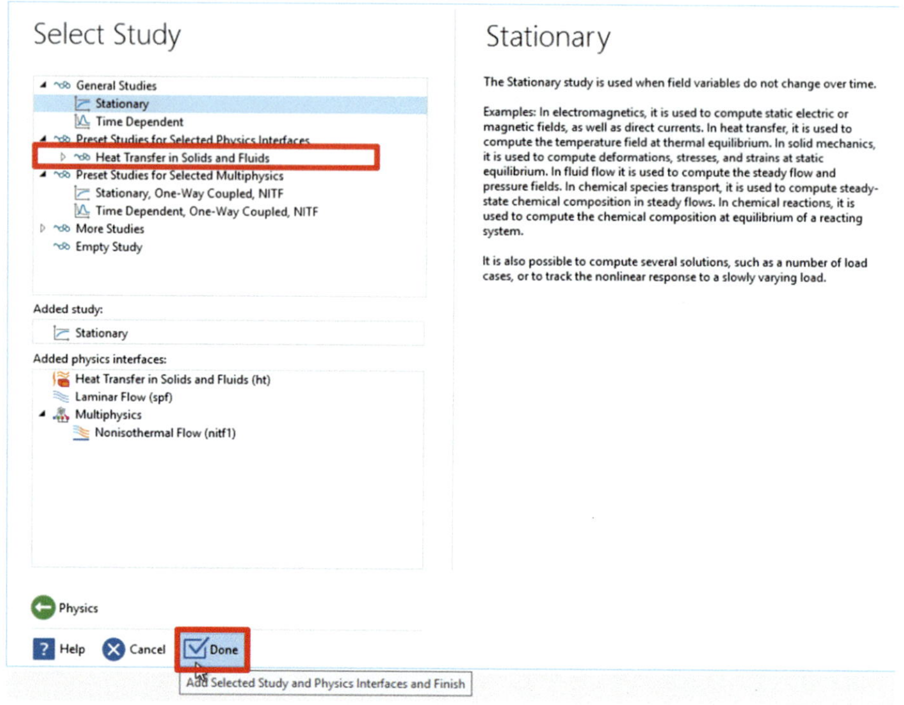

Fig. 2.6   Select study, Stationary

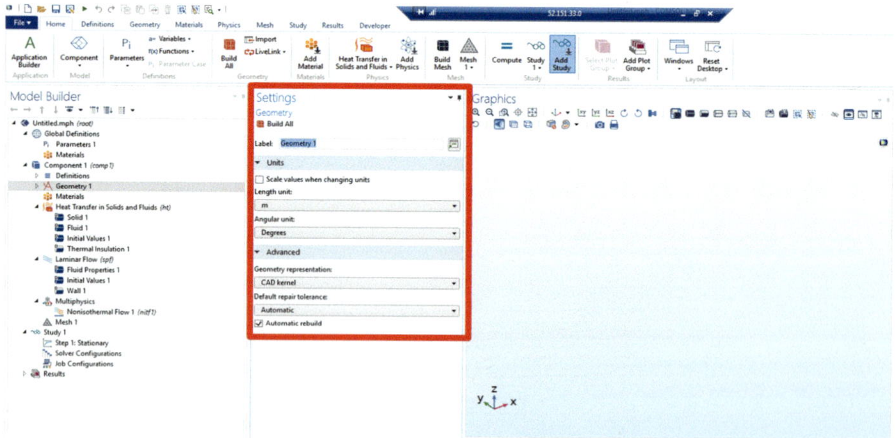

Fig. 2.7   Model Builder

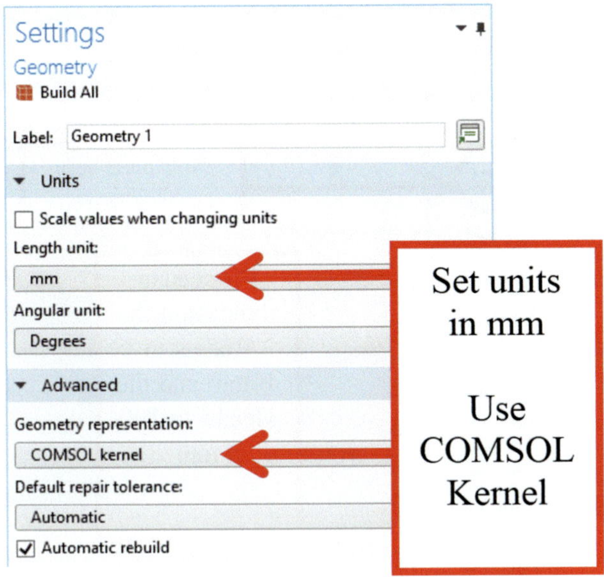

Fig. 2.8 Geometrical unit set

2.4 Building the Geometry

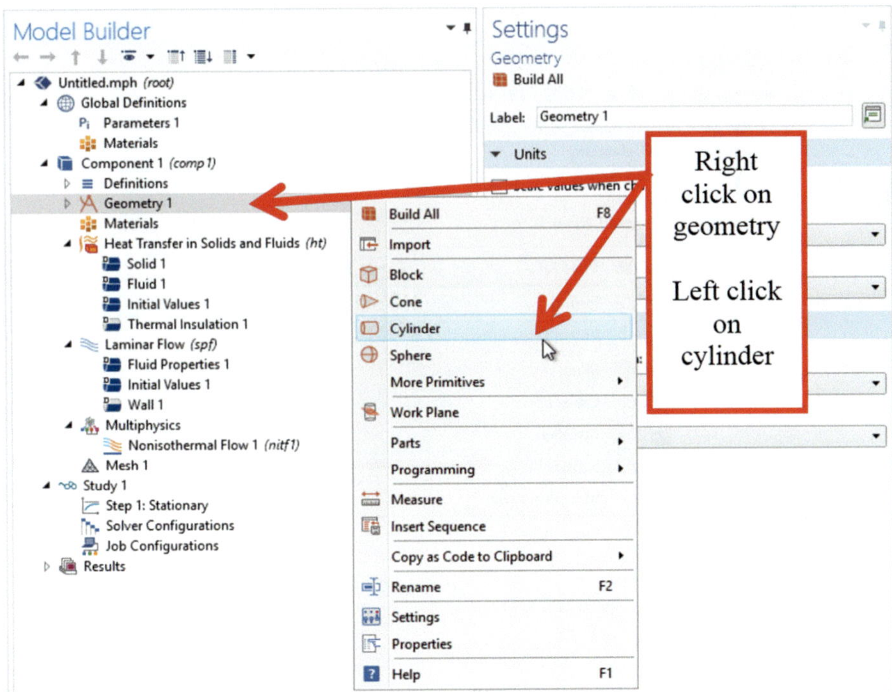

Fig. 2.9 Geometry

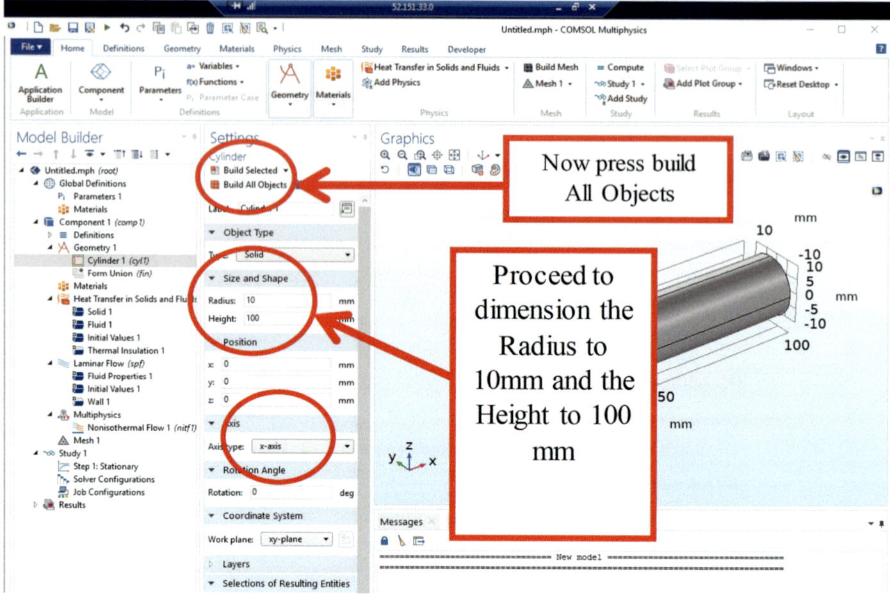

Fig. 2.10   Create Cylinder—Position, Axis

Let us make a sphere by a similar procedure:

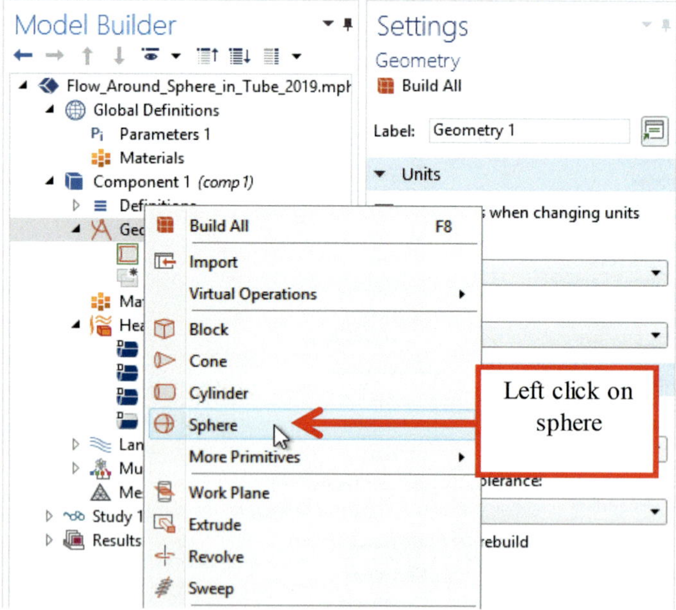

Fig. 2.11   Create a sphere

The sphere has a radius of 5 mm and is located at 50 mm on the *x*-axis.

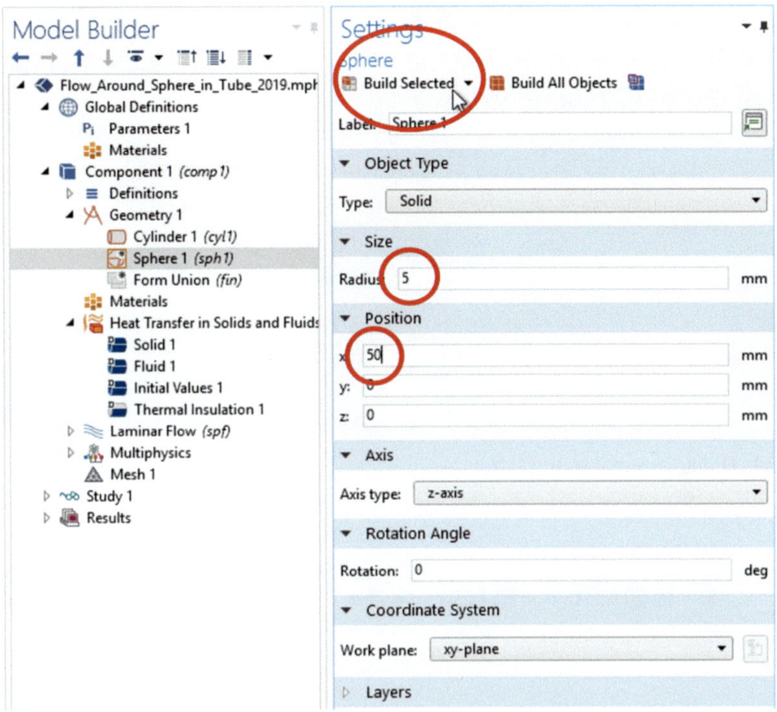

Fig. 2.12 Dimensions and location of sphere

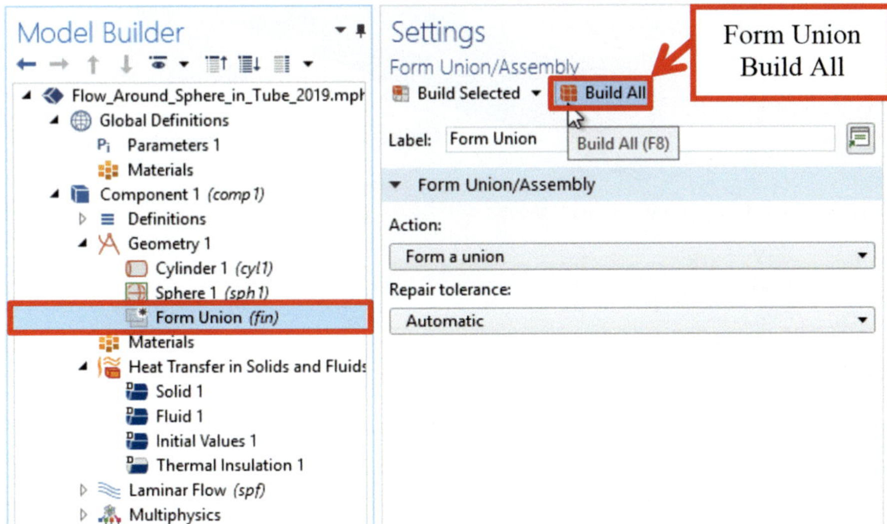

Fig. 2.13 Forming a single body

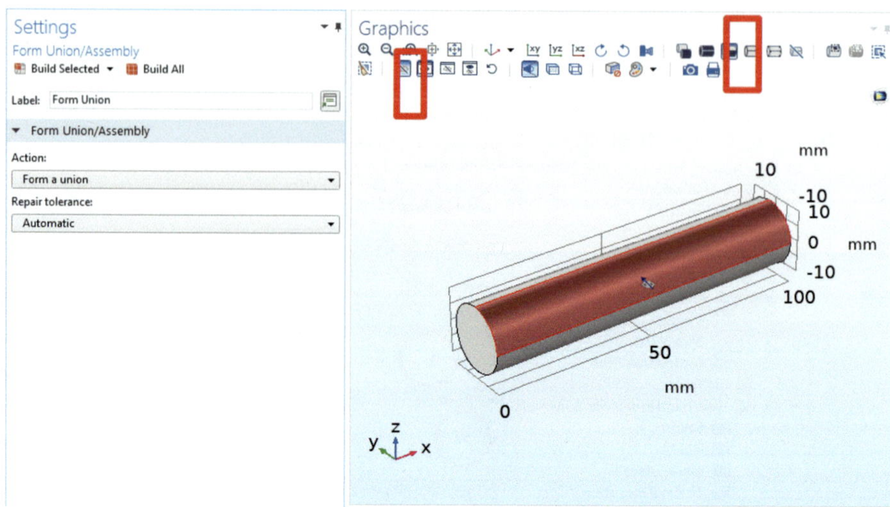

Fig. 2.14 Hiding external wall

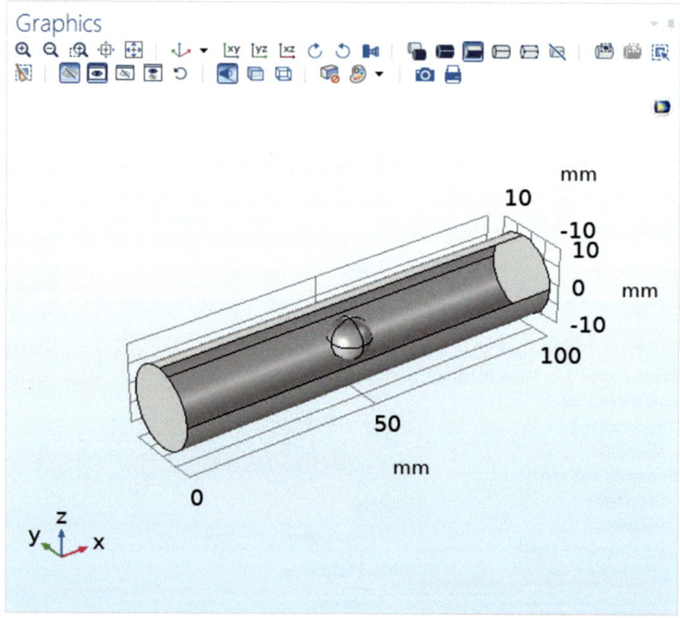

Fig. 2.15 Visualizing interior elements

2.5   Selecting the Materials

Move to materials and select water.

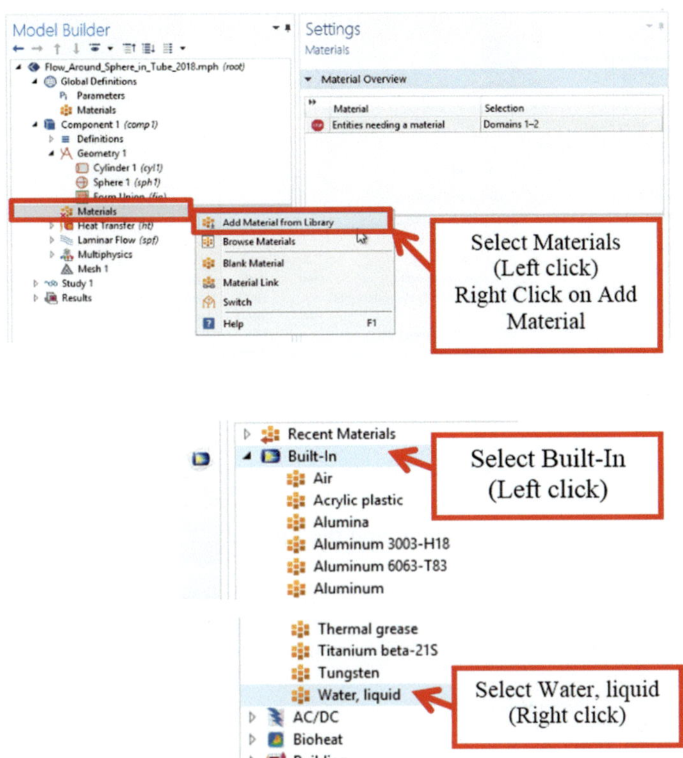

Fig. 2.16   Select Materials—water

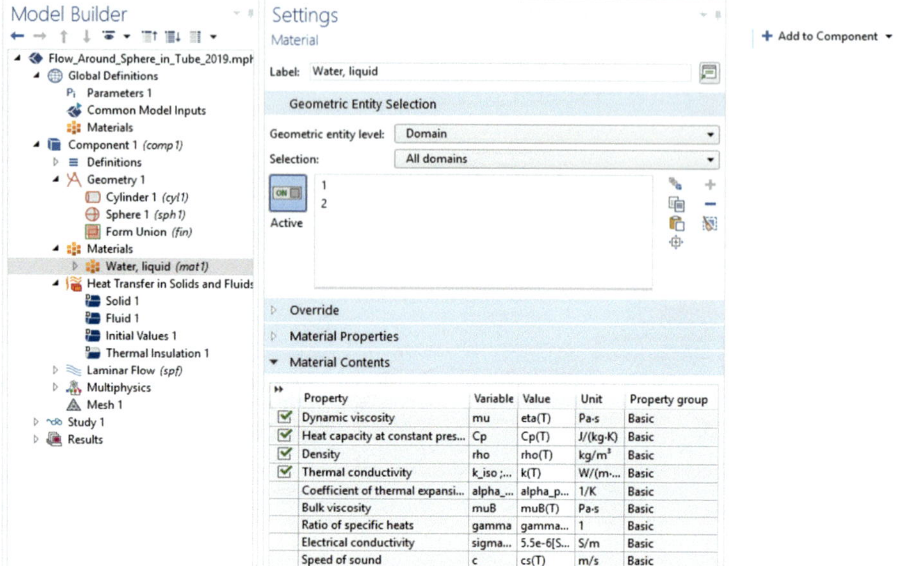

Fig. 2.17 Properties of water in Comsol's database

Need to add copper for the sphere and the tube.

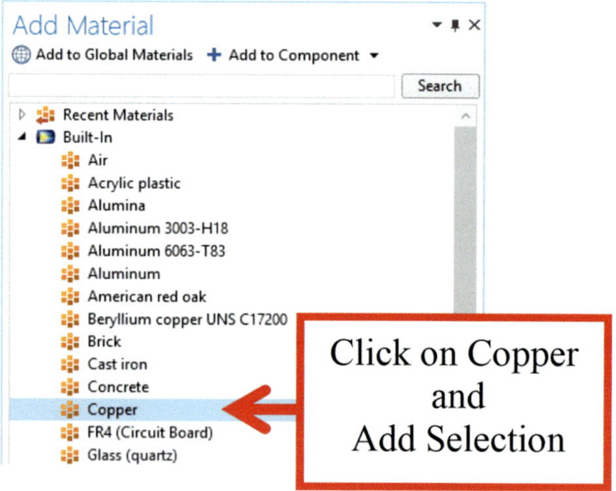

Fig. 2.18 Add Selection—copper

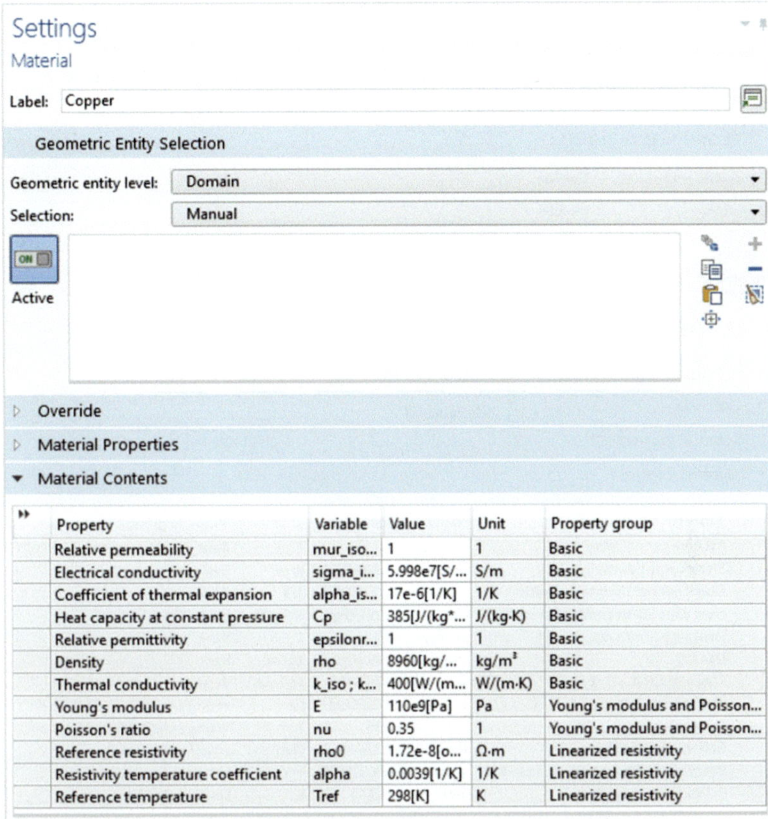

Fig. 2.19 Copper properties in Comsol's database

Water is currently assigned to both domains; it is necessary to correct that and assign copper to the ball (second domain).

Click on the notepad and on the selection line type 2.

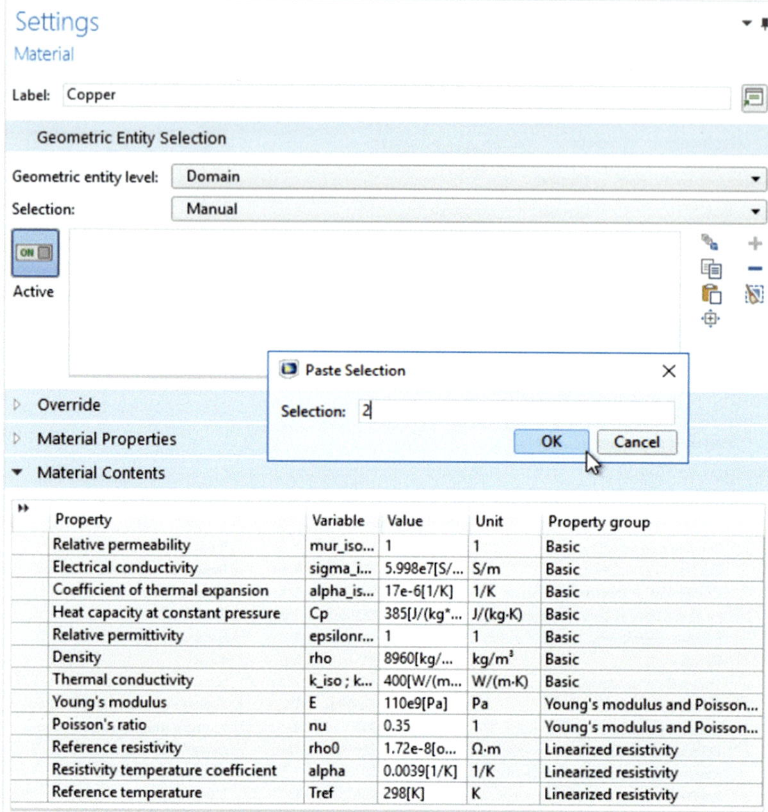

Fig. 2.20 Copper sphere domain and properties

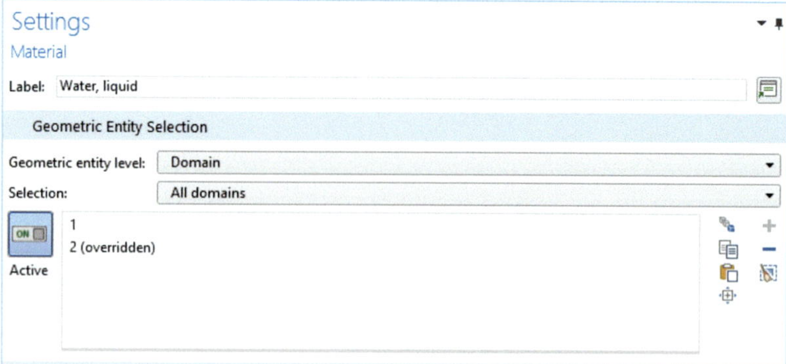

Fig. 2.21 Corrected assignment of materials to each domain

Leave the heat transfer for both domains.

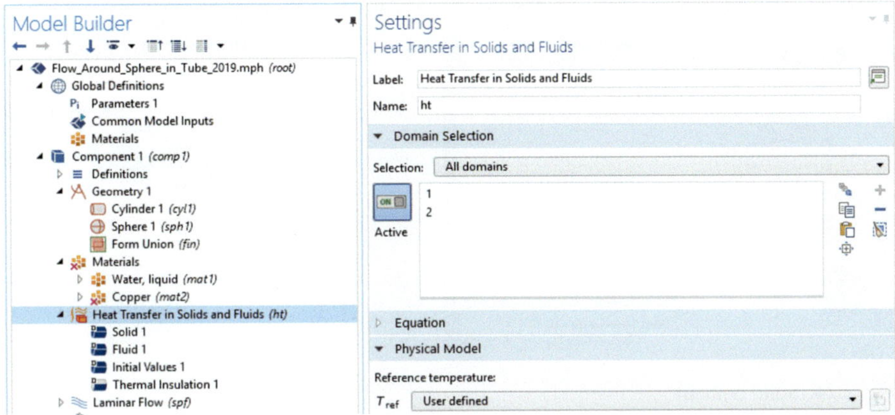

Fig. 2.22   Heat transfer physics on both domains

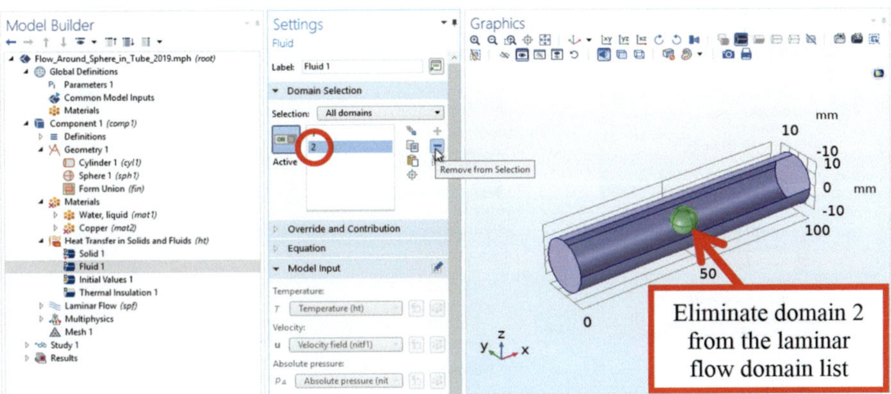

Fig. 2.23   Domain 1

These selections have set the stage for domains 1 and 2 whereby there is heat transfer, but only for domain 1 there is laminar fluid flow.

The study is a steady-state analysis and consequently boundary conditions are needed. If the problem is time-dependent, initial conditions are needed.

There is a solid domain which is the sphere; this is domain 2.

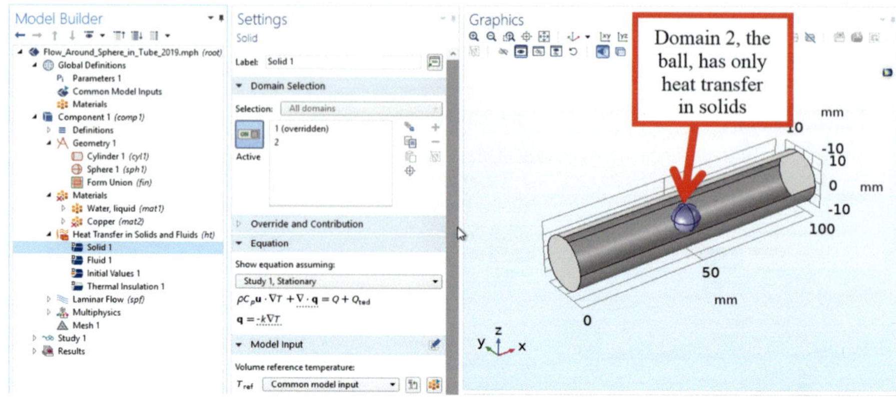

Fig. 2.24 Solid domain

Modify setting so that laminar flow (spf) is active only on domain 1, the cylinder.

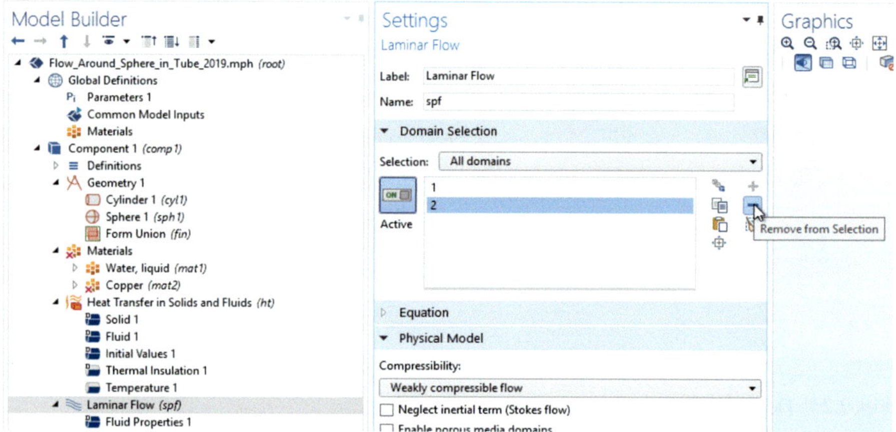

Fig. 2.25 Fluid domain

Verify that solid 1 is active only on domain 2, the sphere.

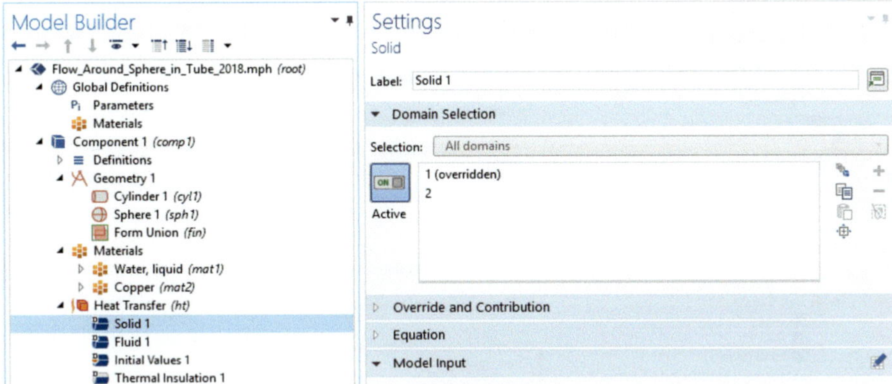

Fig. 2.26 Overriding default setup for solid domain

Initial values stay the same as there are no time-dependent variables.

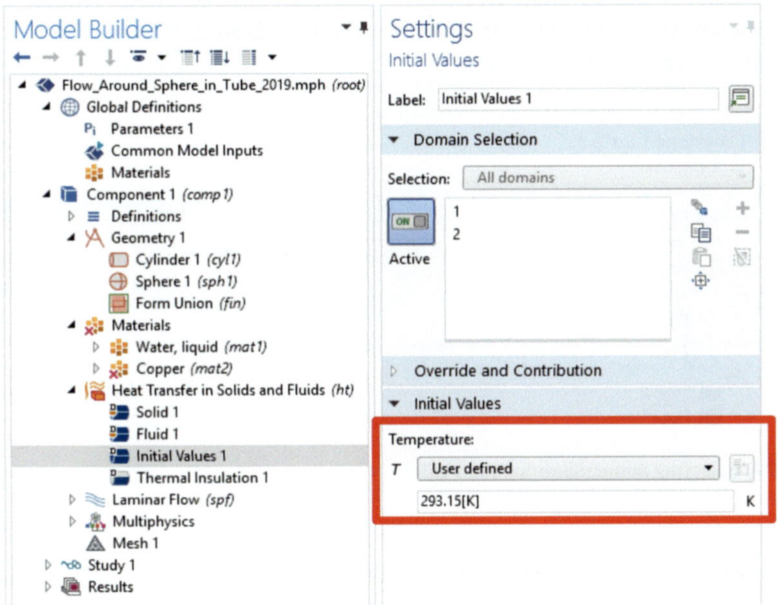

Fig. 2.27 Initial values for the temperatures of the problem

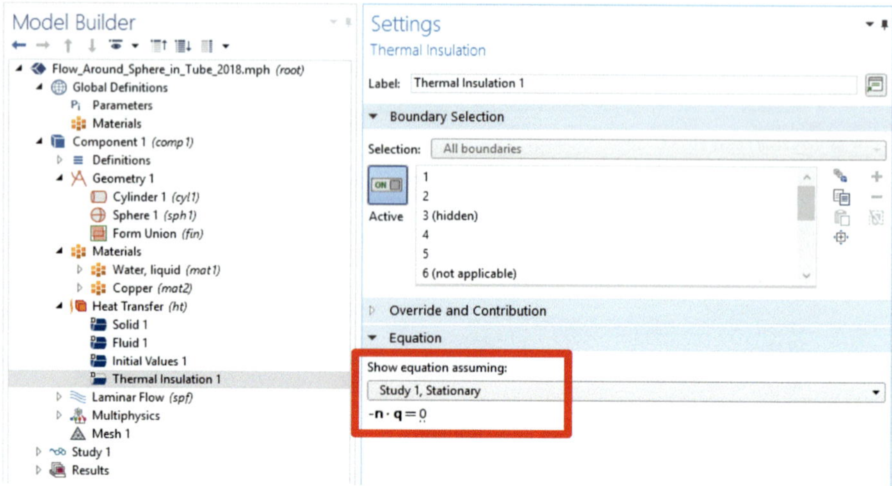

Fig. 2.28 There is no heat flow across the boundary (adiabatic)

The steady-state boundary conditions for each mode are entered by right-clicking on the Heat Transfer (ht) selection.

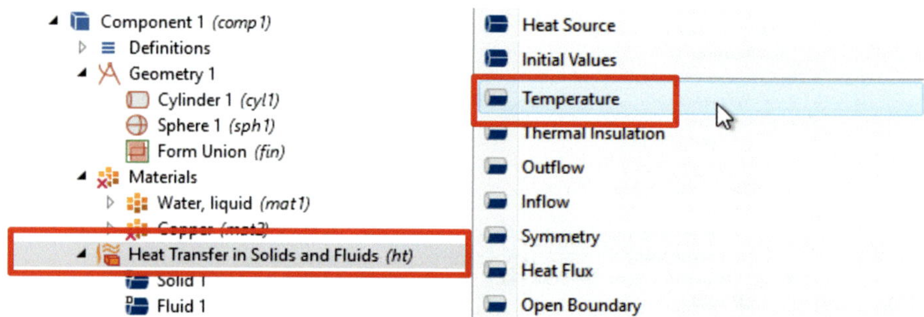

Fig. 2.29 Selecting temperature for solid ball

In order to make the problem more interesting, select only one leaf (or section) on the surface of the sphere. This makes the problem non-axisymmetric.

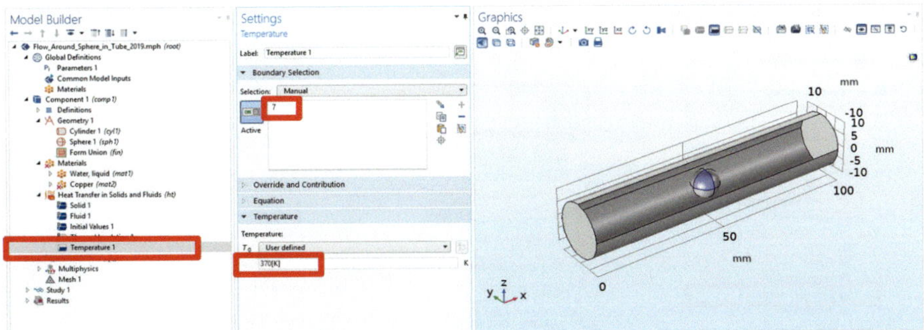

Fig. 2.30 Temperature assignment to solid body

Now the fluid flow has to be described on the Inlet and the Outlet.

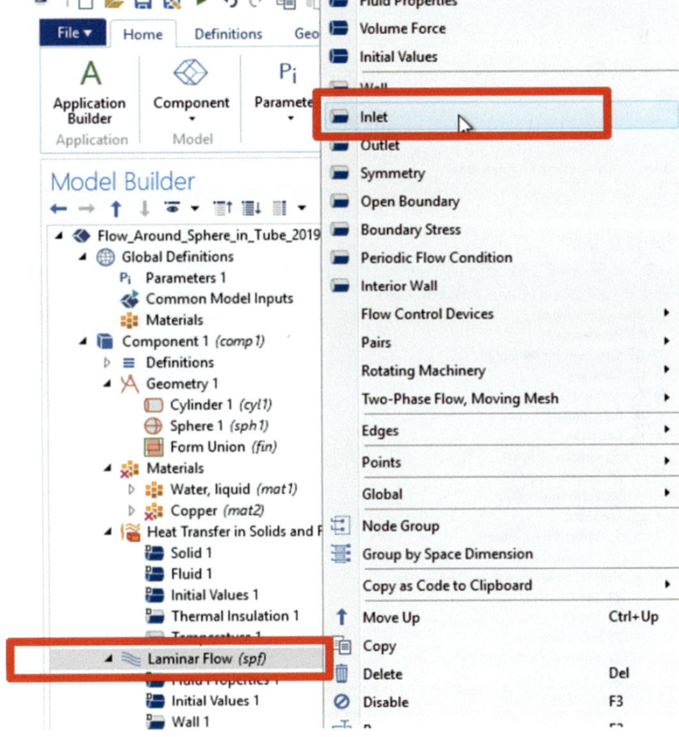

Fig. 2.31 Laminar flow boundary condition on flow

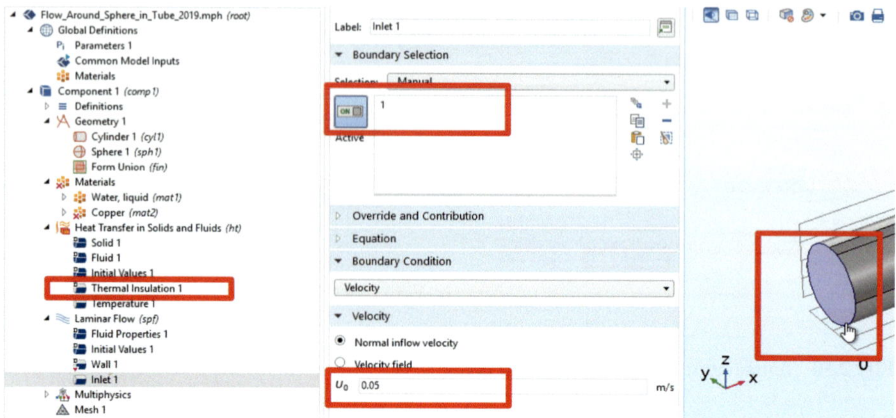

Fig. 2.32 Boundary conditions for entry surface and velocity

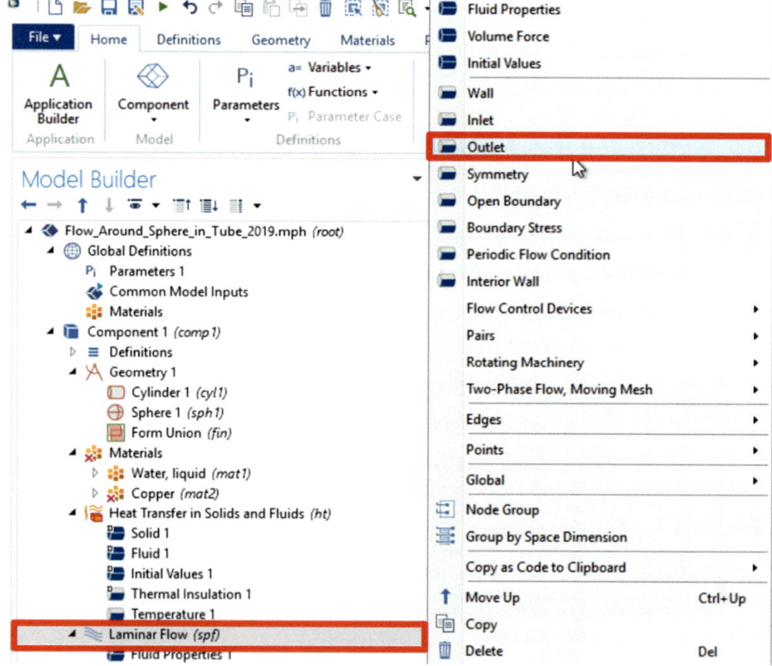

Fig. 2.33 Exit condition given as a pressure condition

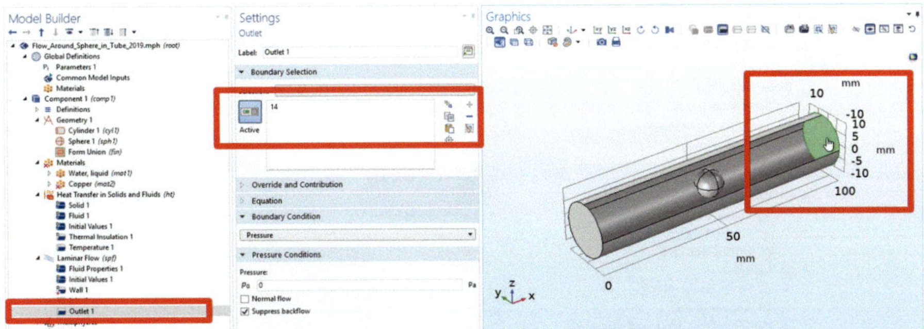

Fig. 2.34 Exit pressure condition

Open the multiphysics menu and confirm that nonisothermal flow (nitf1) is active on domain 1 (the tube).

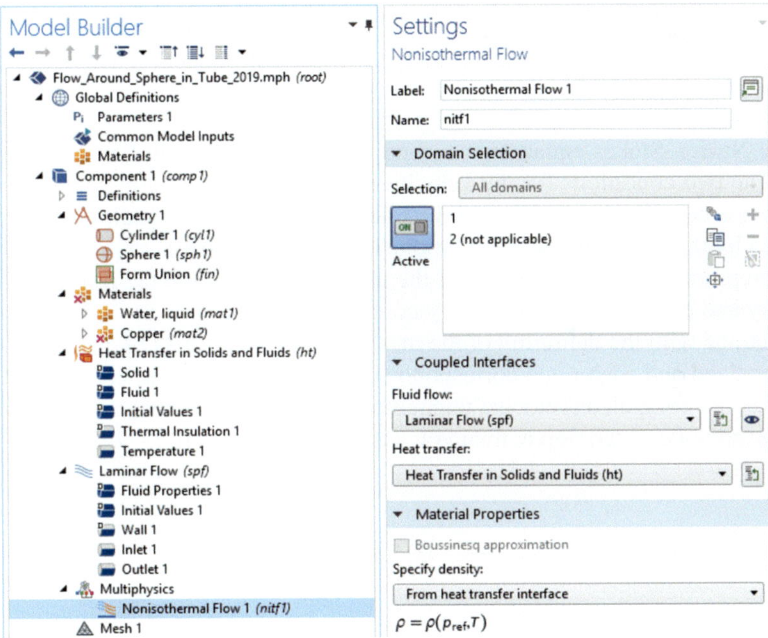

Fig. 2.35 Nonisothermal flow active on domain 1 (the tube)

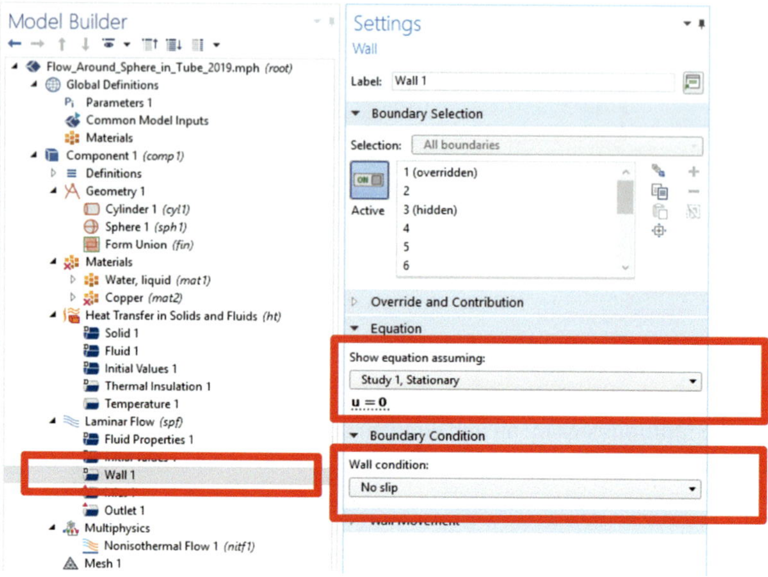

Fig. 2.36 No-slip at the wall

The Navier–Stokes equation is second order in velocity and first order in pressure. The two conditions for the velocity are given by the no-slip condition at the wall that is contained in Wall 1 and the Inlet 1 condition. The condition on Pressure is given by the outlet pressure condition. The following step is to proceed with the solution process that is comprised of the meshing process and the resolution of the velocity and pressure values for every node on the mesh.

Continue with the definition of the mesh. This is the process by which the space is discretized and each node corresponds to location in space where all dependent variables are computed (velocity, temperature, pressure). Consequently, if you have a thousand nodes to compute, then you will need to handle 1000 equations.

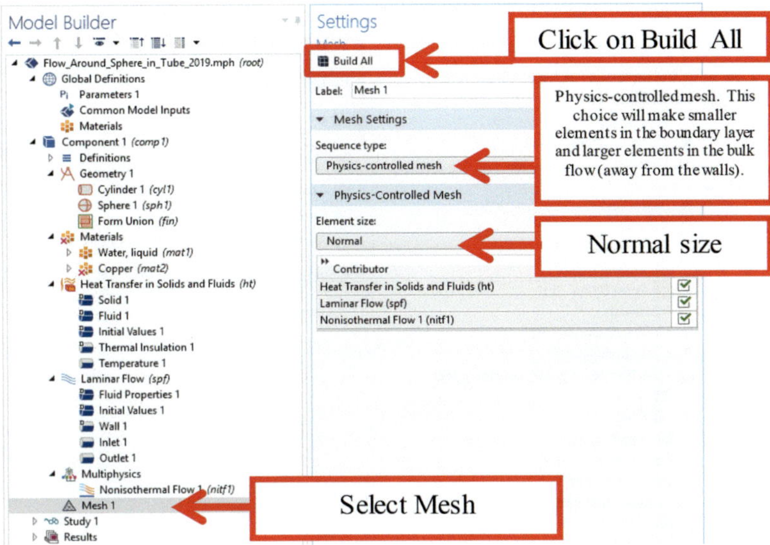

Fig. 2.37   Build mesh

Now your domain has been discretized.

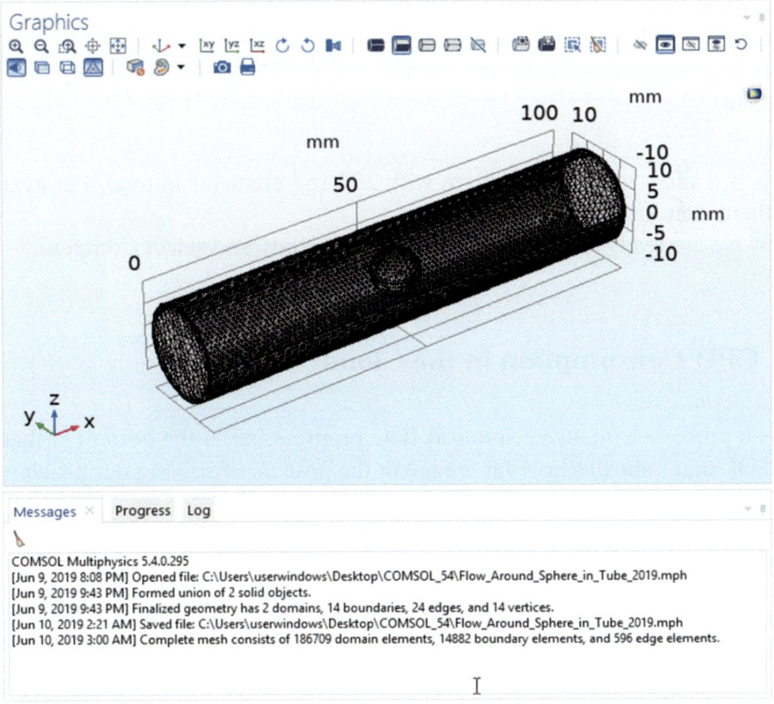

Fig. 2.38   Mesh

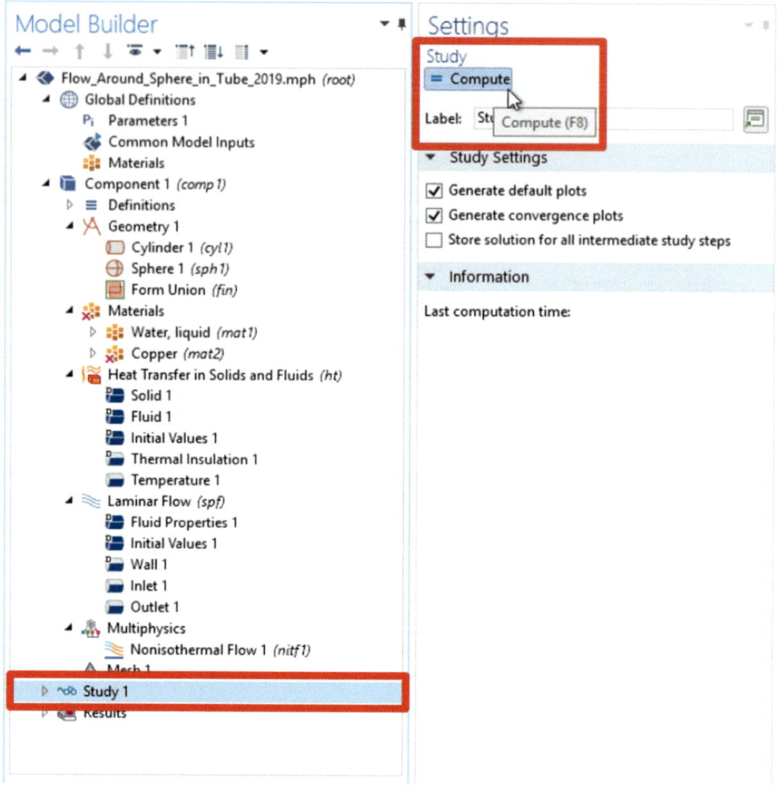

Fig. 2.39 Study 1

This is a very small size problem with 202,187 elements in total. For every element there is an equation to be solved.

Now we are ready to solve our model. Go to study and select Compute.

2.6   CPU Consumption in the Cloud

This is a processor-intensive solution. The progress bar at the bottom is that from COMSOL that indicates how far we are in the process of solving our problem.

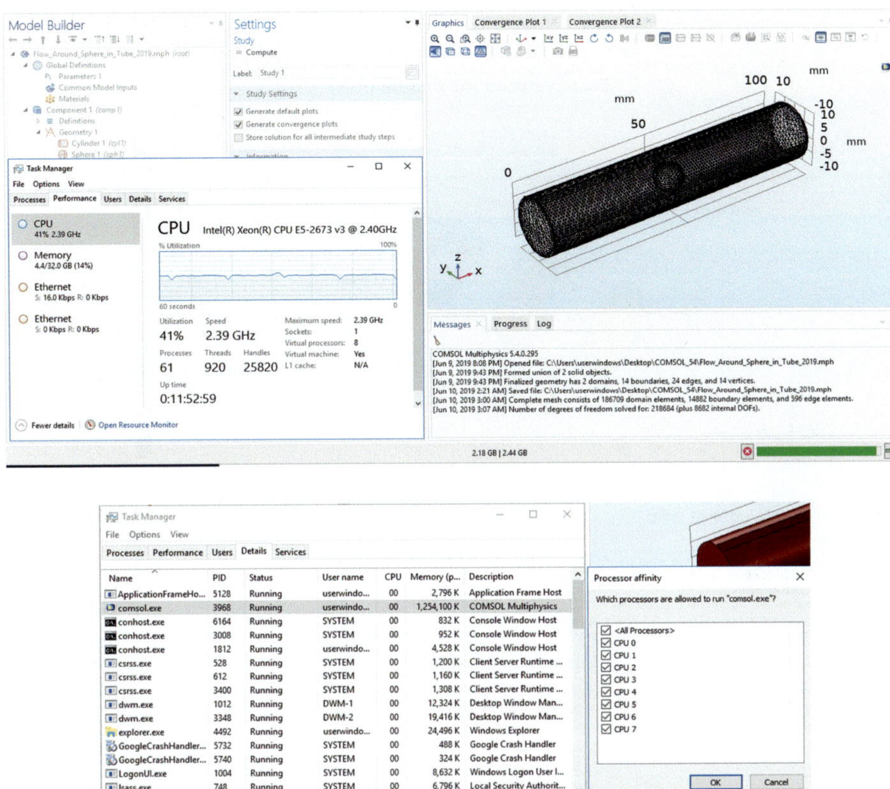

Fig. 2.40 Processor and memory comsuption in a D8s v3 VM (8 vcpus, 32 GB memory)

According to COMSOL's documentation (https://www.comsol.com/support/knowledgebase/1096/), only the actual physical processors are used in the computation (number of virtual processors divided by 2), in this case four physical processors.

The solution for the velocity profile renders a slice plot where the velocity is higher around the inner sphere. This makes sense since the cross-sectional area is smaller, and in order to maintain the flow rate, the fluid must accelerate as water is practically incompressible at the operating pressures. The lowest velocity is $u = 0$ that occurs at the wall, as expected by the application of the no-slip condition.

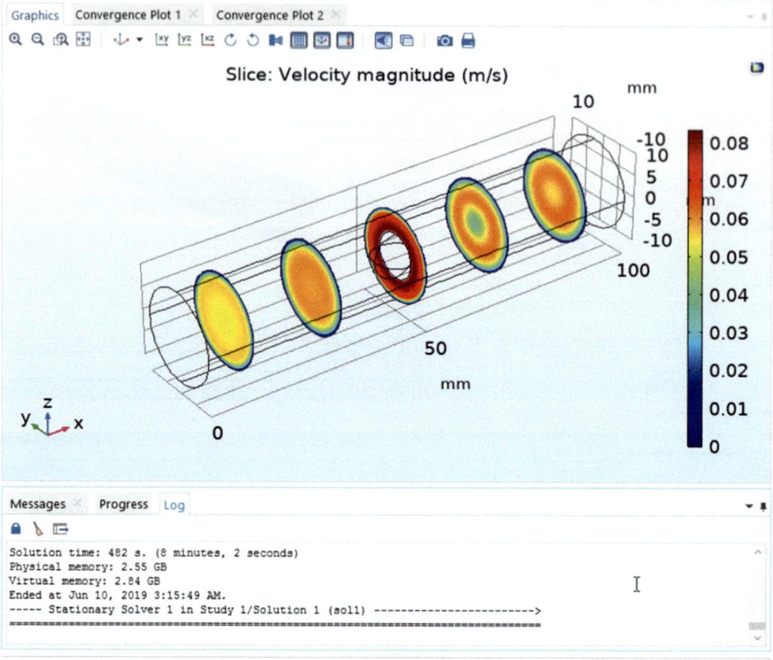

Fig. 2.41 Velocity profile

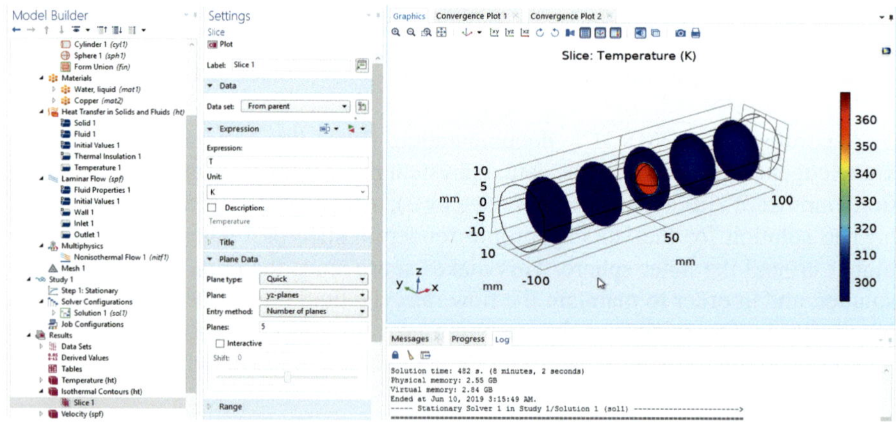

Fig. 2.42 Isothermal contours

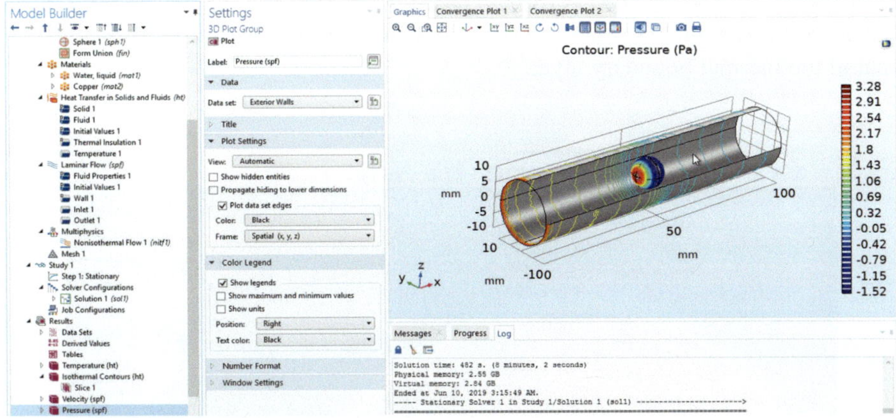

Fig. 2.43 Pressure profile

2.7   Boundary Layers

The momentum boundary layer forms on the surface of the sphere. Notice that in finite elements, the surface is approximated by small linear segments.

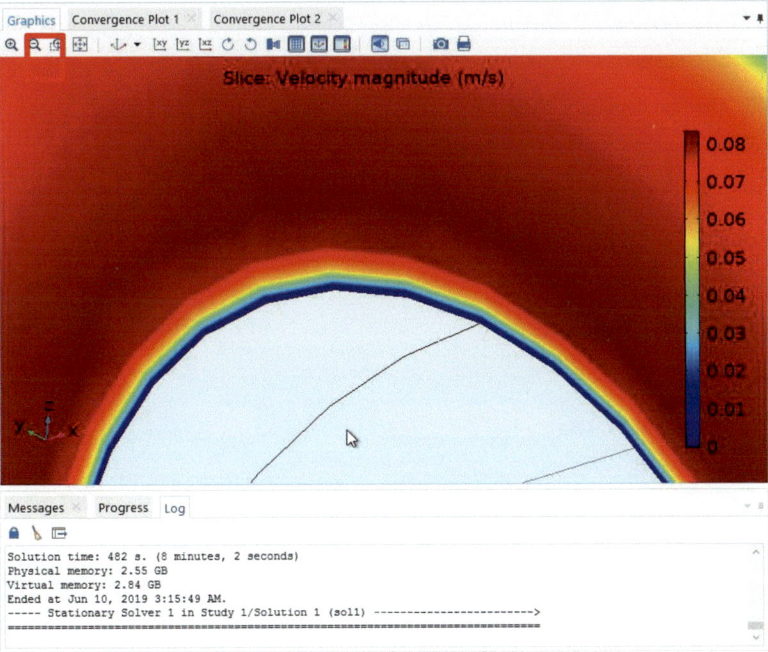

Fig. 2.44 Momentum boundary layer

A thermal boundary layer is formed near the surface (remove transparency for a better view). The thickness of the momentum boundary layer does not coincide with that of the thermal boundary layer.

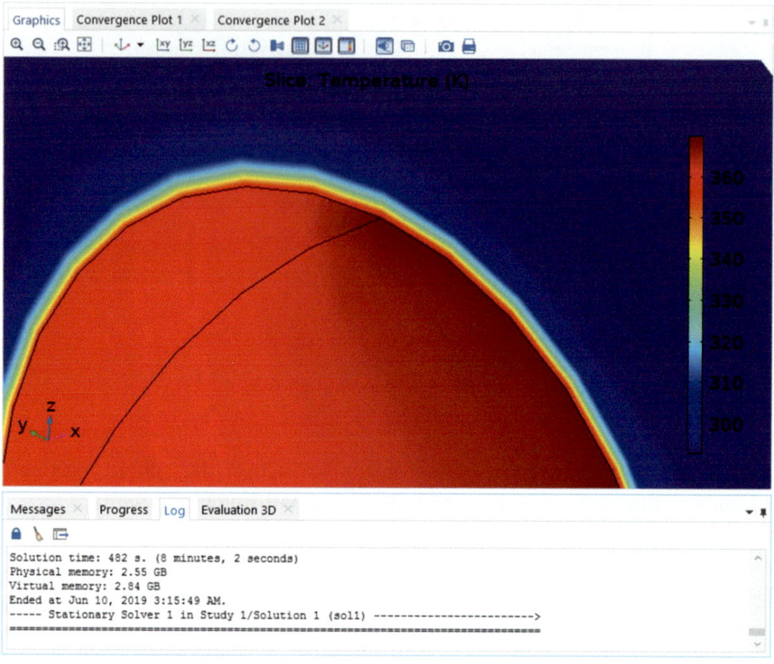

Fig. 2.45 Thermal boundary layer

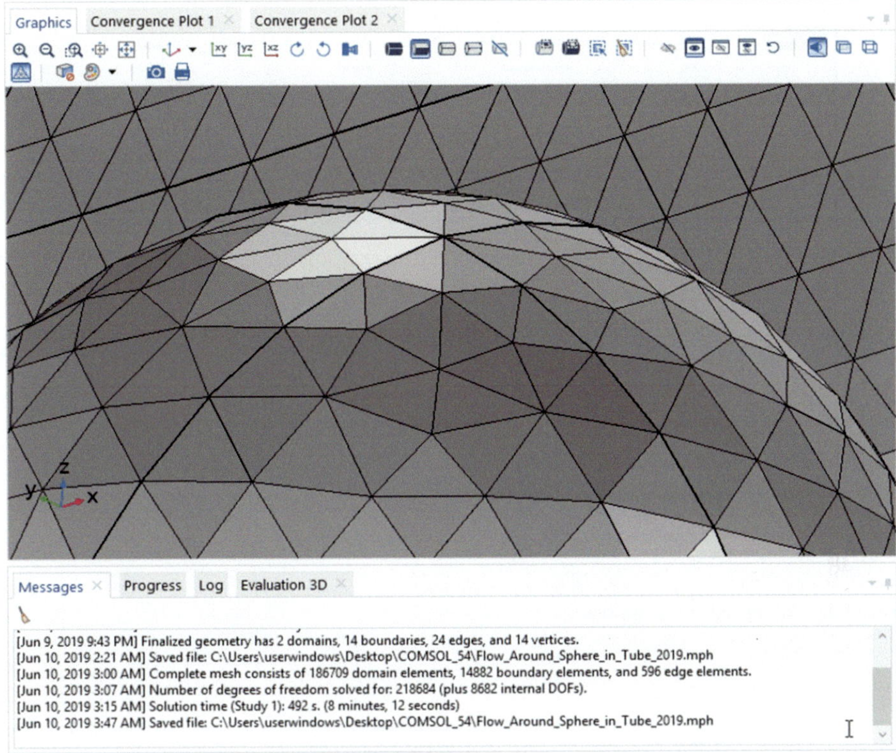

Fig. 2.46 Surface mesh

2.8   Advanced Features: Batch Processing

SAVE YOUR WORK. CONTINUE WITH A COPY OF WHAT HAS BEEN DONE SO FAR

We implement now other features.

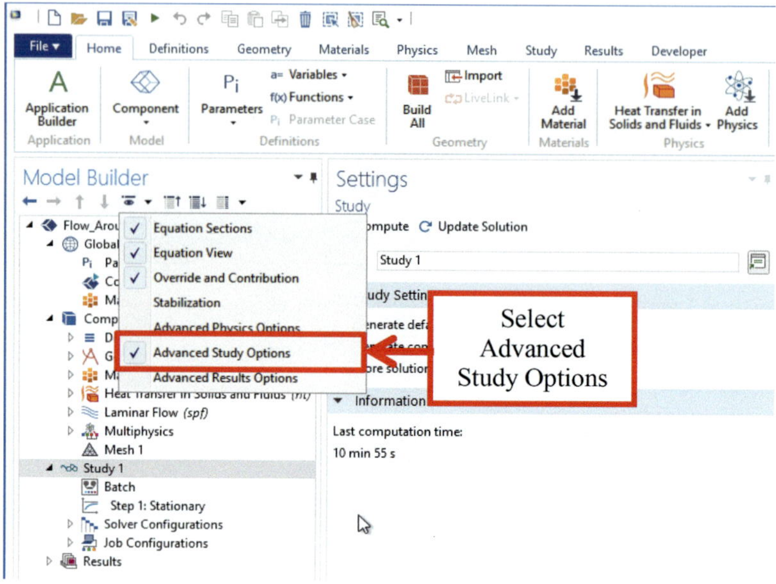

Fig. 2.47 Advanced study options

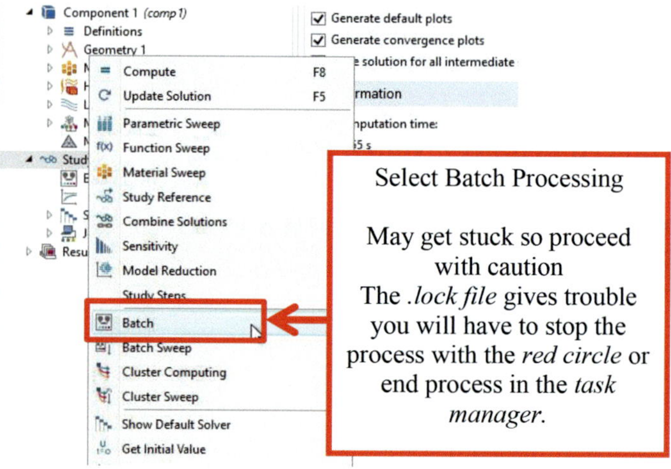

Fig. 2.48 Batch processing

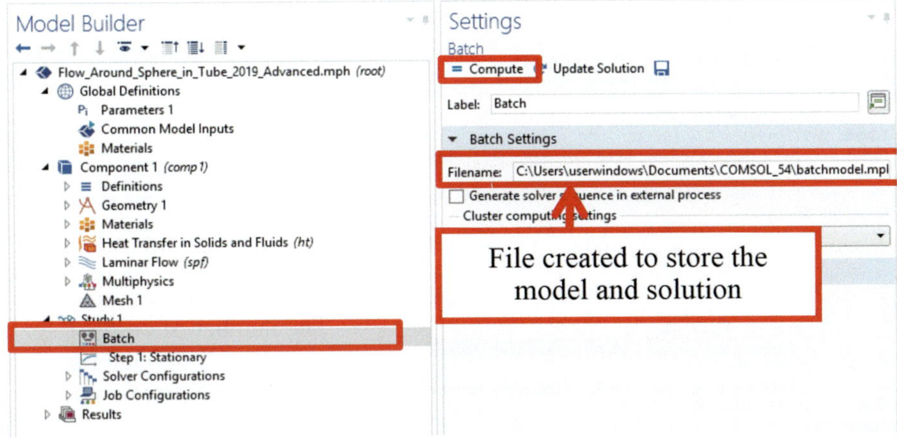

Fig. 2.49 Running the study

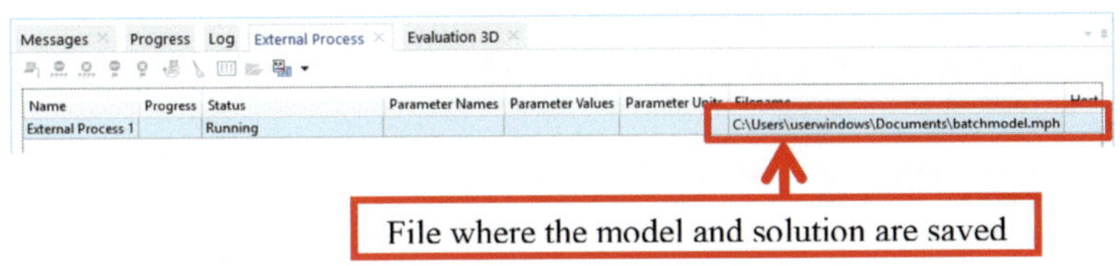

Fig. 2.50 Solutions

Once the COMSOL is finished, your results are in the batchmodel file.

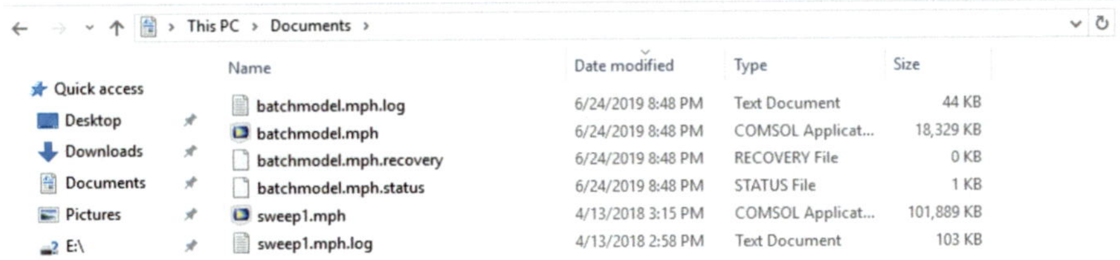

Fig. 2.51 Files with results

There you have your results in a separate file. There is log generated that is useful when problems arise.

```
batchmodel.mph.log - Notepad
File  Edit  Format  View  Help
*********************************************
***COMSOL 5.4.0.295 progress output file***
*********************************************
Mon Jun 24 20:38:46 UTC 2019
COMSOL Multiphysics 5.4 (Build: 295) starting in batch mode
Opening file: C:\Users\userwindows\Documents\batchmodel.mph
Open time: 31 s.
The input file name C:\Users\userwindows\Documents\batchmodel.mph will be used as output file name.
Running: Batch 1
<---- Compile Equations: Stationary in Study 1/Solution 1 (sol1) ---------------
Started at Jun 24, 2019 8:39:17 PM.
Geometry shape order: Linear
Running on Intel(R) Xeon(R) CPU E5-2673 v4 at 2.30 GHz.
Using 1 socket with 4 cores in total on workstation1.
Available memory: 32.77 GB.
            Current Progress:    0 % - Compiling equations
Memory: 772/772 966/966
            Current Progress:    5 % - Compiling equations
Memory: 758/777 949/970
Time: 5 s.
Physical memory: 783 MB
Virtual memory: 1004 MB
Ended at Jun 24, 2019 8:39:22 PM.
```



```
Pseudo time-stepping CFL-ratio:
1
---------- Current Progress: 100 % - Sensitivity
Memory: 1805/2205 2217/2619
Solution time: 537 s. (8 minutes, 57 seconds)
Physical memory: 2.16 GB
Virtual memory: 2.61 GB
Ended at Jun 24, 2019 8:48:21 PM.
----- Stationary Solver 1 in Study 1/Solution 1 (sol1) ---------------------->
Run time: 544 s.
Saving model: C:\Users\userwindows\Documents\batchmodel.mph
Save time: 0 s.
Total time: 575 s.
---------- Current Progress: 100 % - Done
Memory: 953/2205 1250/2619
```

Fig. 2.52 Log file

The log is long; consequently, we only include the beginning and the end. There is a total time that includes the time to open the file. Open the files generated and review your answer.

2.9 Advanced Processing: Parametric Sweep

Go to the Global tab and open definitions. Let us add a pair of new parameters.

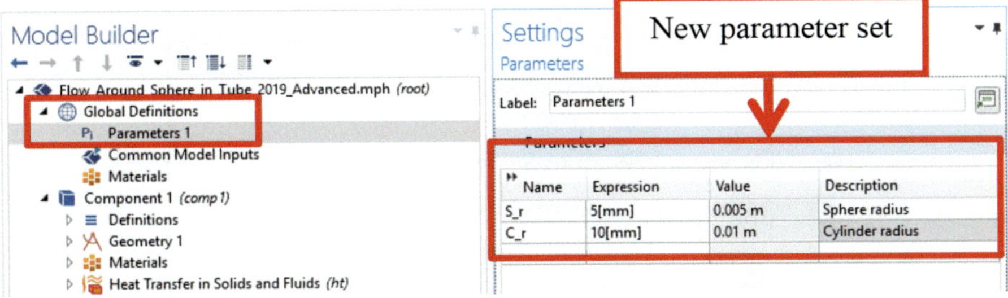

Fig. 2.53 Parameters

Add the parameters in the definition of the geometry.

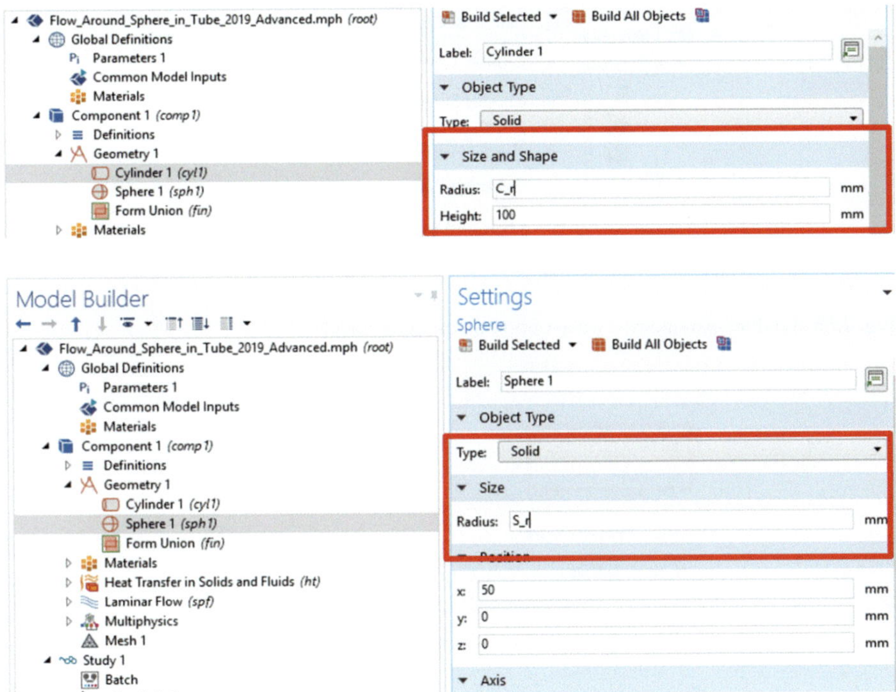

Fig. 2.54 Change the geometry with parameters C_r and S_r

Now select the study Tab and proceed to start the computations, find on the study tab the advanced option for parameter sweep, and compute.

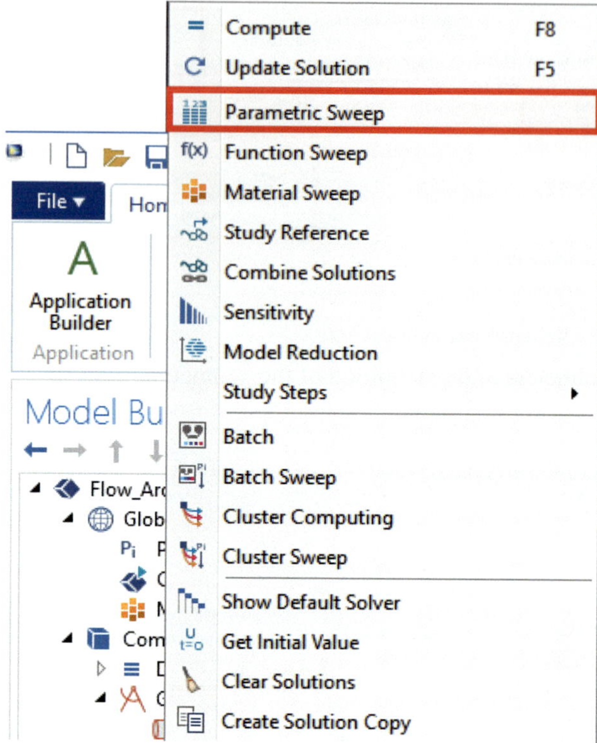

Fig. 2.55 Find the parametric sweep option on the study menu

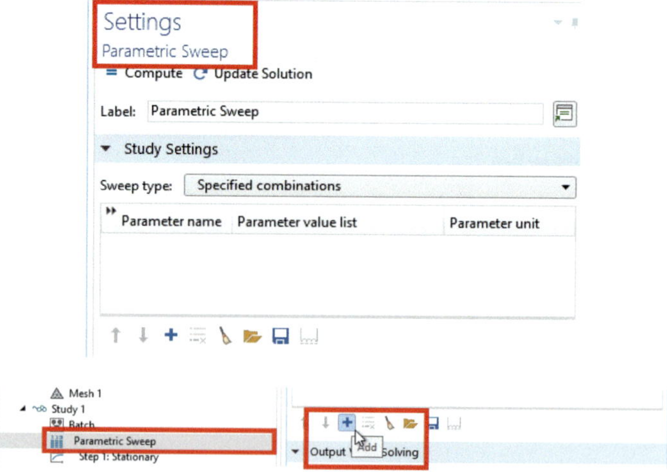

Fig. 2.56 Select Parametric Sweep

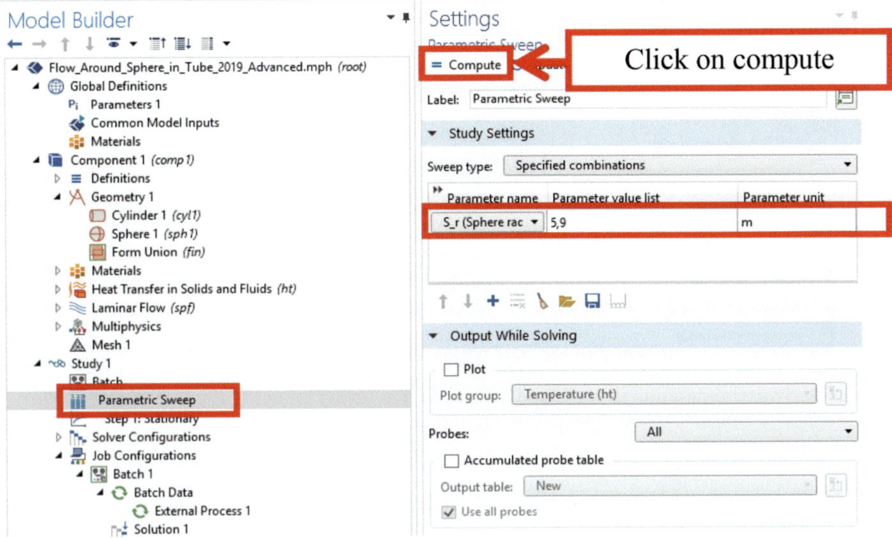

Fig. 2.57 Study—Compute

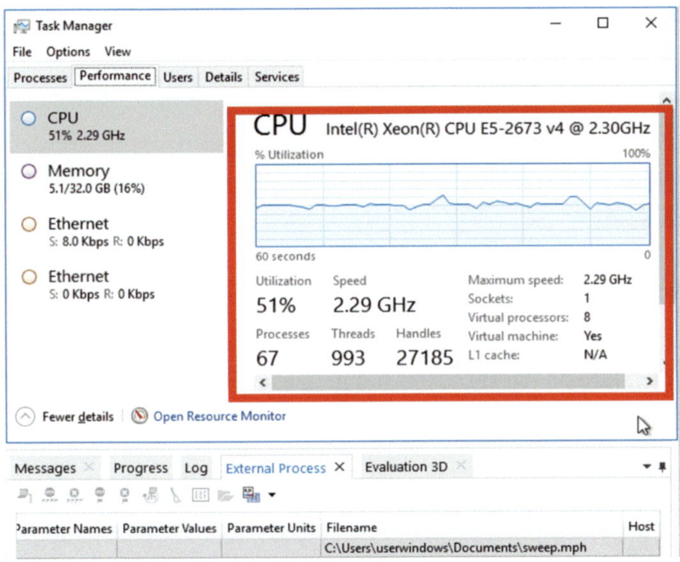

Fig. 2.58 CPU usage in parametric sweep

The processors run close to 100% capacity, in version 5.4; this is because COMSOL is using only the number of physical cores. Thus, the modeling engine is CPU throttled and with sufficient RAM left to work without incident. This is contrary to the traditional concern of memory size rather than CPU capacity bottlenecking.

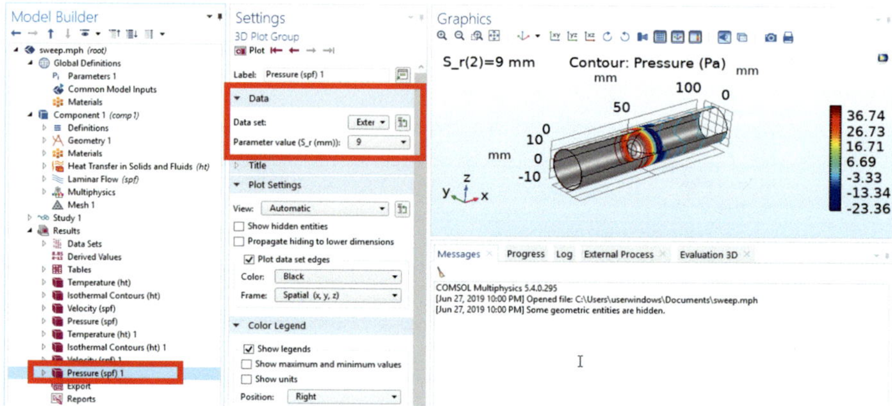

Fig. 2.59 Sweep simulation is generated and saved

Fig. 2.60 Parameter sweep results

The temperature on the surface is hot in one leaf or section of the sphere.

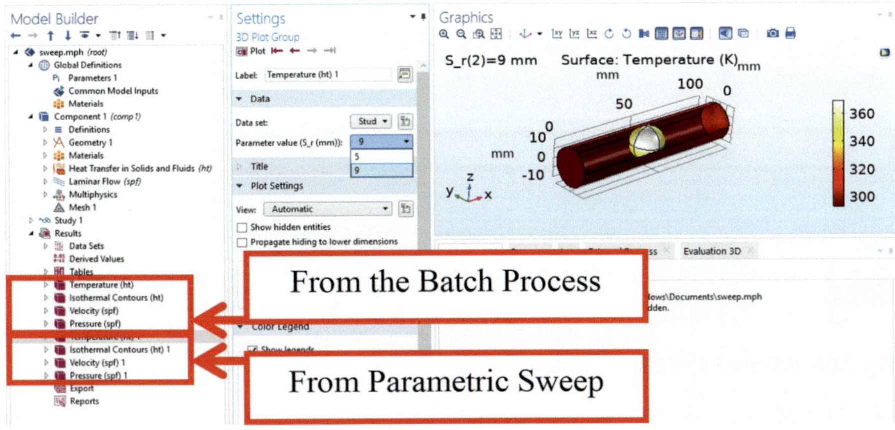

Fig. 2.61 Two sets of results, batch and parametric sweep

The idea is to write a script and let the workstation in Azure run free. In larger applications, one would prefer a larger capacity machine in order to avoid delays particularly at early stages of testing where larger models and multiple tests in a sweep are performed. Consequently prior to do a batch run, a larger machine will be selected.

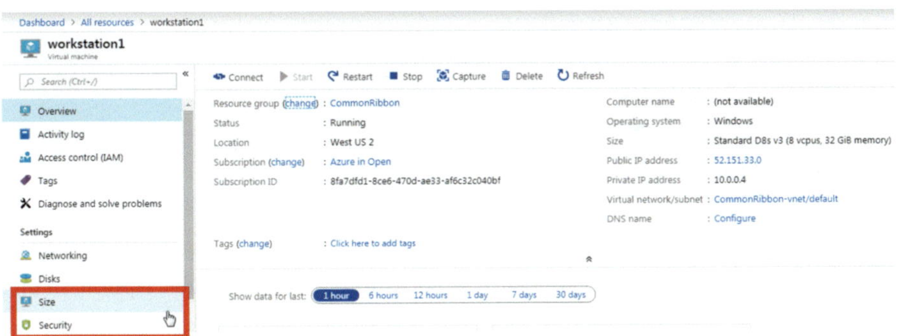

Fig. 2.62 Option to resize the virtual machine

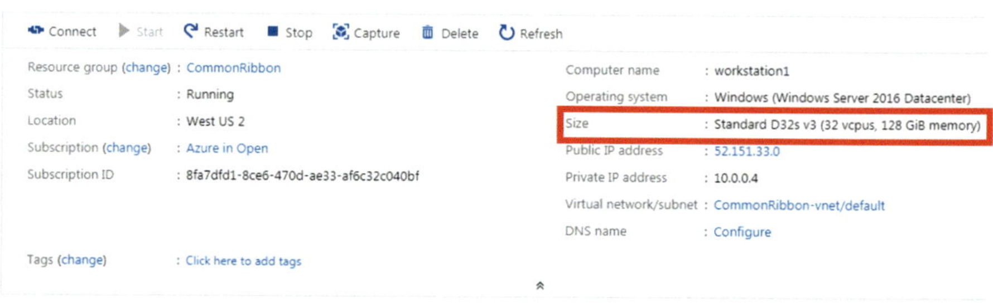

Fig. 2.63 Selecting a 32 vcpus machine

The VM has to be stopped and deallocated prior to proceeding with resizing the VM. However, if that is not done, then the system will proceed with the resizing and restarting the machine and it will take longer, and in the end, a restart with stop and reallocation is needed.

Fig. 2.64 Configuration of larger VM

An Azure account starts with only 10 cores. More cores require a request to technical services to increase the number of cores available for your subscription, in this case 50 cores.

It is important to notice that there are substantial time savings in this upgrade that would take longer if a new machine or part had to be bought.

Generally, batch process bears the idea of programming a task and let it run. Later come back for the results, particularly over weekends and holidays. In notepad create this file and save in the same location where COMSOL is installed (where the Flow_Around_Sphere_in_Tube_2019_Advanced.mph is located).

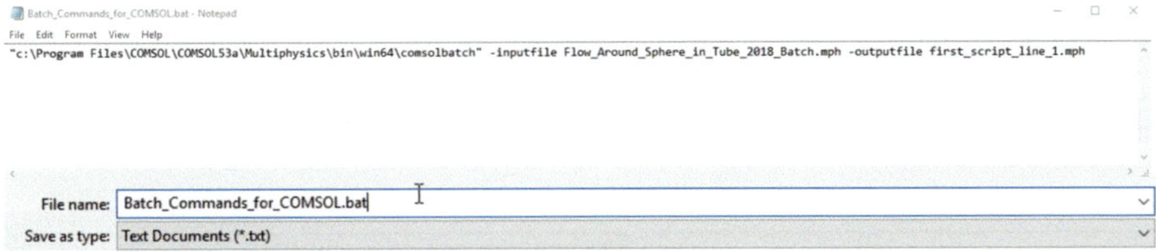

Fig. 2.65 Batch file to run COMSOL simulations

Using Full Screen for the remote VM, press and hold the Windows Key and press
the letter R.

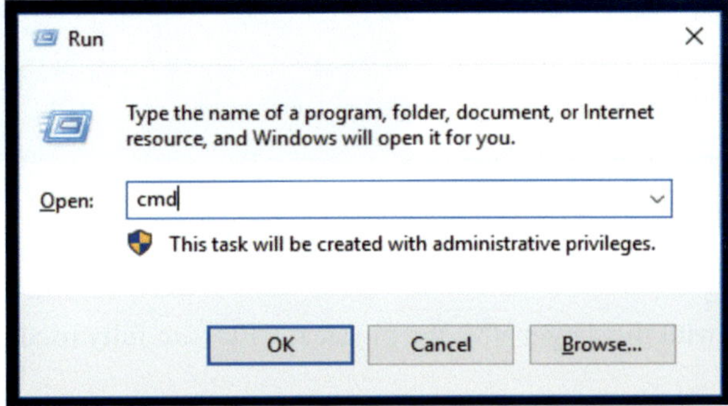

Fig. 2.66 Run command to call the command screen

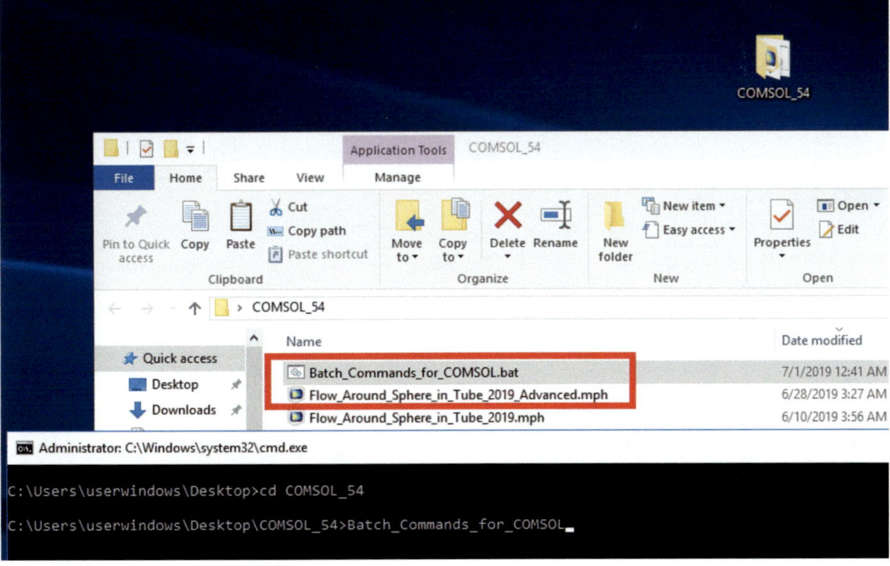

Fig. 2.67 Command screen to call Batch file (.bat)

The script file is run to see how well it works. In a computer with 32 virtual processors, the CPU capacity is fully saturated (only the actual number of physical cores is used, which equals one-half of the total of virtual CPUs) and there is plenty of the memory available.

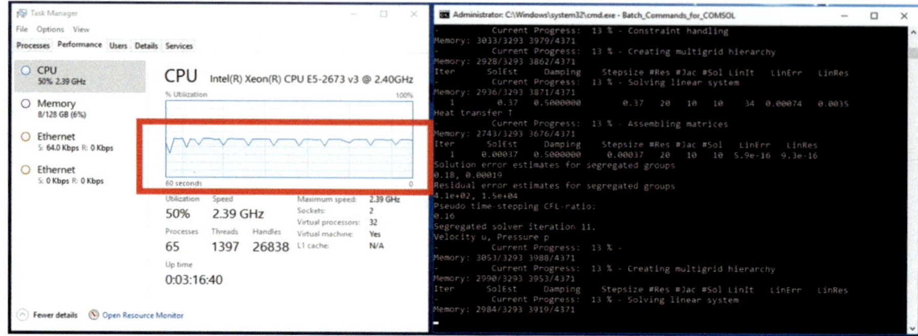

Fig. 2.68 Utilization of 32 vCPU workstation

Notice that with this larger VM, the physical CPUs are fully used.

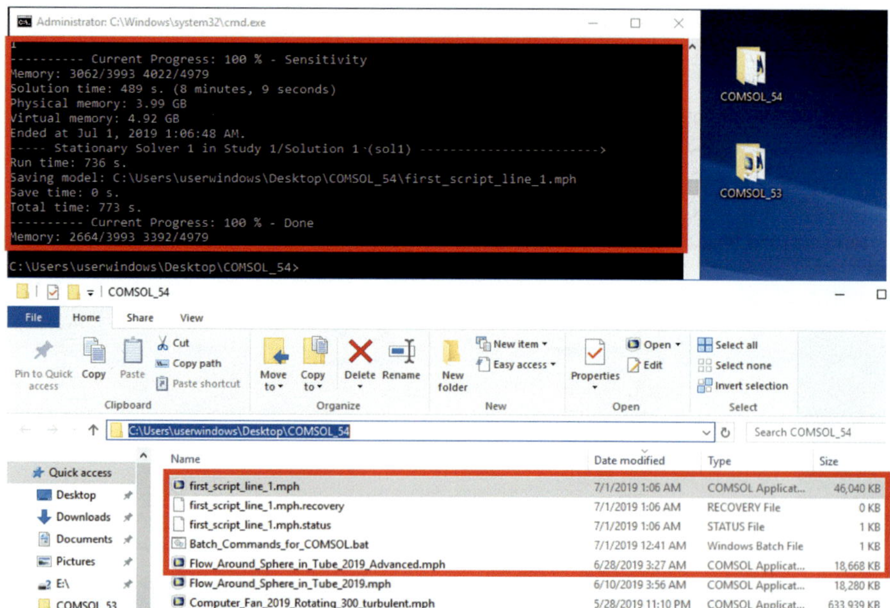

Fig. 2.69 Total running time at 32 vCPUs

At the end of the process, the .mph needed is created.

It is prudent to see the extension of the files being used. Go to the File explorer and tick the box to show File name extensions.

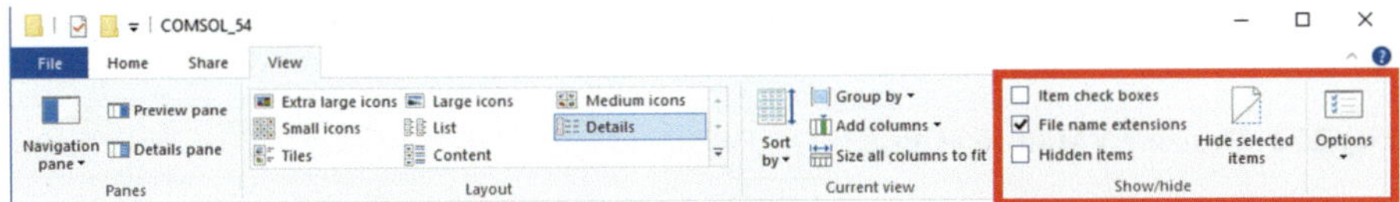

Fig. 2.70 File extensions

Proceed to stop and deallocate the VM. Resize to 16 virtual CPUs and proceed to rerun the .bat file.

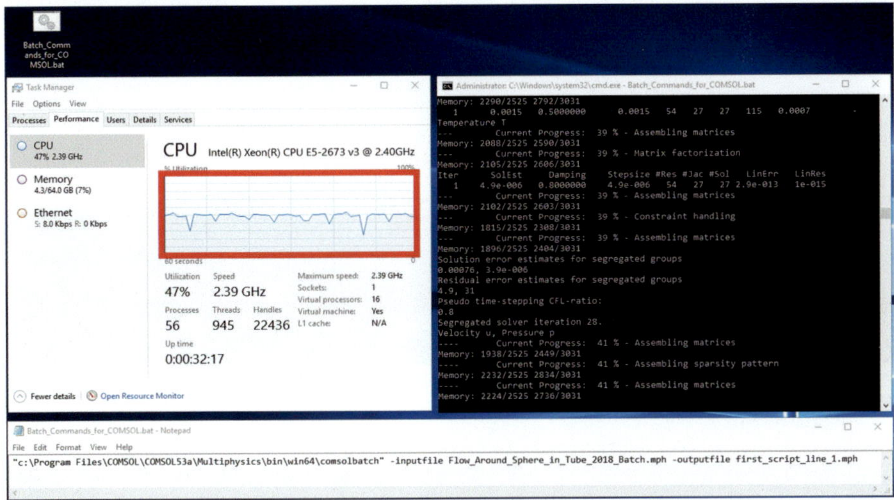

Fig. 2.71 Usage of only physical CPUs

The RAM capacity remains unstressed and the vCPU usage reaches about 50% and the workstation is not freezing.

Fig. 2.72 Resizing VM to 16 vCPUs

Fig. 2.73 Resizing VM to 16 vCPUs verification

Fig. 2.74 Running the batch script

Fig. 2.75 Total running time at 16 vCPUs (D16s_v3)

Rerunning again in batch with eight virtual processors and take note of the total processing time for comparison (Table 2.1).

```
---------- Current Progress: 100 % - Sensitivity
Memory: 1686/2650 2121/3141
Solution time: 1179 s. (19 minutes, 39 seconds)
Physical memory: 2.58 GB
Virtual memory: 3.14 GB
Ended at Jul 1, 2019 4:29:46 AM.
----- Stationary Solver 1 in Study 1/Solution 1 (sol1) -------------------------->
Run time: 1739 s.
Saving model: C:\Users\userwindows\Desktop\COMSOL_54\first_script_line_1.mph
Save time: 0 s.
Total time: 1778 s.
---------- Current Progress: 100 % - Done
Memory: 1387/2650 1759/3141

C:\Users\userwindows\Desktop\COMSOL_54>
```

Fig. 2.76 Total running time at eight vCPUs (D8s_v3)

The 16 processor is a middle-of-the-road choice where some speed is gained but not at the highest cost.

Table 2.1 Computational times for increasing size virtual machines

Number of processors	Total time (s)	Cost per month (US $)	Cost per second (US $)	Total cost (US $)	Extra money spent from base case (US $)	Time saved from base case (s)	Cost of seconds saved (US $)
8	1778	559.49	2.1585×10^{-4}	0.3838	Base	Base	Base
16	1123	1118.98	4.3171×10^{-4}	0.4848	0.1010	655	0.2828
32	773	2237.95	8.6341×10^{-4}	0.6674	0.2836	1005	0.8677

1 month = 30 days of 24 h = 2,592,000 s

2.10   Case of the Turbulent Flow Fan: A More Elaborate Problem

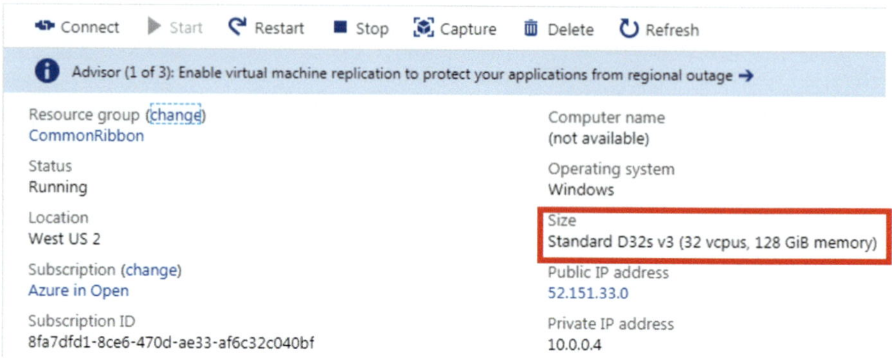

Fig. 2.77 Larger VM for a large model

A sufficiently large machine for any concerned reader.

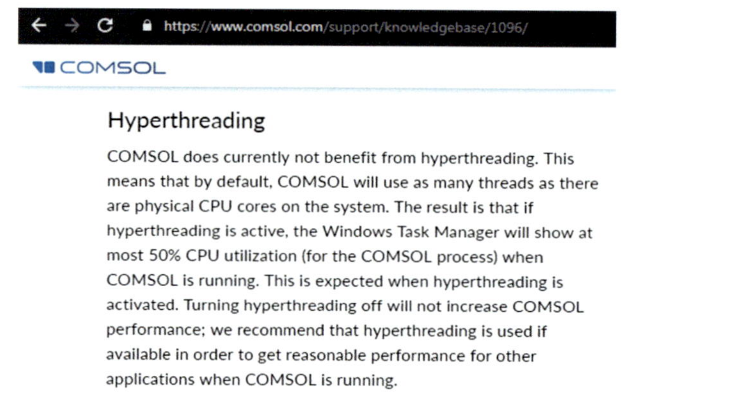

https://www.comsol.com/support/knowledgebase/1096/

Fig. 2.78 Hyperthreading (link consulted June 30, 2019, https://www.comsol.com/support/ knowledgebase/1096/)

The VM starts without recognizing the 32 vCPUs; only 16 physical cores are recognized.

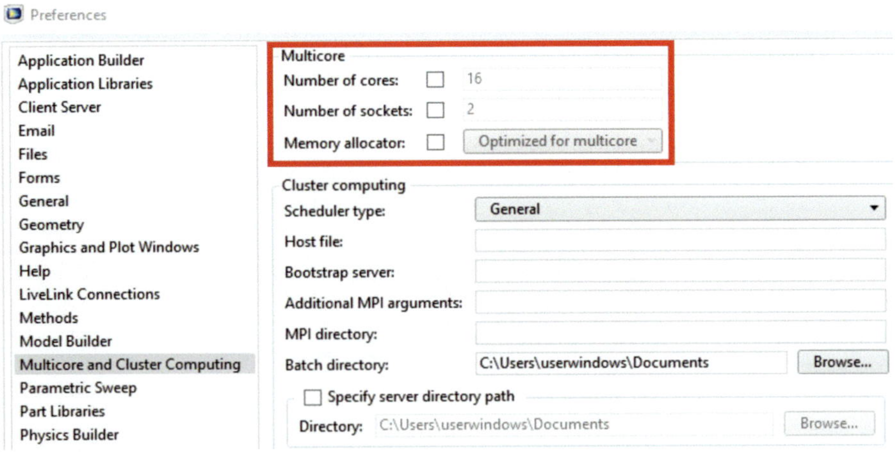

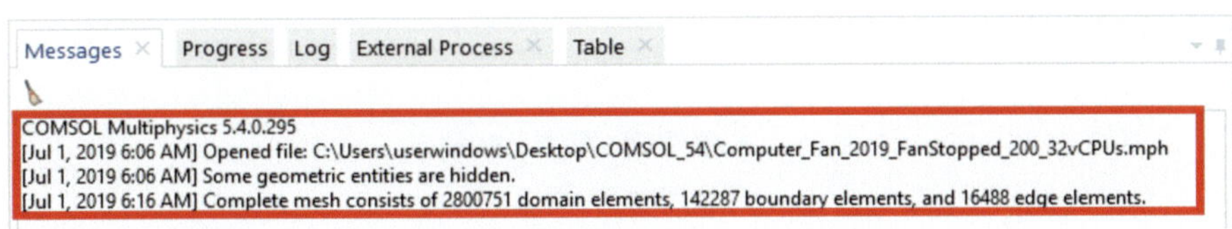

Fig. 2.79 Multithreading is not recognized

Sixteen physical cores are recognized, not 32 vCPUs. Using normal-size mesh, this is a larger problem with more than two million elements (two million equations). Sequence type: Physics-controlled mesh, Element size: Fine).

Fig. 2.80 Larger model with more than two million elements

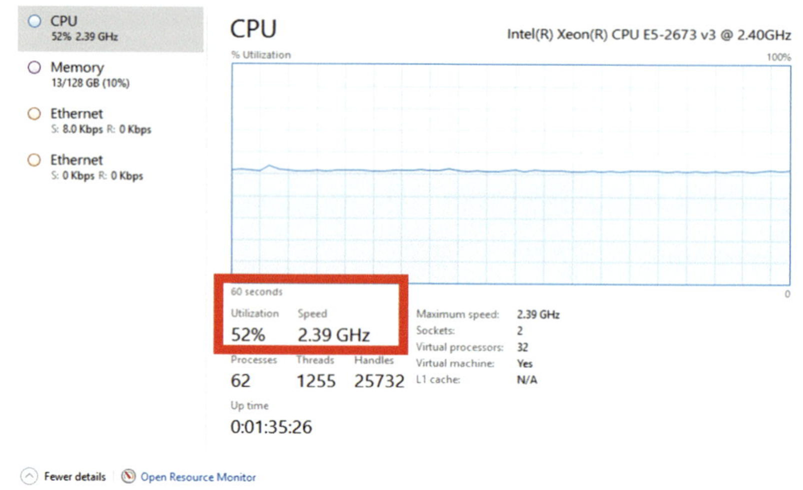

Fig. 2.81 Low use of vCPUs and full use of physical CPUs

There is fan motion in this first case. Not over 50% of the CPU capacity is used, because COMSOL cannot use hyperthreading. This means that the physical cores (the ones used by COMSOL) are fully used. The system is not RAM limited.

There are 32 virtual cores and 16 actual cores. COMSOL uses the actual cores, not the virtual cores.

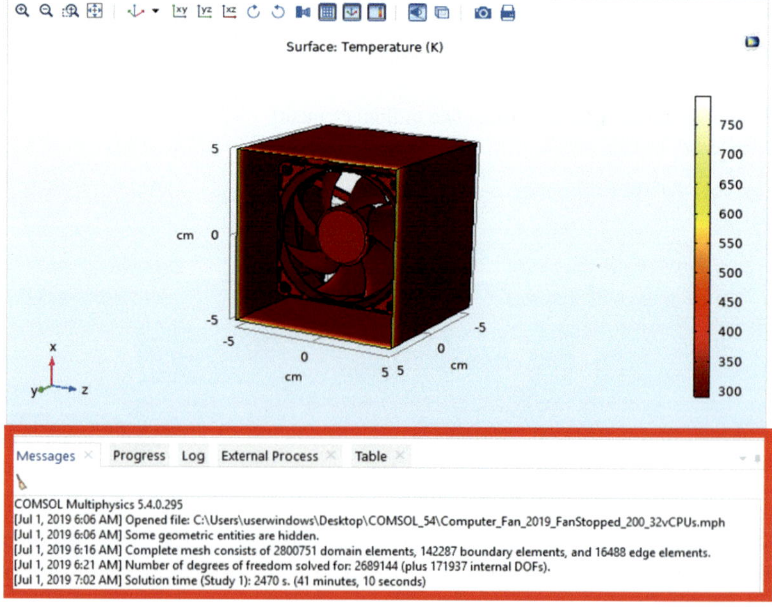

Fig. 2.82 Computing time with 16 cores

Now the settings of COMSOL are changed and made to use the 32 cores and see what happens. This is because the technology sometimes advances faster than current documentation.

Disregard the guideline for the use of only the 16 physical cores and force the system to use the 32 vCPUs. Leave the Sockets at default.

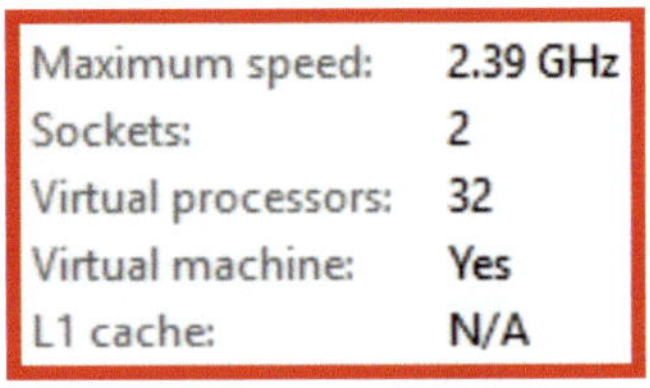

Fig. 2.83 Number of virtual CPUs

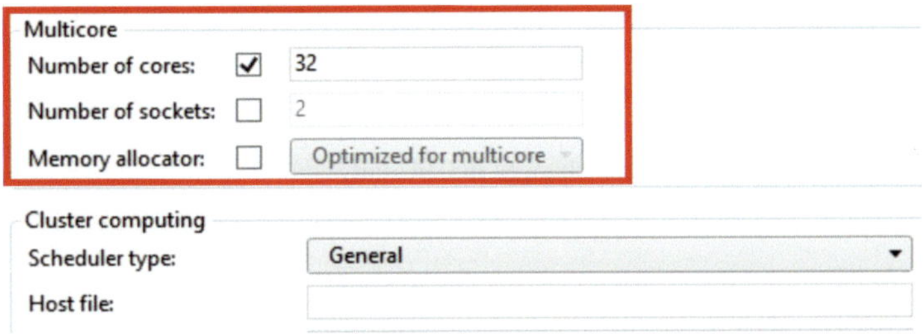

Fig. 2.84 Specifying the number of CPUs

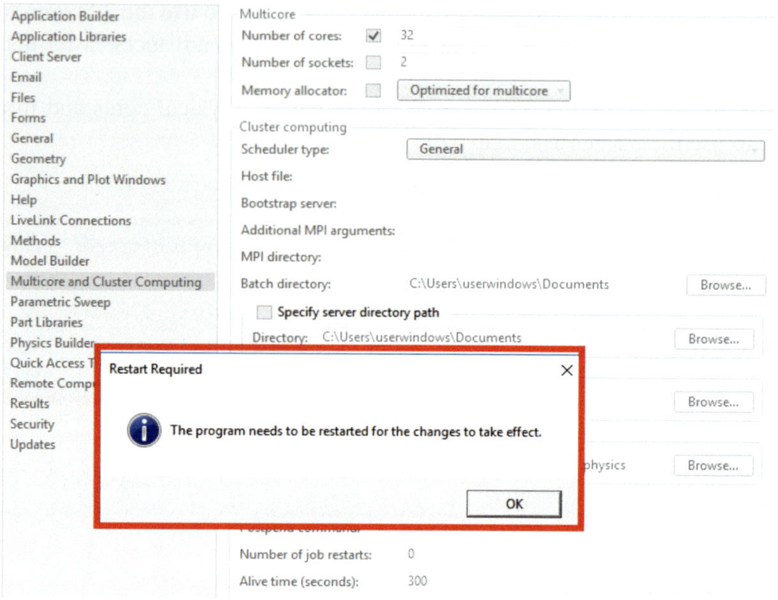

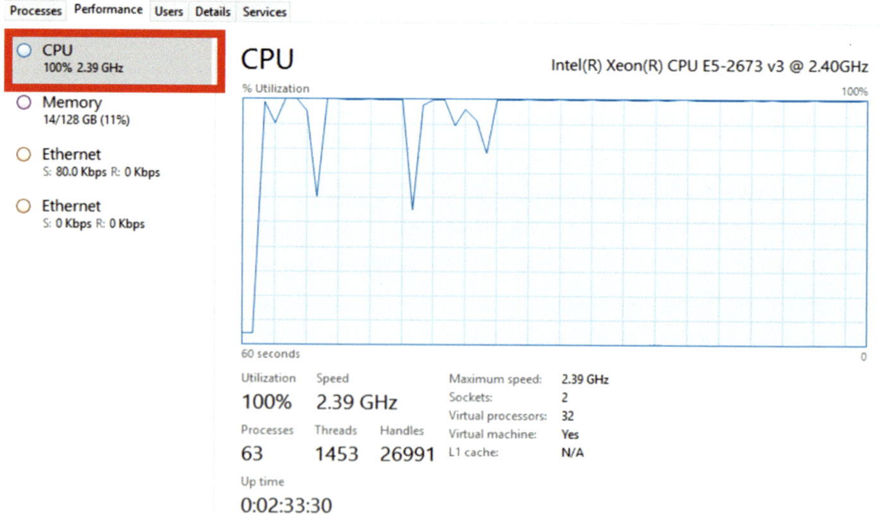

Fig. 2.85 Restart COMSOL to make changes in CPU tally

Fig. 2.86 All 32 vCPUs received a larger workload

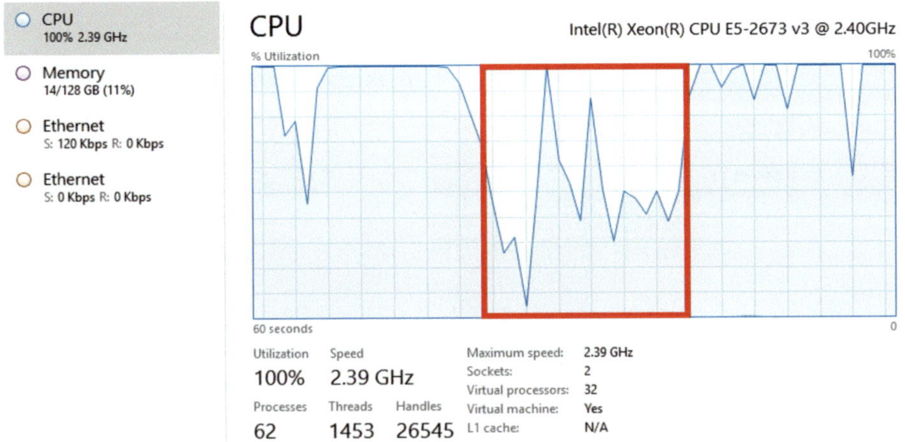

Fig. 2.87 Period of low usage of the 32 vCPUs

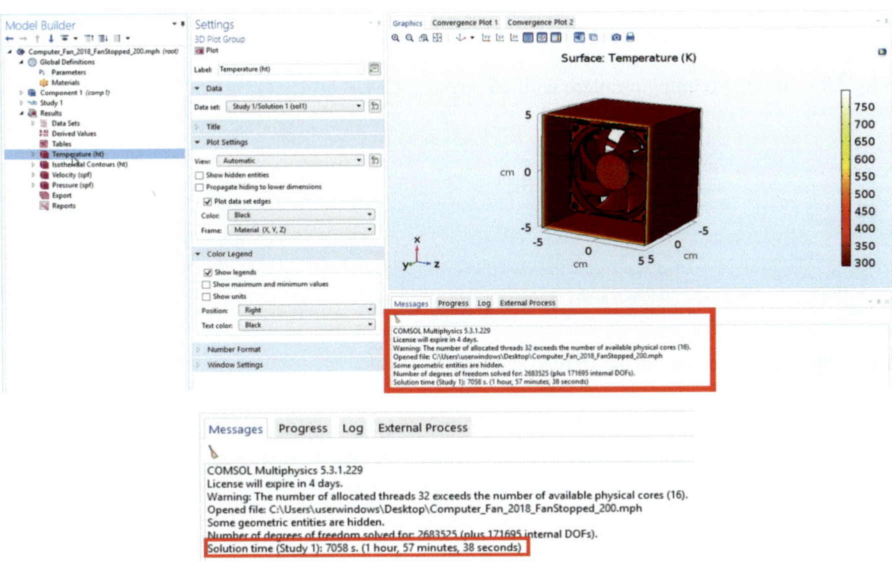

Fig. 2.88 Results using 32 vCPUs

Even though there are a larger number of processors involved, the actual computation time grew by about a half hour. Consequently, it would seem to indicate that the path to follow is to use only the physical cores.

The problem is more complex when a rotating axis is added. Still the difficulty is limited as the flow is laminar.

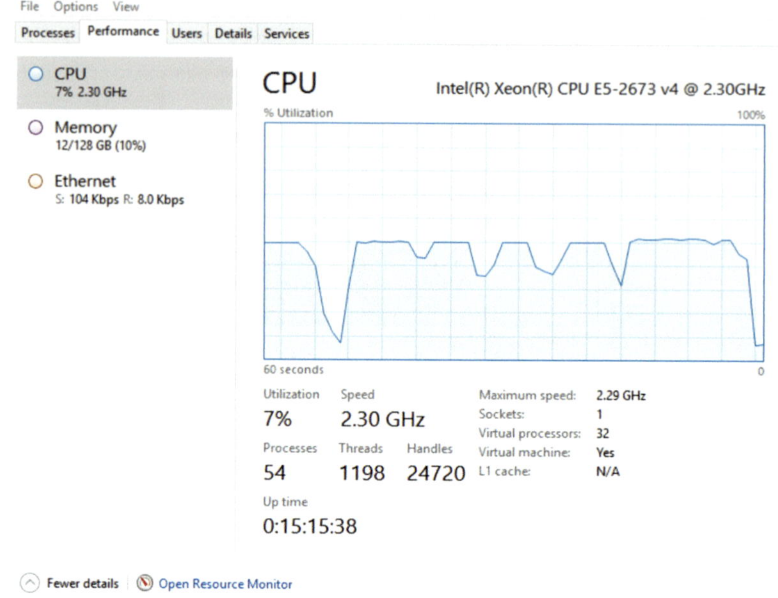

Fig. 2.89 Solving rotating problem with 16 physical CPUs

Fig. 2.90 Configuration set for 16 physical CPUs

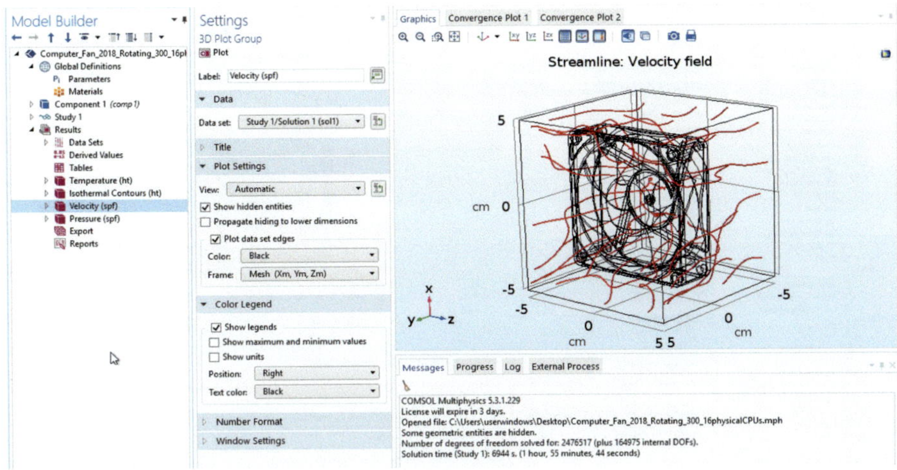

Fig. 2.91 Running rotating laminar fan with 16 physical CPUs

The solution of the problem can be repeated with the 32 vCPUs to check if yet again the time consumed is much longer.

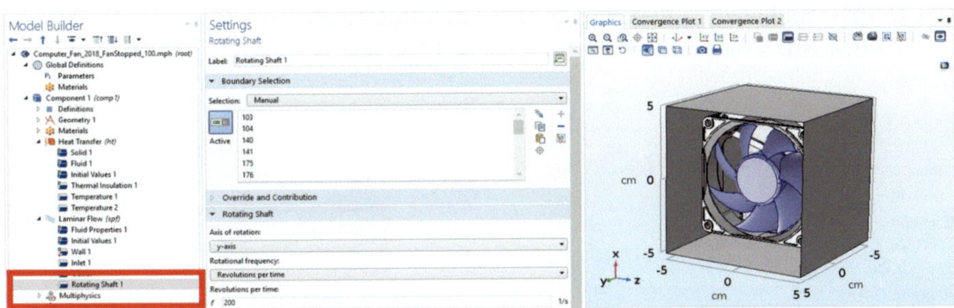

Fig. 2.92 Adding rotating shaft with laminar model 32 vCPUs

Fig. 2.93 Rotating shaft with laminar model 32 vCPUs

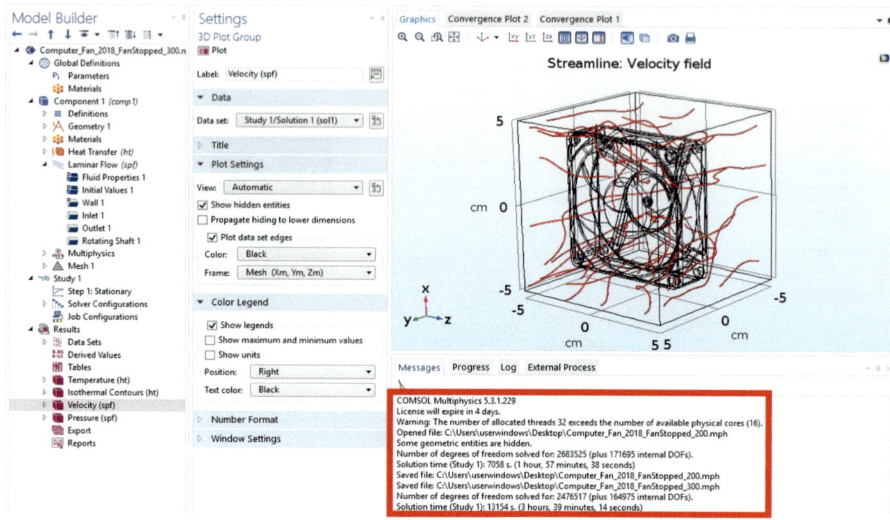

Fig. 2.94 Time consumed with 32 vCPUs

The time increased from 1 h 55 min to 3 h 39 min.

An experiment that is needed is to increase the number of physical cores to 32 (64 vCPUs) and check the behavior of the solution time.

There is still a larger machine with 64 vCPUs; this is 32 physical CPUs.

Fig. 2.95 Increasing to 64 vCPUs (32 physical CPUs)

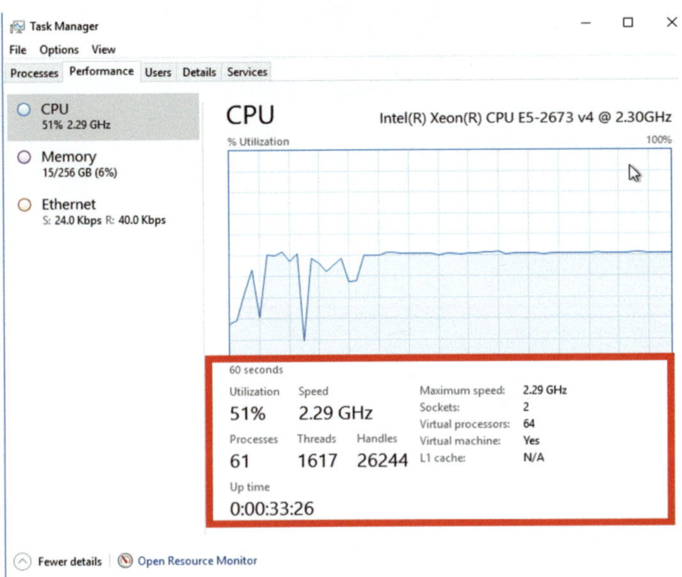

Fig. 2.96 Increasing to 64 vCPUs (32 physical CPUs)

Fig. 2.97 Processor usage at 32 physical CPUs

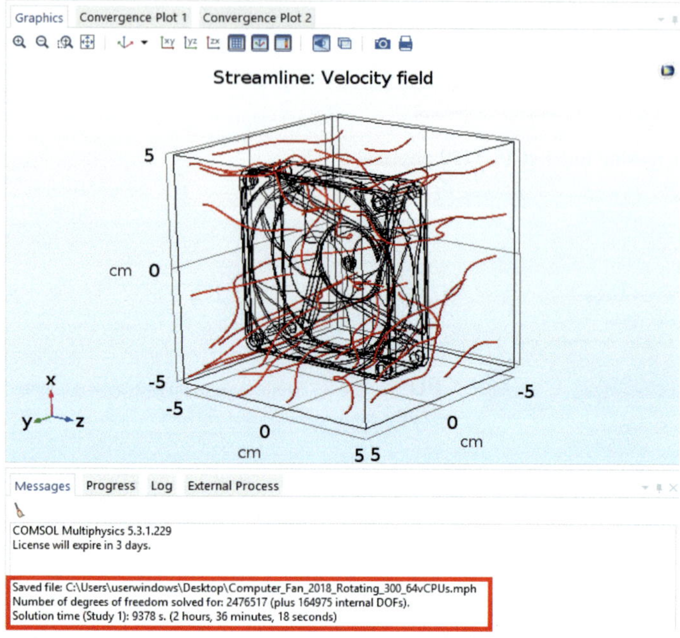

Fig. 2.98 Processing time at 32 physical CPUs

Forcing 64 vCPUs to be seen by COMSOL.

Fig. 2.99 Forcing COMSOL to see 64 vCPUs

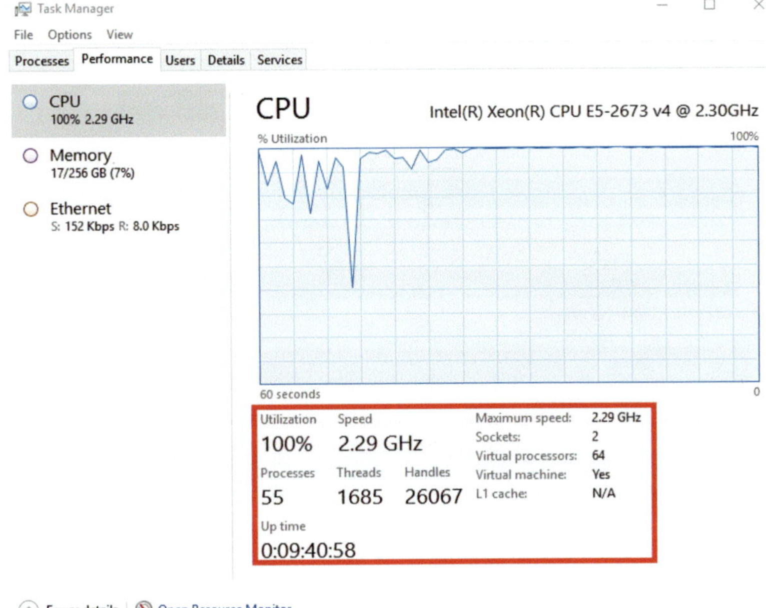

Fig. 2.100 CPU consumption with COMSOL making use of 64 vCPUs

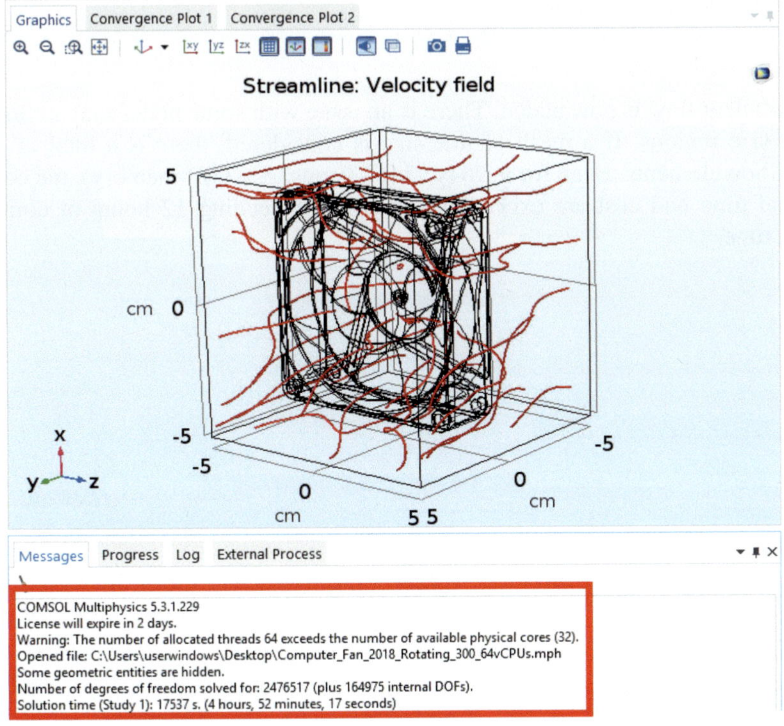

Fig. 2.101 Processing time with 64 vCPUs

The following is a detail of the expenditures during the setup and use the VM and different VM sizes.

Fig. 2.102 Principal expenditures

Turbulent flow is now added. There is an issue with some nodes that are local in small size regions. If a mesh of fine size is considered, there is a total of about 43 million elements. Even for a D64s v3 (64 vcpus, 256 GiB memory), the computational time and cost are exceedingly high by exceeding 12 hours of computational time.

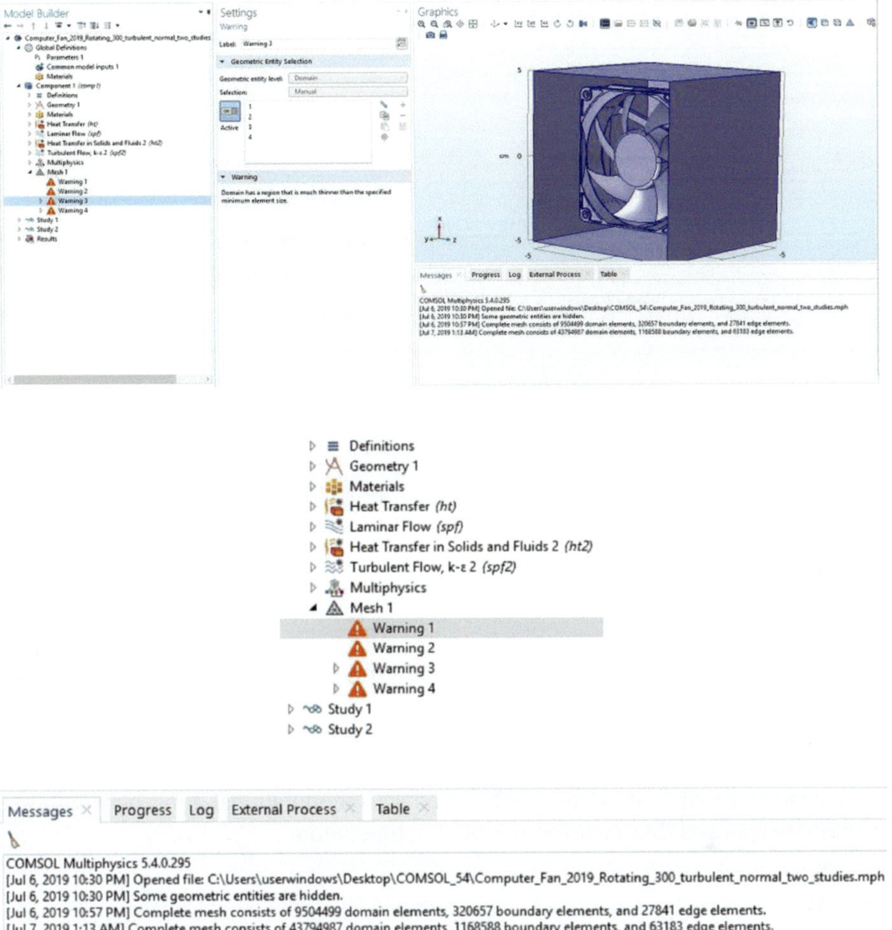

Fig. 2.103 Number of elements for a fine-size mesh

A large model of this nature with a fine mesh uses the entire memory capacity of a 256 GB memory workstation when doing the computations for turbulent flow. This fact should be taken into account when defining mesh sizes. In fact, if looked closely, the consumption of memory reaches the 500 GB mark as indicated by the consumption of physical and virtual memories.

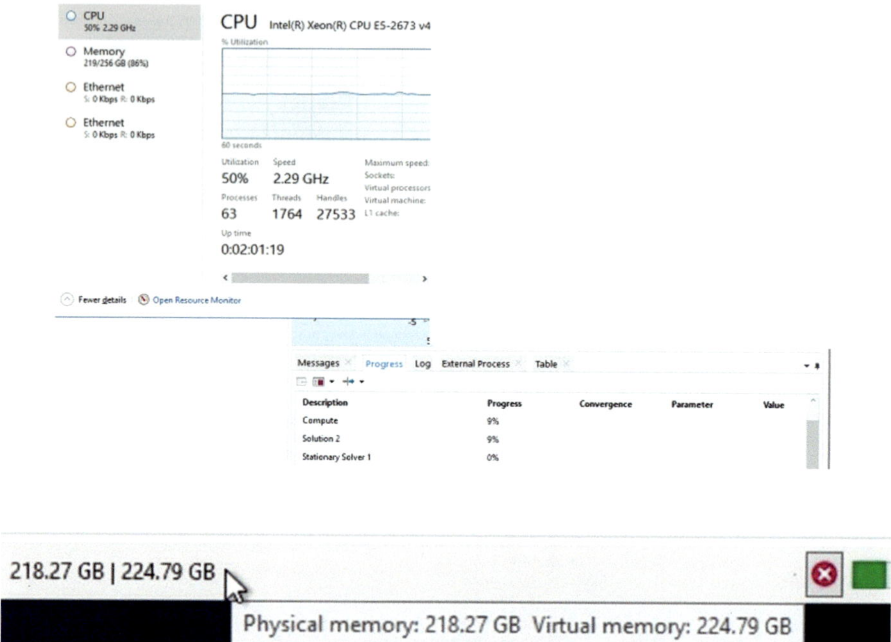

Fig. 2.104 Massive memory requirements for a fine-size mesh

A coarser mesh is used in order to reduce computational time consumption, memory usage, and thus cost for an initial computation.

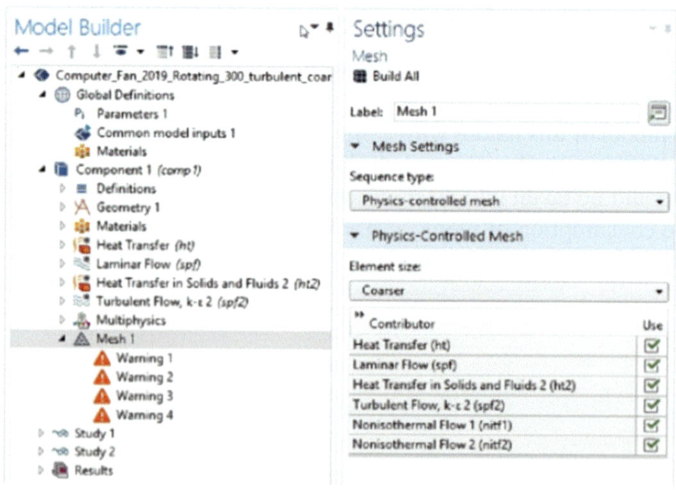

Fig. 2.105 Coarser mesh settings for turbulent-flow computations

First, the laminar flow study is run to obtain an initial solution that serves as an initial condition for the turbulent flow computations.

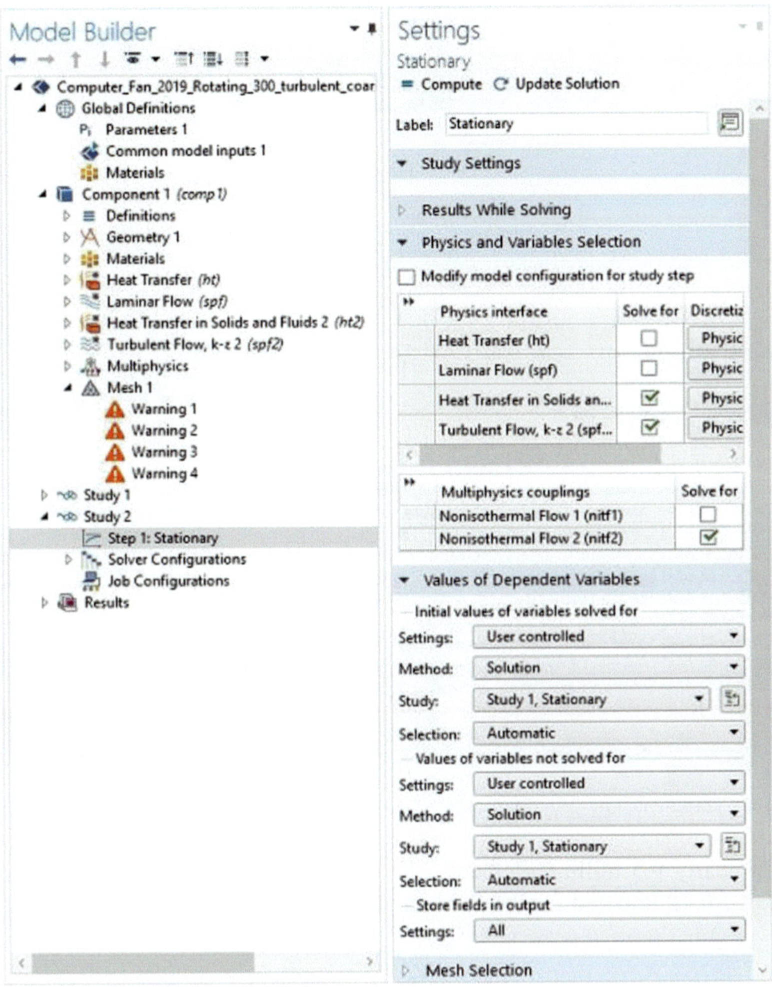

Fig. 2.106 Initial conditions for turbulent-flow computations

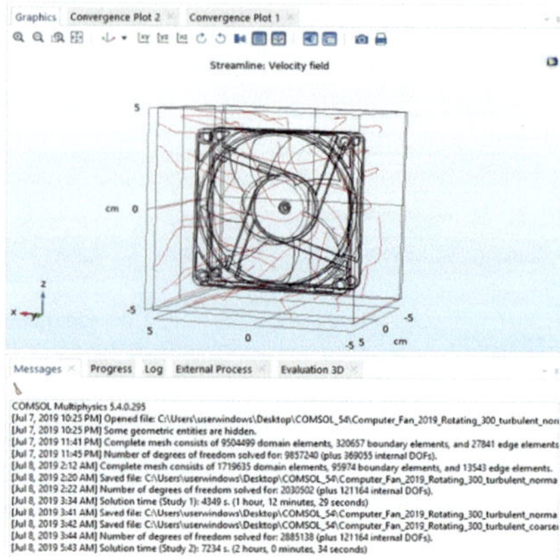

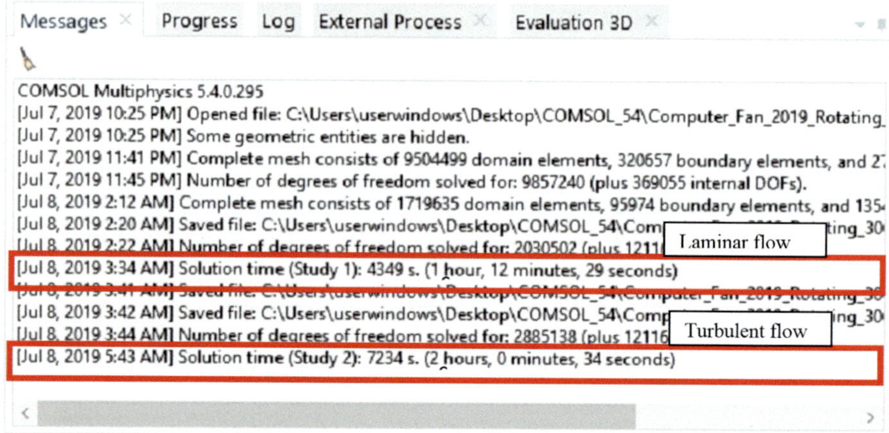

Fig. 2.107 Turbulent-flow velocity profile

With nearly 9.5 million elements, the computational times are for the laminar flow part 1 h 12 min and for the turbulent flow part 2 h 0 min.

Important Equations: Laminar Flow (spf)
Appendix

Equation 1. Navier–Stokes Equation

$$\rho\left(\mathbf{u}\cdot\nabla\right)\mathbf{u} = \nabla\cdot\left[-\rho\mathbf{I} + \mu\left(\nabla\mathbf{u} + \left(\nabla\mathbf{u}\right)^{\mathrm{T}}\right) - \frac{2}{3}\mu\left(\nabla\cdot\mathbf{u}\right)\mathbf{I}\right] + F$$

Equation 2. Continuity Equation

$$\nabla \cdot \left(\rho \mathbf{u} \right) = 0$$

Inlet

Equation 3. Inlet Velocity Equation

$$\mathbf{u} = -U_0 \mathbf{n}$$

Outlet

Equation 4. Pressure Equation

$$\left[-\rho \mathbf{I} + \mu \left(\nabla \mathbf{u} + \left(\nabla \mathbf{u} \right)^{\mathrm{T}} \right) - \frac{2}{3} \mu \left(\nabla \cdot \mathbf{u} \right) \mathbf{I} \right] \mathbf{n} = -\hat{\rho}_0 \mathbf{n} \; ; \qquad -\hat{\rho}_0 \le \rho_0$$

Important Equations: Heat Transfer in Solids (ht)

Equation 5. Heat Transfer in Solids Equation

$$\rho C_p \mathbf{u} \cdot \nabla T = \nabla \cdot \left(k \nabla T \right) + Q$$

Equation 6. Thermal Insulation Equation

$$-n \cdot \left(-k \nabla T \right) = 0$$

Equation 7. Heat Transfer in Fluids

$$\rho C_p \mathbf{u} \cdot \nabla T = \nabla \cdot \left(k \nabla T \right) + Q + Q_{vd} + Q_p$$

Further Readings

Data, A., & Rakesh, V. (2009). *Cambridge texts in biomedical engineering: An introduction to modeling of transport processes* (p. 532). Cambridge, UK: Cambridge University Press.

Pryor, R. W. (2011). *Multiphysics modeling using COMSOL v.4. A first principles approach* (p. 700). Sudbury, MA 01776, Mercury Learning and Information.

Zimmermann, W. B. (2006). *Multiphysics modeling with finite element methods* (p. 432). Singapore: World Scientific.

Chapter 3
Building a Prototype

3.1 The Solidworks Environment

1. Proceed to start Solidworks

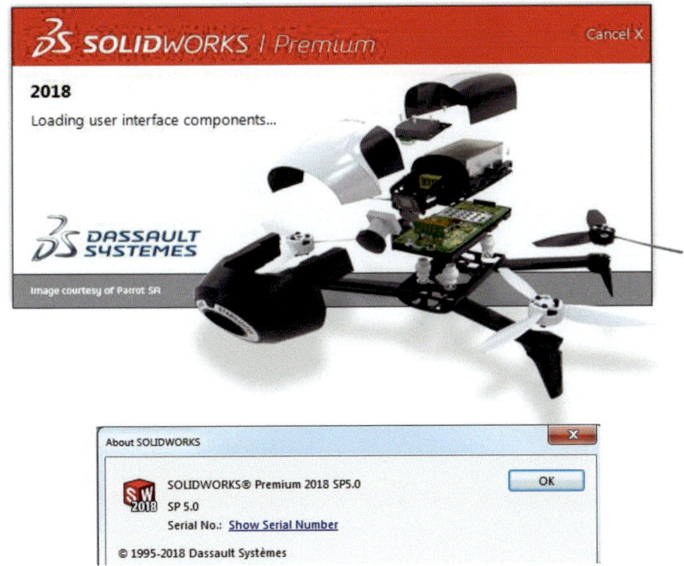

Fig. 3.1 Opening screen for Solidworks

© Springer Nature Switzerland AG 2020

B. A. Stradi-Granados, *Cloud Computing for Engineering Applications*,

https://doi.org/10.1007/978-3-030-40445-1_3

2.   Start to begin a new workfile

Fig. 3.2 On the first screen, select New

3.   From the menu select to create a new part

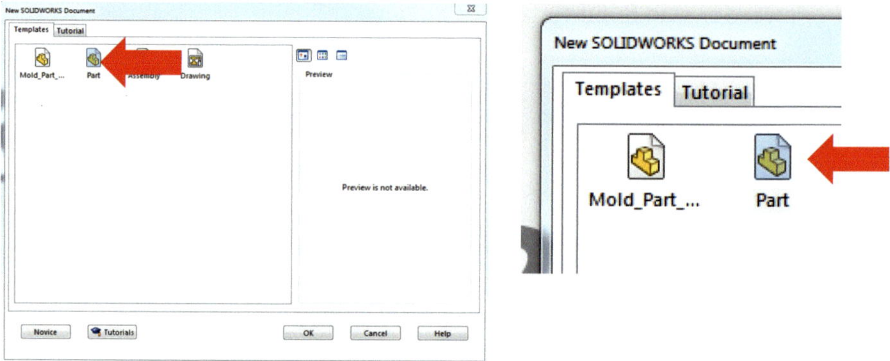

Fig. 3.3 Select Part from the Menu

4.   This is the working canvas for Solidworks. There are a number of features that will be covered on the road in more detail.

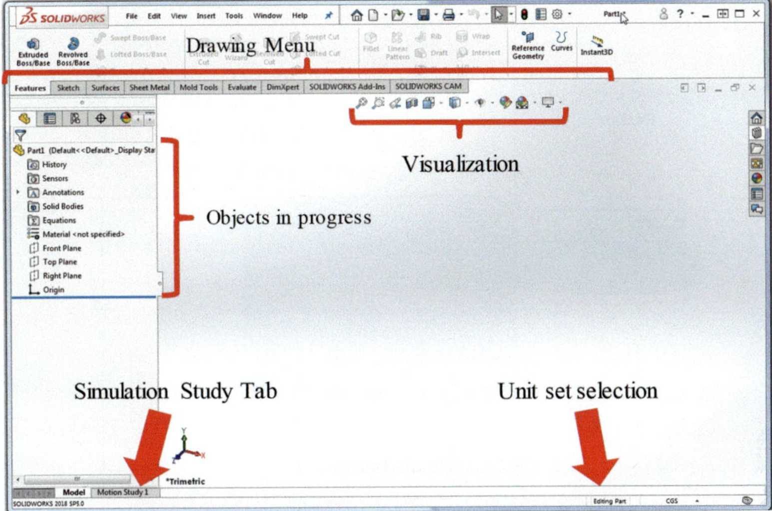

Fig. 3.4 Working canvas for Solidworks

5.   We have to carry out a series of operations to make our frame as a first step.

These operations are:	Create a frame in 2D
	Make necessary holes
	Draw internal features
	Detail geometry
	Reproduce symmetries
	Generate object

6. In order to work in order, it is important to know the elements to be drawn, this is the first element to be created.

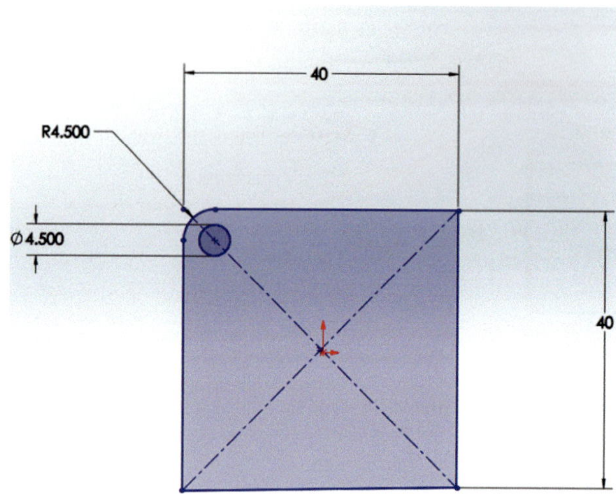

Fig. 3.5 First sketch to create

7. Initiate the drawing on the top plane.
   7a. Right-click on the Top Plane and click on the eye to show the plane.
   7b. The top plane turns light blue.
   7c. Right-click again and select the positioning Normal To.
   7d. Obtain your working plane perpendicular to line of sight.

7a,b   **7c**

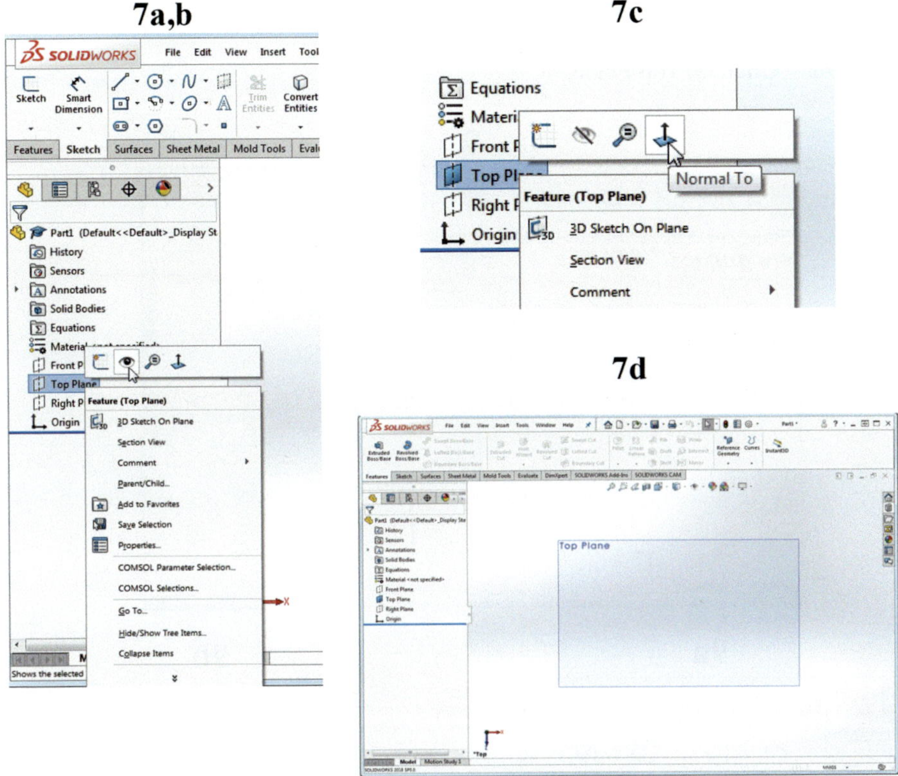

Fig. 3.6 Visualizing the drawing canvas

8.   Set the units set to your engineering or scientific specification using the sub-menu at the bottom.

  8a.   In this case, it is MMGS

8a

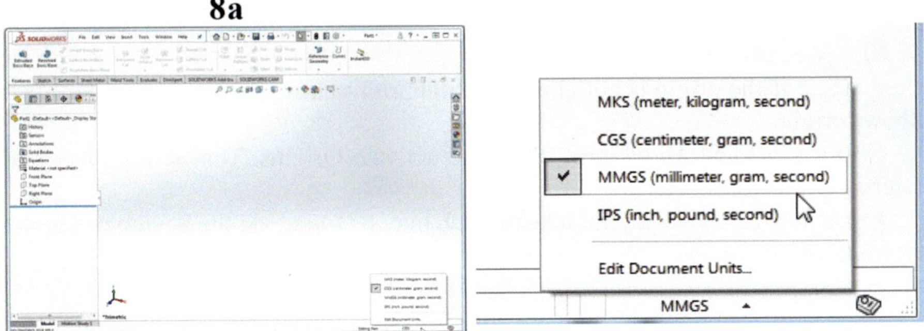

Fig. 3.7 Selecting measurement units

9.   This is essential, start with a Sketch.
   9a.   Click on Sketch.
   9b.   Click on Top Plane.

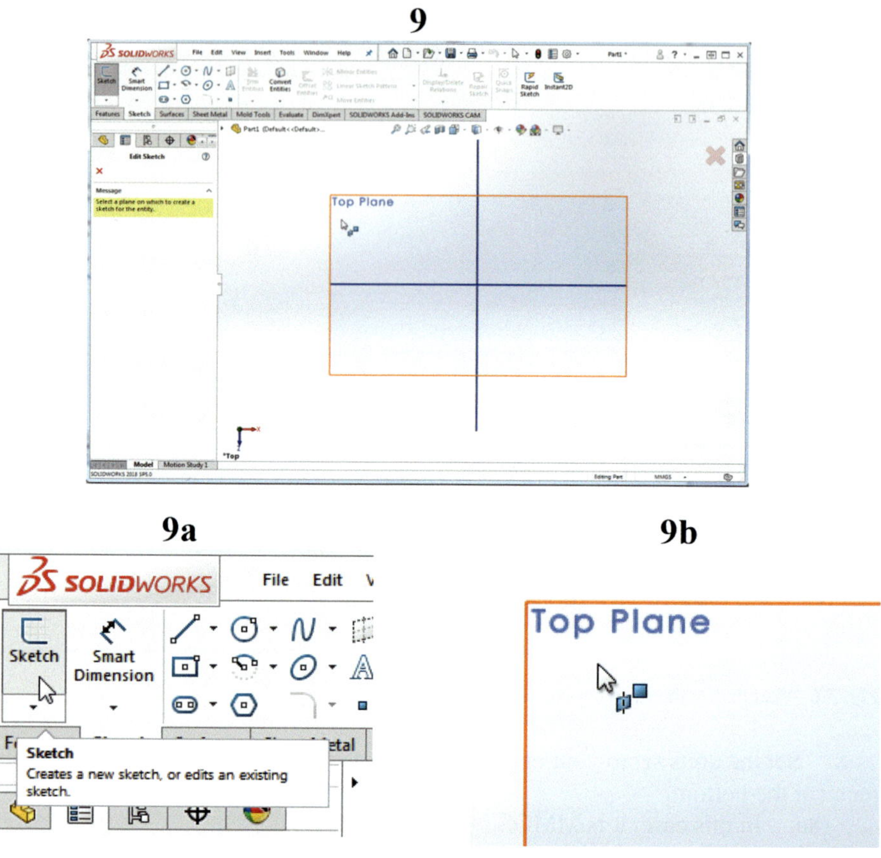

Fig. 3.8   Creating a sketch plane

10.   Visualize your origin.

10a.   If the origin is not already visible, right-click, choose the eye to select Shows origin.

10b.   In View, Hide/Show, Select to activate Origins. This action places an asterisk at the origin.

A grid may be activated, the selection tab may be found via the Search cell at the top right of the screen.

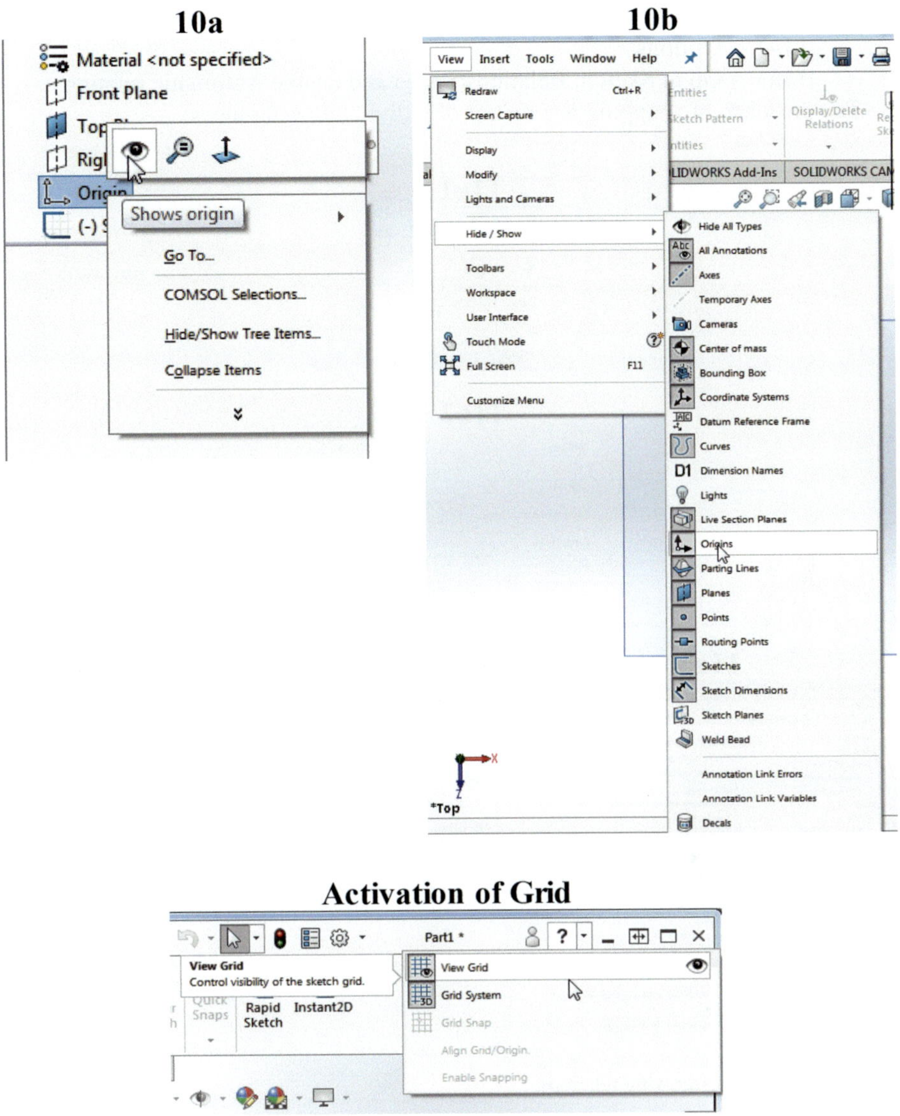

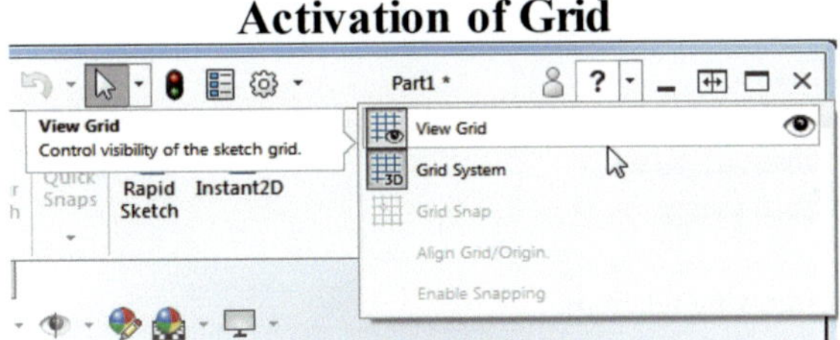

Fig. 3.9 Grid menu

11.   Draw a square with 40-mm side length and round the corners to 4.50-mm radius.

This is the first geometry drawn; consequently, we would like to get it right the first time.

11a.   Make sure that when drawing the characteristics of the figures remain proportional upon a change in size. This means if you draw a square and change the dimensions of one side, the other sides should follow suit.

11a1.   Go to Tools
11a2.   Options
11a3.   Go to Sketch, Relations/Snaps and enable Automatic relations.
The presence of snapping is a matter of individual preferences.

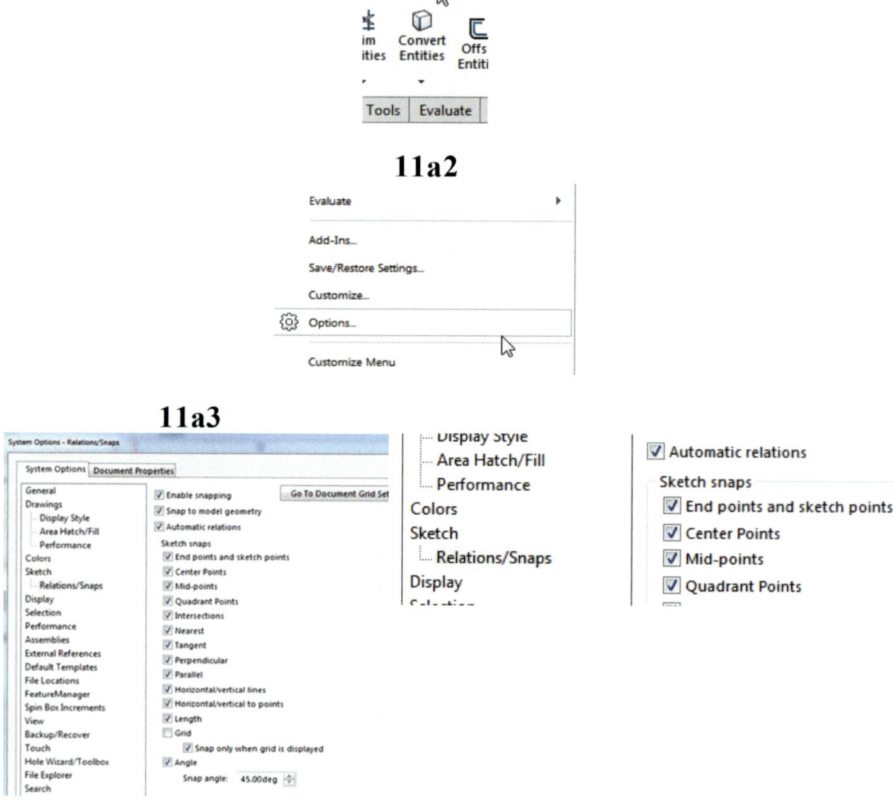

Fig. 3.10 Preserving drawing relations

12.   The problem in the past was that a square would not keep its shape when one side was modified and it required modifying all four sides.

12a.   Left-click at the origin. This is step of approximating the desired position.

12b.   Go directly to Center Rectangle and click on that selection.

12c.   Place the pen at the origin and drag outwards to a measurement close to 40 mm.

12d.   Refine the side measurements by clicking on Smart Dimensions.

12e. Click on the size numbers in blue and type in the right-side measurement.

12f. Refine the side measurements by clicking on Smart Dimensions and make the side length 40 mm.

12g. Repeat with the other side.

12h. The first geometry is now ready.

12a. Left-click at the origin. This is step of approximating the desired position.

Fig. 3.11 Origin location

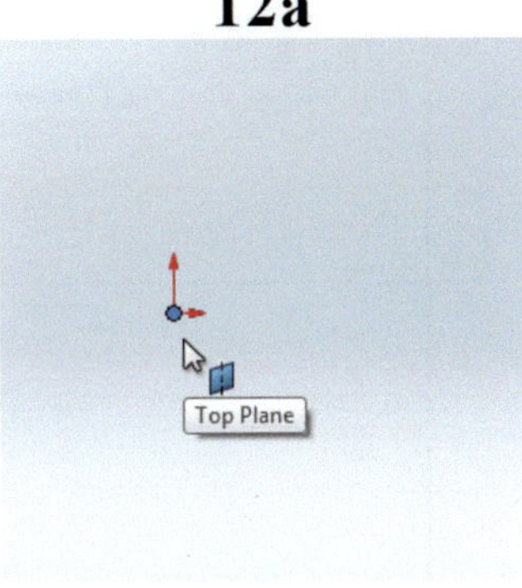

12b. Go directly to Center Rectangle and click on that selection.

12c. Place the pencil at the origin and drag outwards to a measurement close to 40 mm.

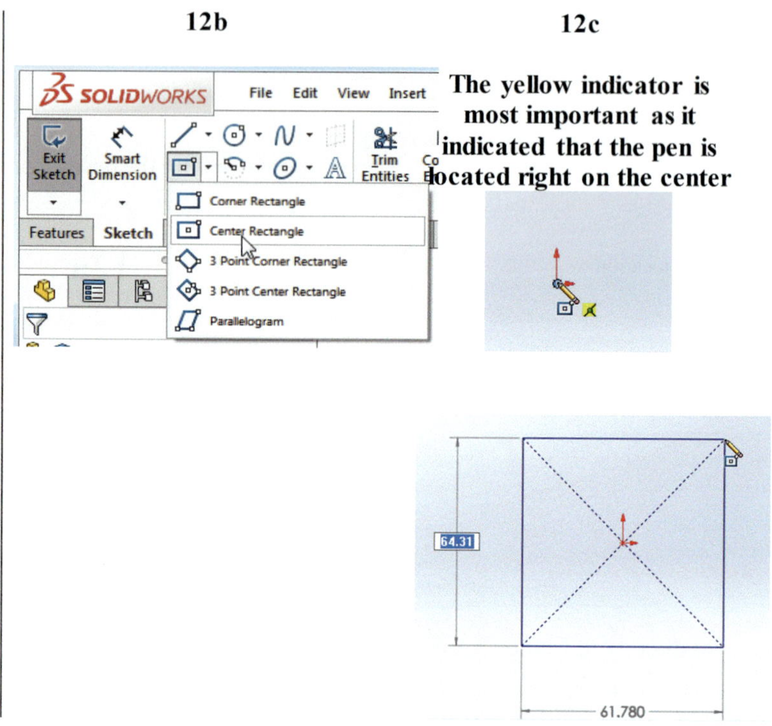

Fig. 3.12 Generating sketch rectangle

12d.   Refine the side measurements by clicking on Smart Dimensions.
12e.   Click on the size numbers in blue and type in the right-side
measurement.

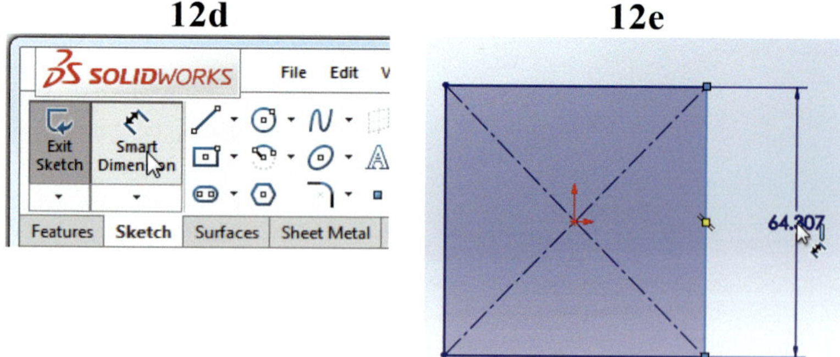

Fig. 3.13 Sizing the sketch rectangle

12f.   Refine the side measurements by clicking on Smart Dimensions and make the side length 40 mm.

12g.   Repeat with the other side.

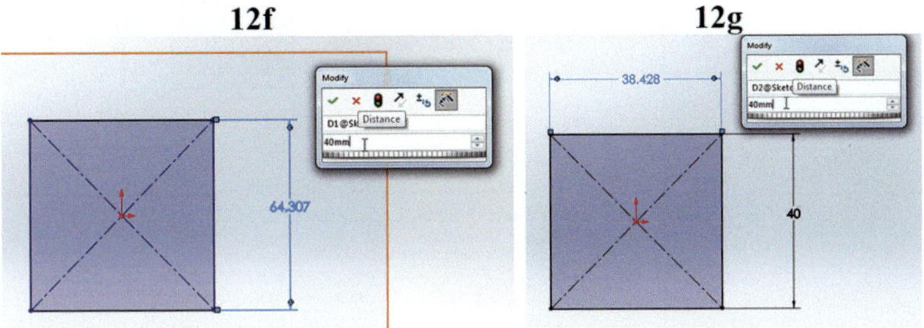

Fig. 3.14 Using Smart Dimension

12h.   The first geometry is now ready.

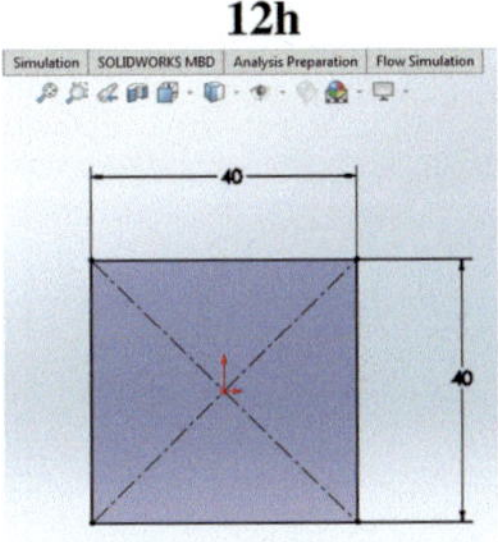

Notes

*   In this case the square is centered and only two sides had to be resized, the other two were automatically adjusted.

*   In the past the figure would loose the parallel feature of the sides when a dimension was modified. This implied separate resizing of all four sides.

*   Similarly sometimes a drawing is not centered exactly. It is possible to center the figure using Smart Dimension or Display/Delete Relations if necessary.

Fig. 3.15 Initial centered square

13.   Resize to fit the screen and review relation for the center of the figure.

13a.   Adjust the size of the square to the size on the screen using the Zoom to Fit tool.

13b.   The center of the figure is fixed to the center of the canvas: the center of each diagonal is coincident with the center of the canvas (two relations), and the centers of the diagonals are coincident at their midpoint (one relation).

13a

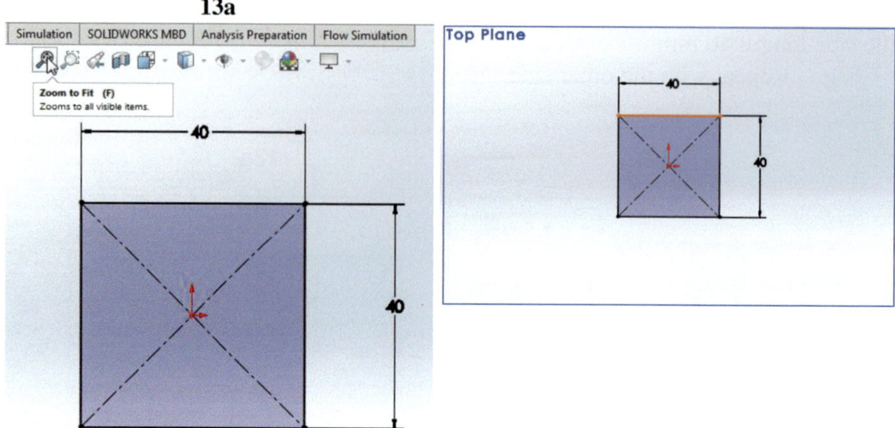

Fig. 3.16 Zoom to Fit

13b

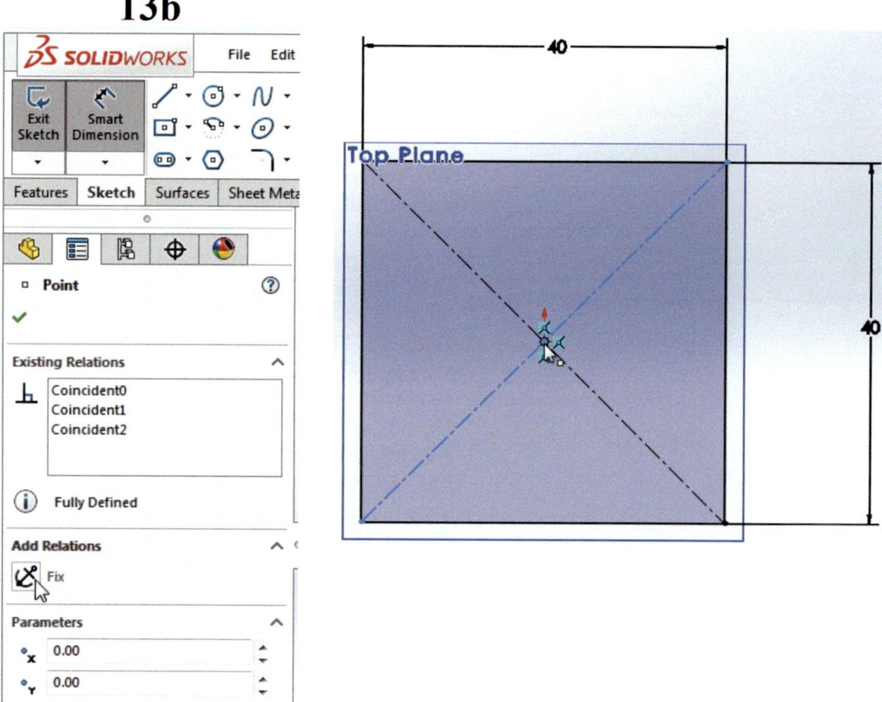

Fig. 3.17 Fixing the center in place

14.   Proceed to round the corners to a radius of 4.50 mm.
   14a.   Select Sketch Fillet
   Specify the radius as 4.5 mm
   14b.   Select two sides: select one side and while holding the Crtl key down select the other; the rounded corner appears. Then click on the Check mark.

14a

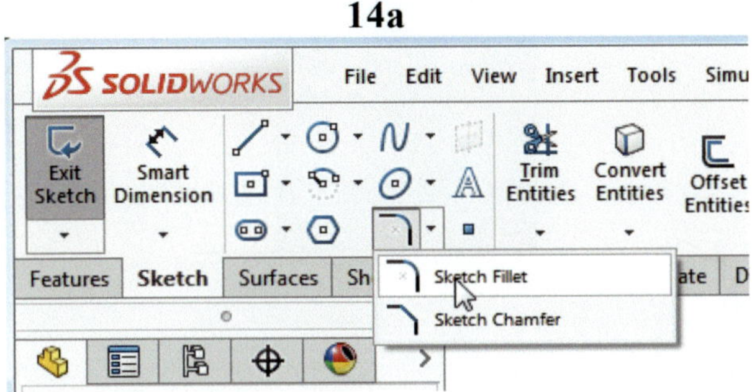

Fig. 3.18   Sketch Fillet

14b

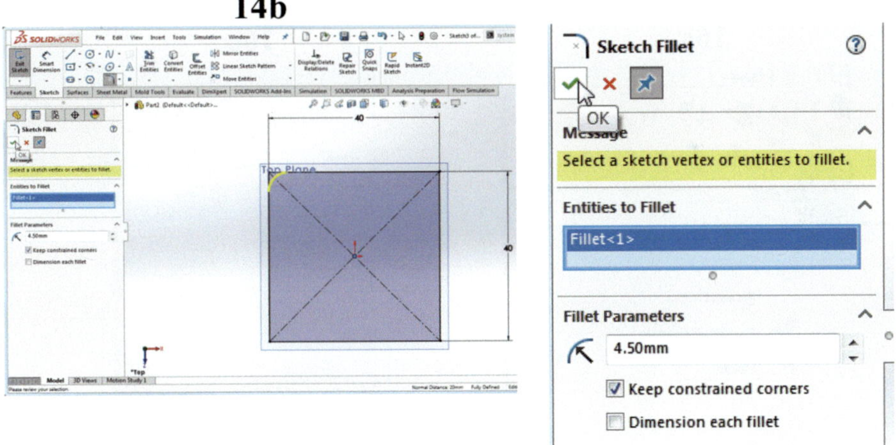

Fig. 3.19   Sketch Fillet at 4.50 mm

15.   The figure should look like the one below.
   15a.   Before advancing, click on the Traffic Light at the top. That action will take the construction process out of the Sketch option.

15a

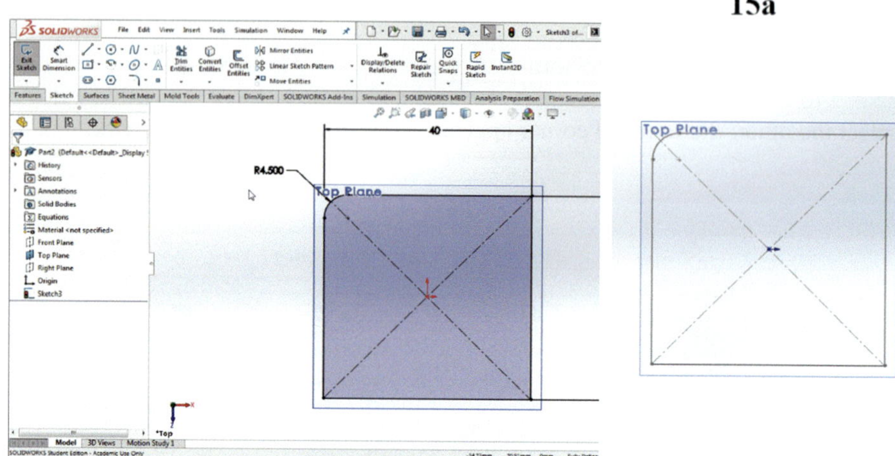

Fig. 3.20 Redrawing out of Sketch Mode

16.   Return to Sketch mode to continue the drawing.
    16a.   On the left-hand menu, right-click on Sketch1 and select Edit Sketch.
    16b.   The drawing should have the lines in blue color and with the indication
of dimensions.

16a   **16b**

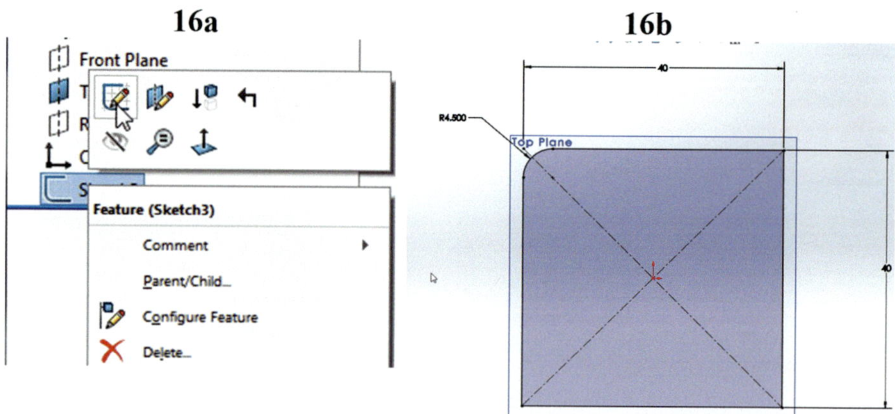

Fig. 3.21 Return to sketch mode (dimensions reappear)

17.   Create a Circle of 4.50 mm in diameter on the top left-hand side corner.
    17a.   Select the circle geometry from the menu.
    17b.   Place the circle at the center of the radius of curvature of the
rounded corner.
    17c.   Size the circle to 45 mm in diameter using Smart Dimension.
    17a.   Select the circle geometry from the menu.

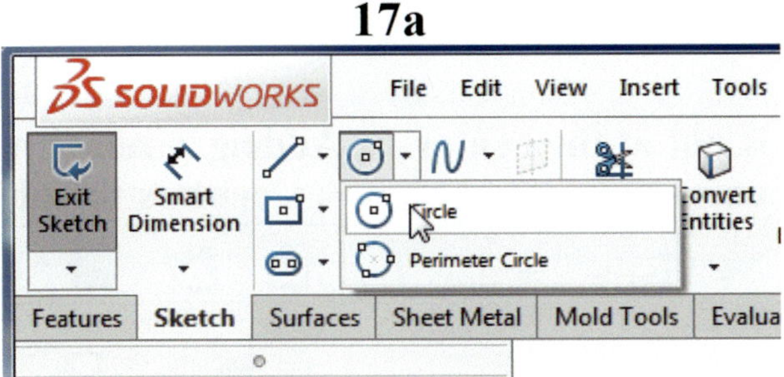

Fig. 3.22 Select circle geometry

You can use the *Scroll Wheel* to magnify (minify) the feature of interest.
Left-click on the Center of Curvature.

Move to the menu, select and draw the circle with 45-mm diameter (or close to that measurement).

17b.   Place the circle at the center of the radius of curvature of the rounded corner.

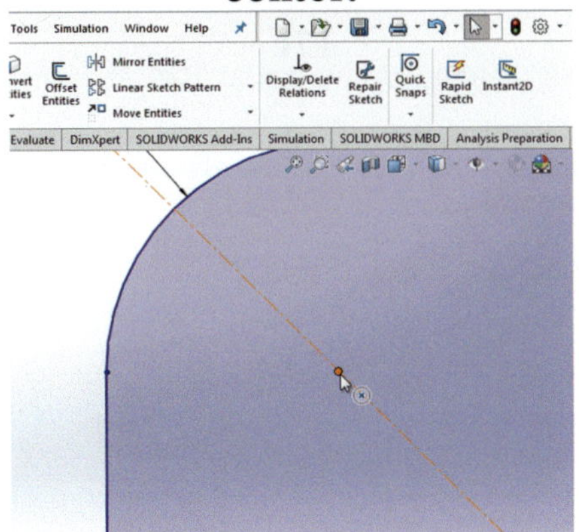

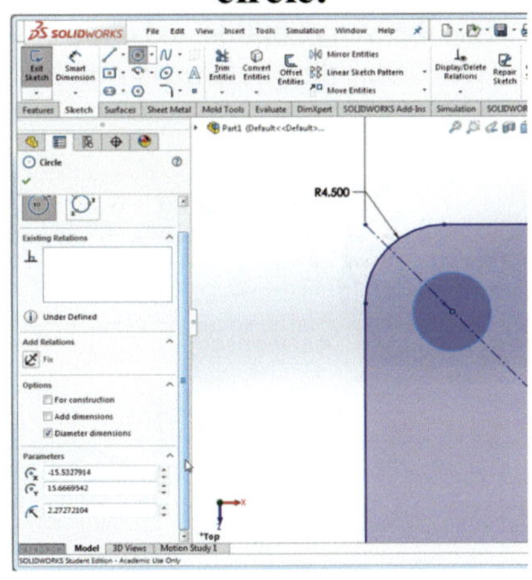

Fig. 3.23 Select circle geometry

The previous procedure is sometimes cumbersome and a more stylized process is preferred.

Draw the circle separated from the center of curvature and add a relation between the two centers: the center of curvature and the center of the circle.

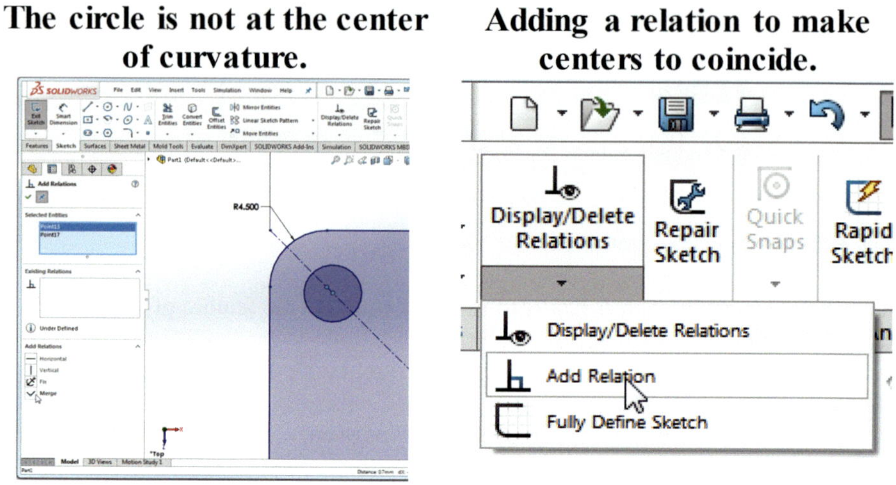

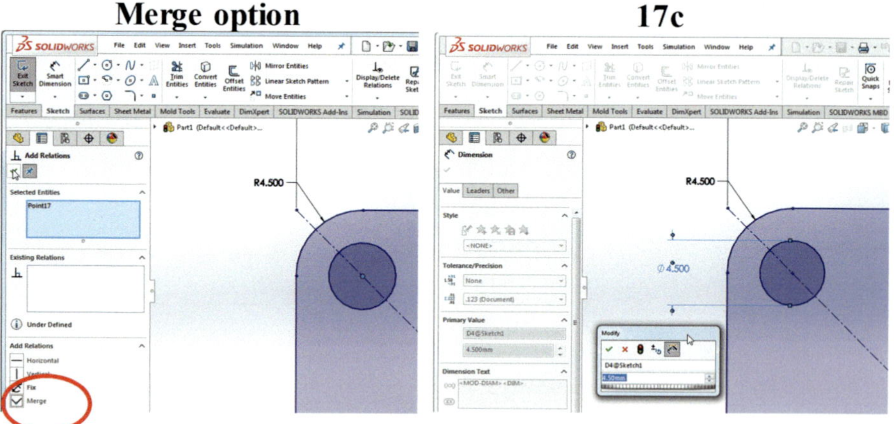

Fig. 3.24 Add relation to overlay centers of curvature

Use the Merge option and click on the check mark.

17c.   Size the circle to 45 mm in diameter using Smart Dimension.

Fig. 3.25 Merge centers and dimension of the circle

This is result.

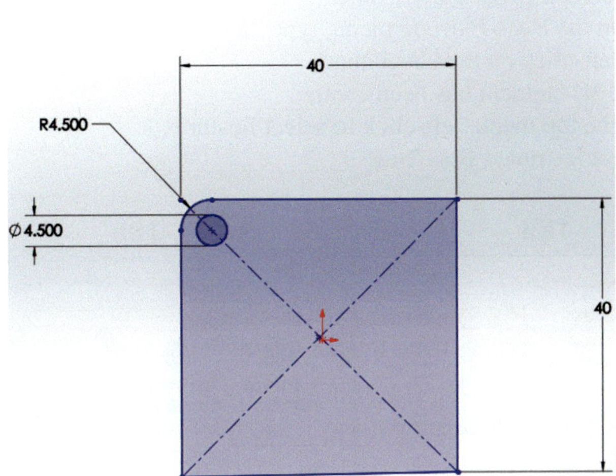

Fig. 3.26 Resulting part with dimensions

It is possible to move the figure around the screen using the combination of simultaneously holding the *Ctrl key* and pressing down on the *Scroll Wheel*.

We have created the first element making use of a few features of Solidworks. The second element to be created is the following.

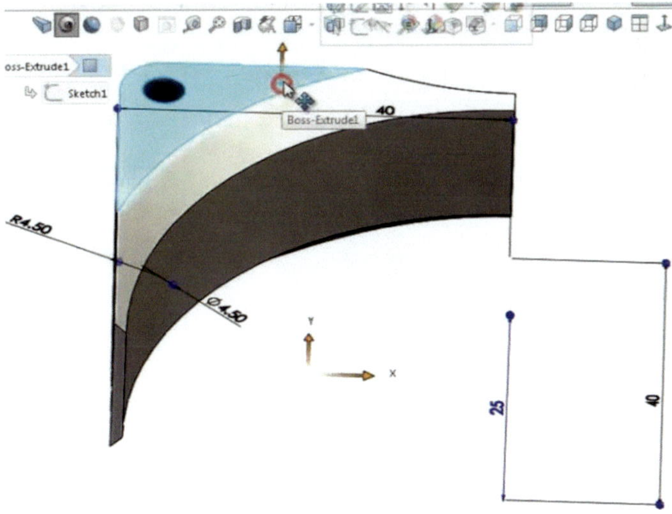

Fig. 3.27 Second element to be created

18.   Create a 3D object with a depth of 25 mm.
    18a.   On the top menu, left-click to select Features
    18b.   Select Extruded Boss/Base
    18c.   On the Boss-Extrude menu, type 25 mm on the Depth option
    18d.   Left-click on the check mark
    18e.   A 3D element has been created
18a.   On the top menu, left-click to select Features
18b.   Select Extruded Boss/Base

18a   **18b**

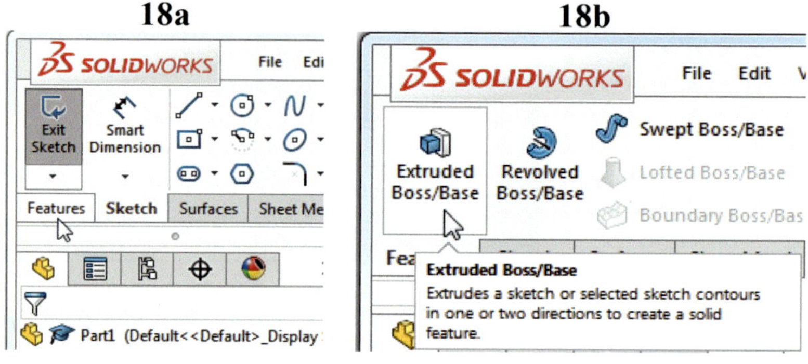

Fig. 3.28 Extruding Boss/Base

18c.   On the Boss-Extrude menu, type 25 mm on the Depth option

18c

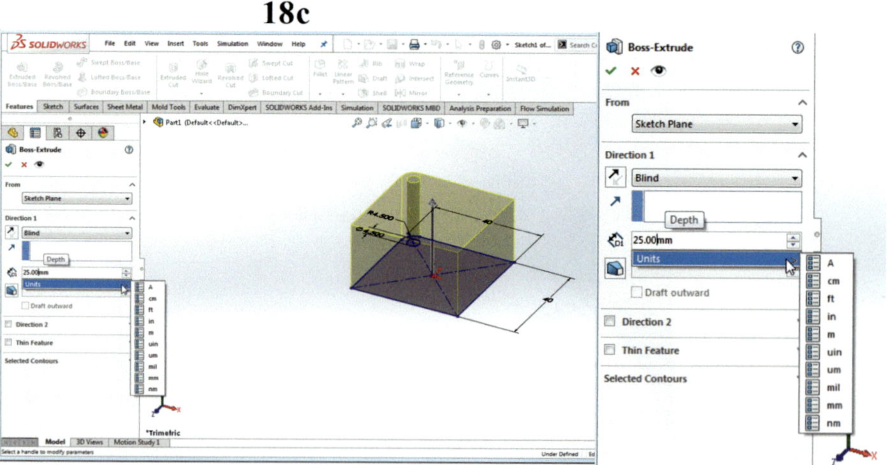

Fig. 3.29 Apply depth of 25 mm

18d.   Left-click on the check mark

18d

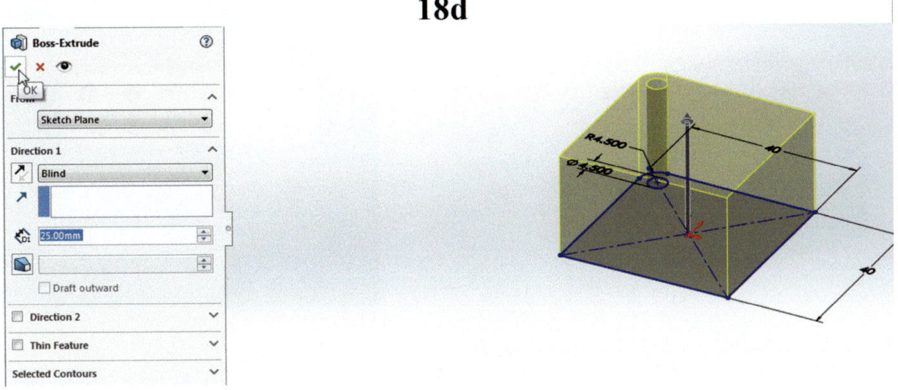

Fig. 3.30 Accept current geometry

18e.   A 3D element has been created

Fig. 3.31 Current
geometry

18e

Next, there is the subject of learning to draw on the side of the cubical shape already created. This is an important topic as the shape drawn on the face of the cubical shape is rotated in later steps.

19.   Create a profile with lines on the side of the cubical shape to form a closed perimeter.

19a.   Left-click on the side of the cubical shape. That side will be the new drawing plane.

19b.   Left-click on the perpendicular orientation (Normal To) so that it becomes the new drawing canvas.

19a.   Left-click on the side of the cubical shape. That side will be the new drawing plane.

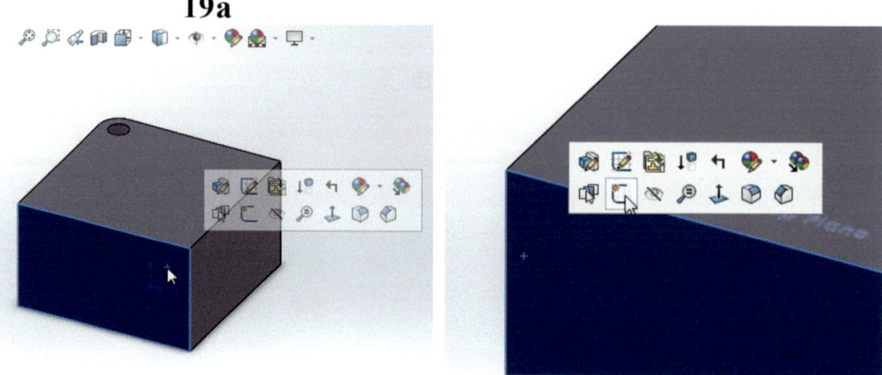

Fig. 3.32 Selecting a side of the volume to draw

19b.   Left-click on the perpendicular orientation (Normal To) so that it becomes the new drawing canvas.

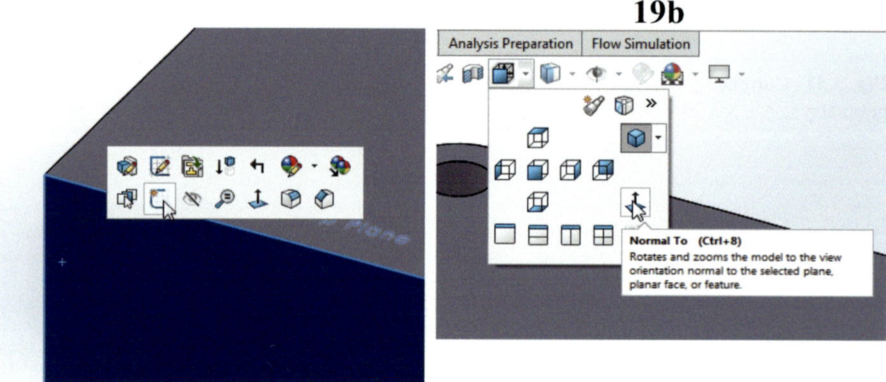

Fig. 3.33 Orientation of plane to draw

Fig. 3.34 Side plane
position for sectioning of
volume

On the new plane, create this new figure

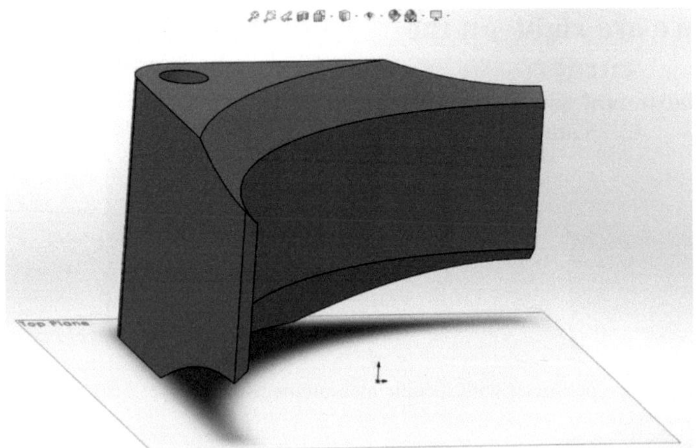

Fig. 3.35 Volume sectioned

20.   Create a closed perimeter using linear segments with measures of 45.75 (bottom horizontal edge), 38.75 (as shown in the figures), 42.75 (top horizontal edge) mm and a vertical segment to the right closing the figure.

20a.   Select to sketch two lines (one at the top and one at the bottom) from the menu and approximate segment dimensions indicated. Notice the relation indicators that appear as you place the pencil over the vertices.

20b.   Once the initial sketch of the geometry is ready, left-click on the check mark (top corner created).

20c.   Refine the side measurements to the indicated values using Smart Dimension.

20d.   Refine the angles to the indicated values using Smart Dimension.

20a.   Select to sketch two lines (one at the top and one at the bottom) from the menu and approximate segment dimensions indicated. Notice the relation indicators that appear as you place the pen over the vertices.

20a
The yellow marker indicates that we are right on the corner
(draw a horizontal line to the left of 42.75 mm).

The yellow marker indicates that we are right at a vertex and that the line is horizontal
(draw a horizontal line to the left of 45.75 mm).

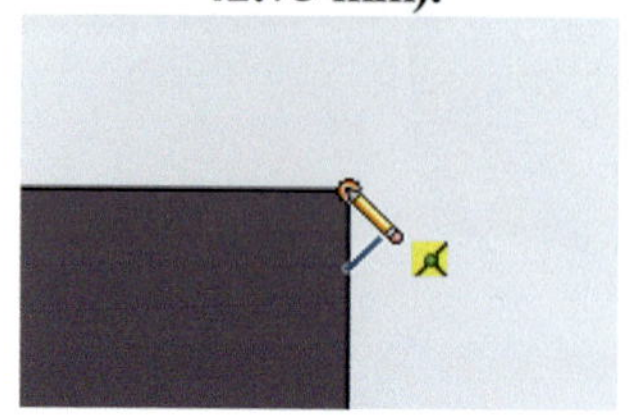

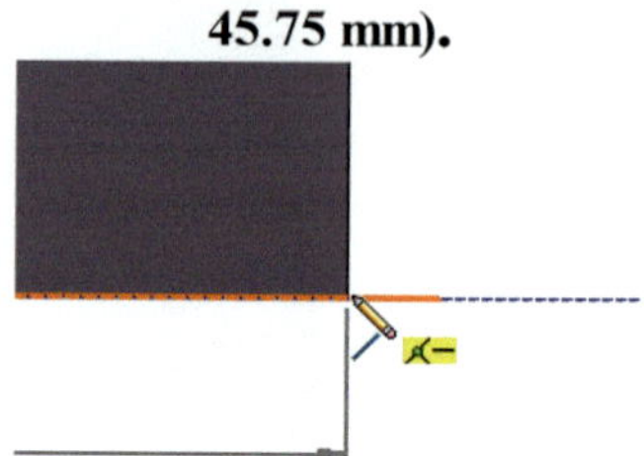

Fig. 3.36 Generate a perimeter with specific measurements

20b.   Once the initial sketch of the geometry is ready, left-click on the check mark.

The yellow markers indicate that we are coincident with the initial point and that the line is vertical.

20b. Top corner created (repeat procedure at the bottom)

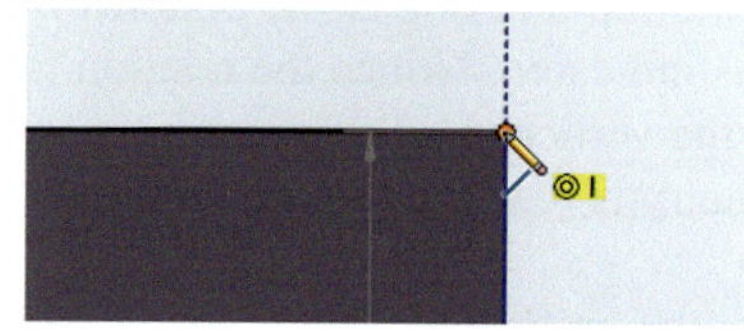

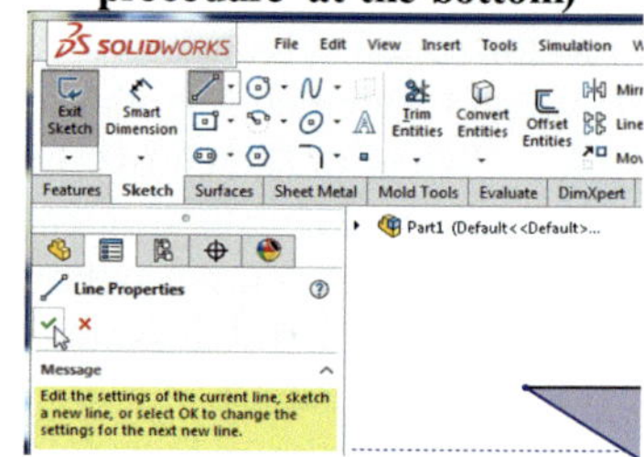

Fig. 3.37 Verify indicator of alignment during construction

In order to create the measurement 38.75 mm, press the Crtl key and while pressing it, left-click on the two vertical elements on the sketch. Then establish the required length between the two elements.

20c.   Refine the side measurements to the indicated values using Smart Dimension.

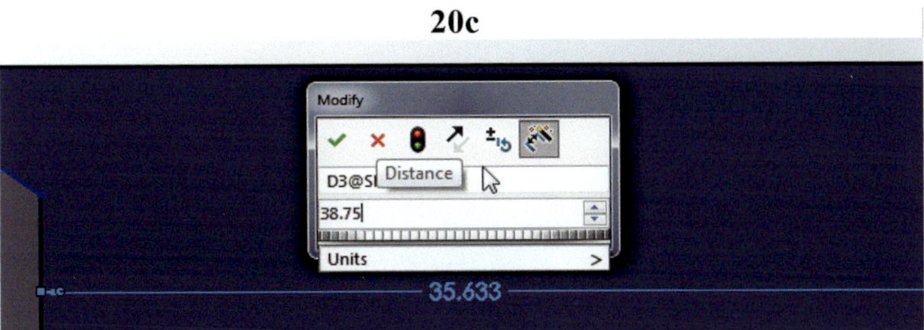

Fig. 3.38 Refine measurements

In order to create the 45° angle, press the Crtl key and while pressing it, left-click on the two sides that share the vertex and type the angle of interest.
    20d.   Refine the angles to the indicated values using Smart Dimension.

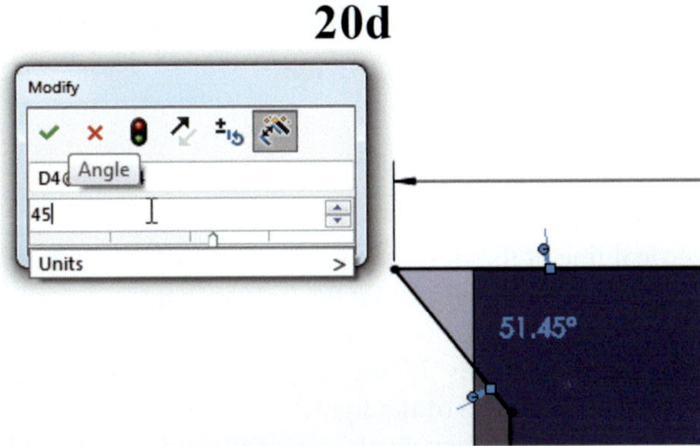

Fig. 3.39 Apply angle of 45° to one corner

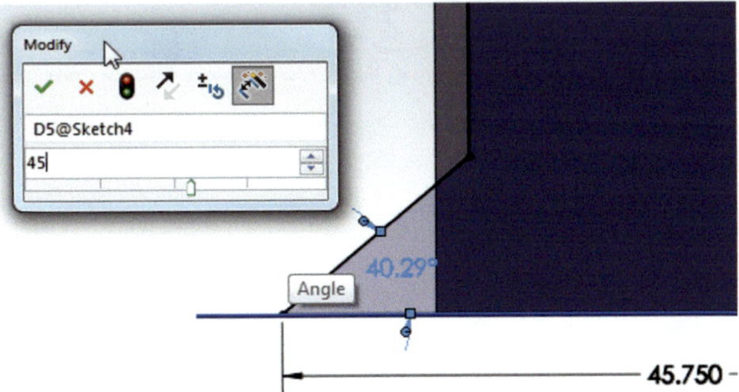

Fig. 3.40 Apply angle of 45° to the second corner

Now, we have the construct in 2D.

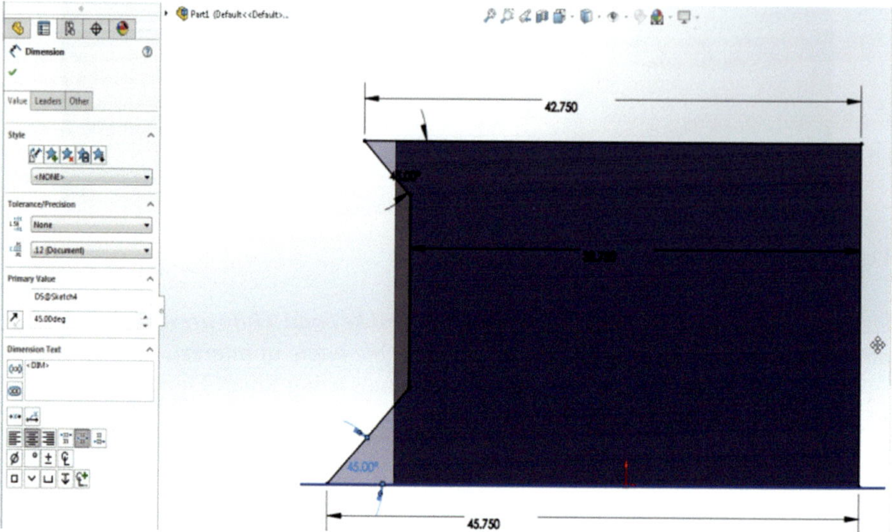

Fig. 3.41 Planar figure to use as mandrel

There is a vertical line at the right-hand side that closes the figure in this plane.

21.   Now rotate the figure and cut out the excess parts.

21a.   Select Revolved Cut.

21b.   Select the Axis of Revolution on the left-hand menu. Click on the vertical line segment at the right edge of the figure.

21c.   Left-click on the green check mark and evidence the 3D cut figure just generated.

21a.   Select Revolved Cut.

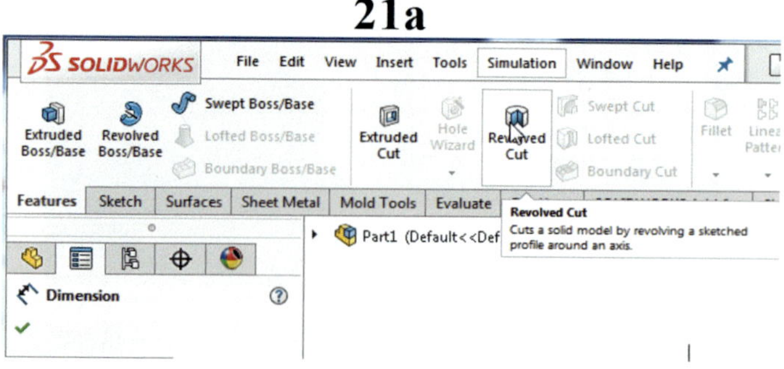

Fig. 3.42 Revolved cut menu

21b.   Select the Axis of Revolution on the left-hand menu. Click on the vertical line segment at the right edge of the figure.

21b

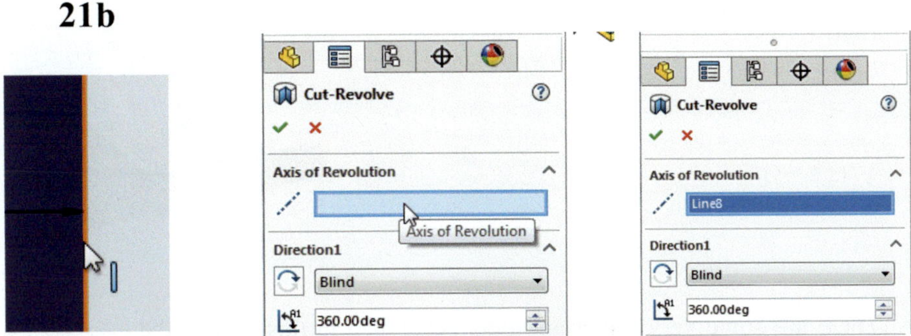

Fig. 3.43 Selecting the axis of revolution

21c.   Left-click on the green check mark and evidence the 3D cut figure just generated.

21c

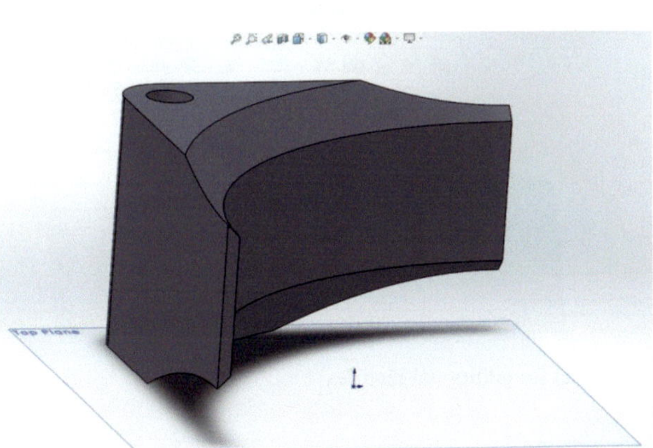

Fig. 3.44 First segment of the three-dimensional casing

We are going to construct the following figure from the previous results.

Fig. 3.45 Eliminating
large portions of extra
material

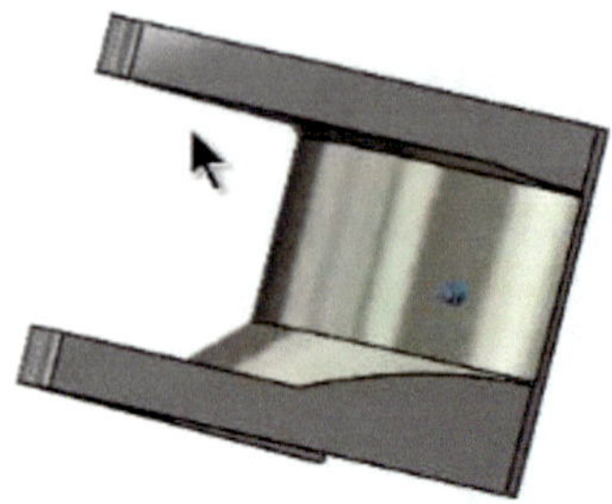

There are some very important tools developed in this part of the exercise by
using Convert Entities.

22.   Create a plane on the area that bears the hole.

   22a.   Left-click on the area.

   22b.   Right-click on the area and choose Sketch.

   22c.   Select the region again and make it perpendicular to your line of sight.

   22a.   Left-click on the area.

   22b.   Right-click on the area and choose Sketch.

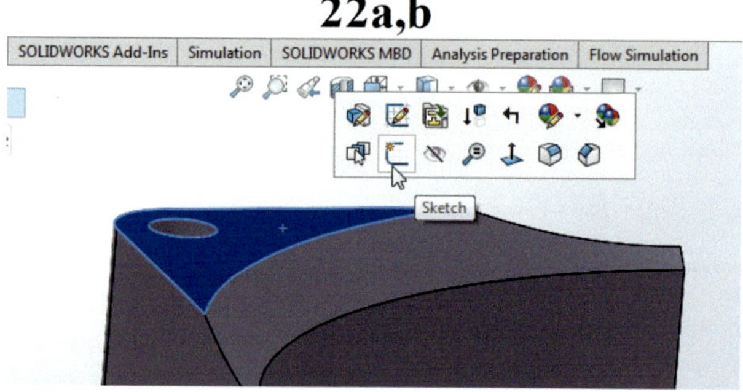

Fig. 3.46 Selecting plane for additional features

22c.   Select the region again and make it perpendicular to your line of sight.

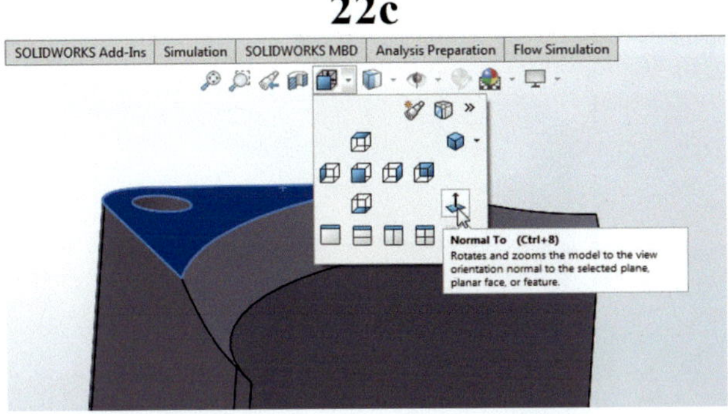

Fig. 3.47 Orientation of plane to draw

23.   Create a thin border using the perimeter of the part already drawn.

23a.   Press and hold the Ctrl key and left-click on the lines or elements of the external border.

23b.   Select Convert Entities. Notice that the black line that is created on the plane, from the projection of elements below the plane. This is a very important feature because it allows using underlying geometry to design additional features for perfect fit.

23c.   Select the external border just created. Using Offset Entities create another line parallel to the border at an internal distance of 1.25 mm. Left-click the check mark.

23d.   Join the two-line profiles at the ends using the line tool. Extrude the profile using the Extruded Boss/Base selection. Make sure to extrude from one side of the object to the other indicated using the selection Up To Surface (need to click on the surface that defines the limit to the extrusion in this case Face<1> on the side opposite to the drawing plane).

23a.   Press and hold the Ctrl key and left-click on the lines or elements of the external border.

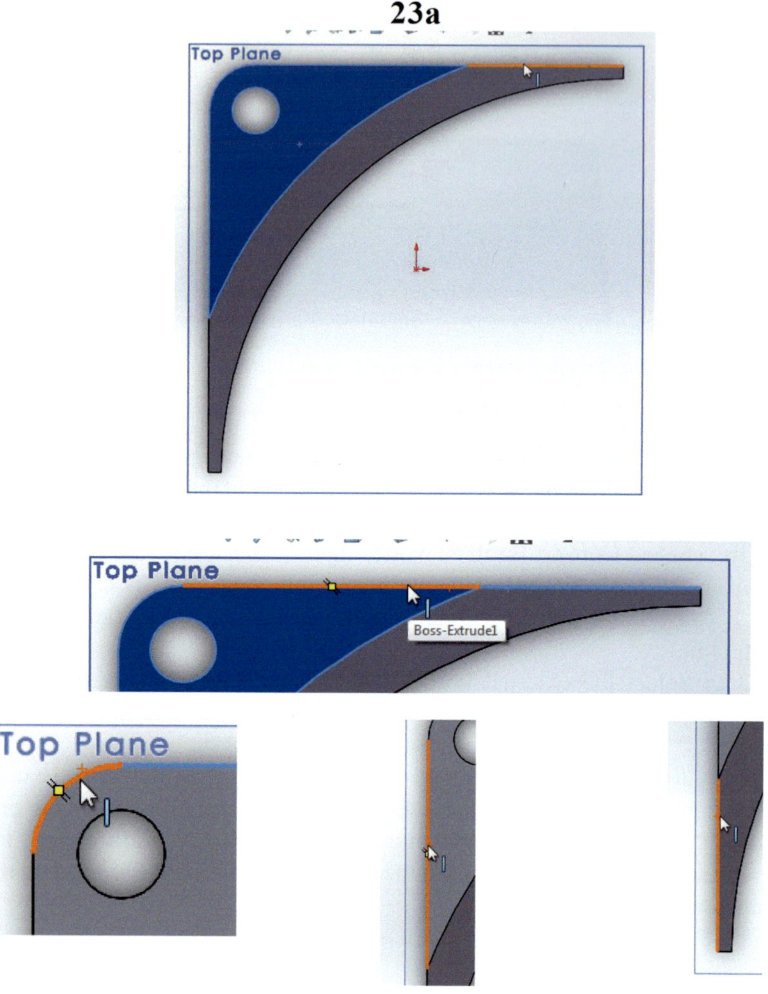

Fig. 3.48 Lines for finer sketch features

The selected external border is highlighted with light blue color.

Fig. 3.49 External border selected for sketching

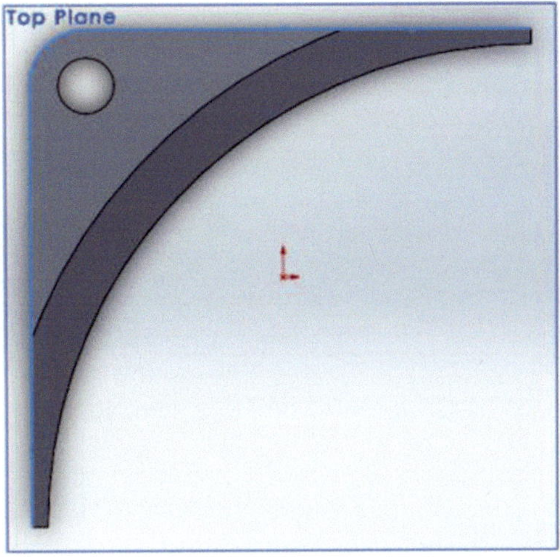

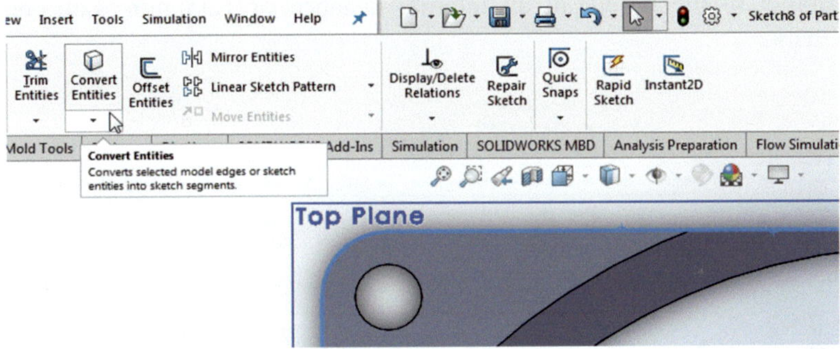

Fig. 3.50 Use Convert Entities

23b.   Select Convert Entities. Notice that the black line that is created on the plane, from the projection of elements below the plane. This is a very important feature because it allows using underlying geometry to design additional features for perfect fit.

Fig. 3.51 Transfer border profile to sketch plane

23c.   Select the external border just created. Using Offset Entities create another line parallel to the border at an internal distance of 1.25 mm. Left-click the check mark.

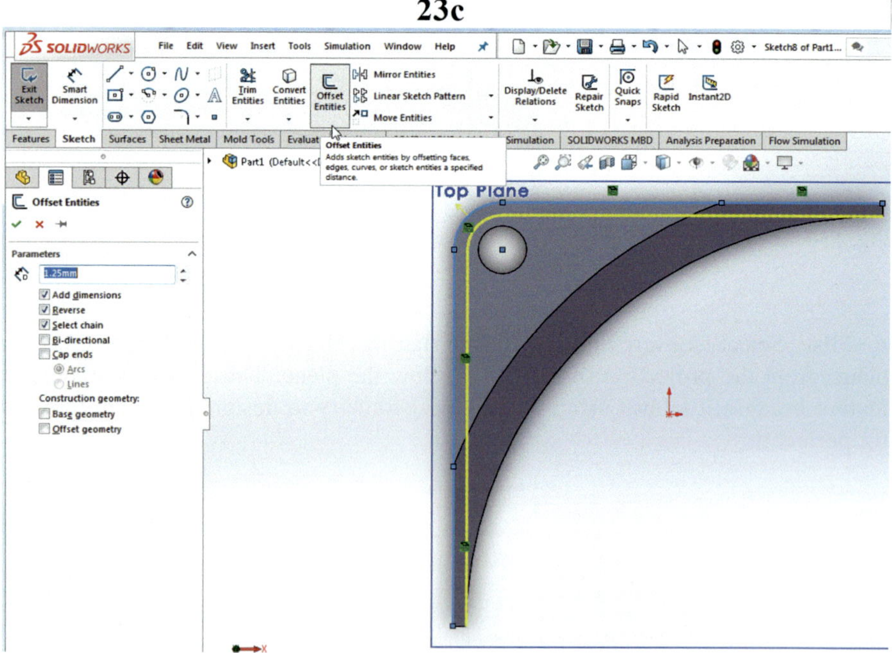

Fig. 3.52 Offset entities

It is important to notice that the points are on a plane that took its guide lines from the elements below the plane and that are not in contact with the sketch plane: basically drawing in the air.

23d.   Join the two-line profiles at the ends using the line tool.

23d

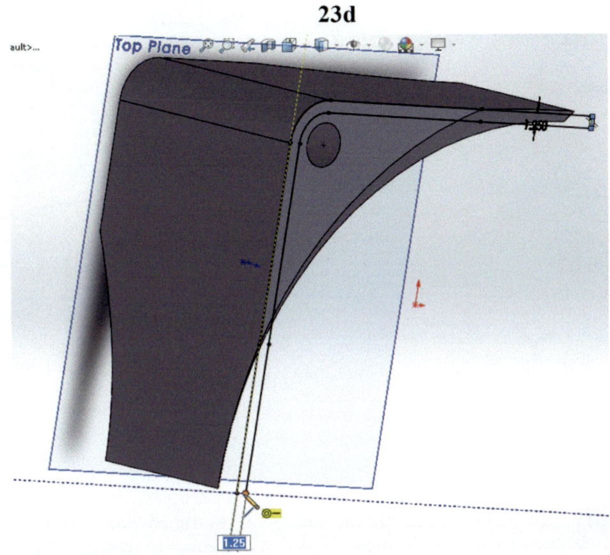

Fig. 3.53 Join the two perimeter line profiles

In the following process to extrude the area between the lines, the selection is Extruded Boss/Base on the top menu, but the choice of Up To Surface on the side menu *requires turning the figure around to choose the face on the opposite side.*

23e.   Extrude the profile using the Extruded Boss/Base selection. Make sure to extrude from *one side of the object to the other side* using the selection Up To Surface (need to click on the surface that defines the limit to the extrusion in this case Face<1> on the side opposite to the drawing plane).

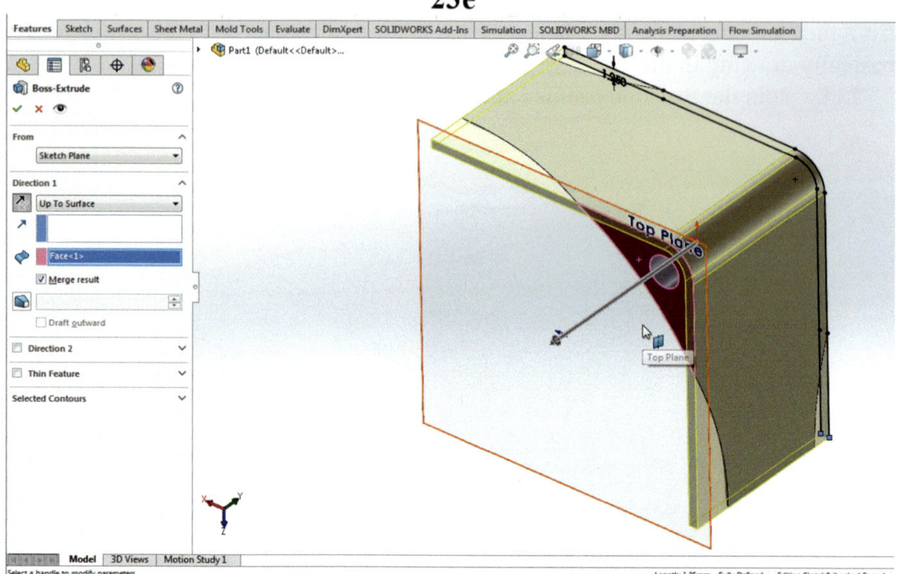

Fig. 3.54   Extrude profile up to surface

The geometry has now rectangular sides. It is necessary to trim more of the object, by now it should be clear that a new geometry is going to be created on an auxiliary sketch to be rotated in order to complete the trim.

Fig. 3.55   Straight edges
of the part

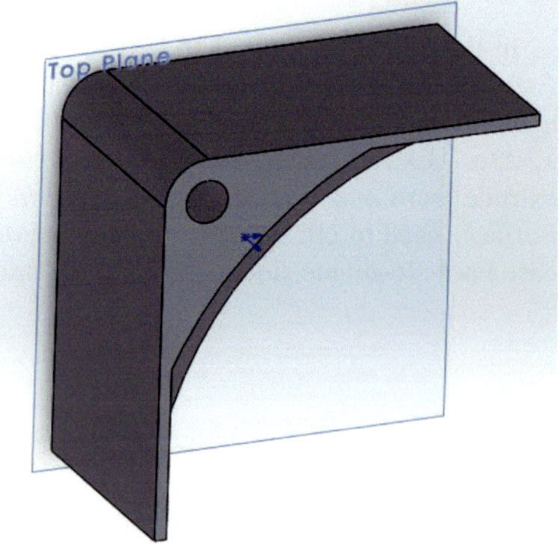

The step is to trim further this 1/4 of the total figure.

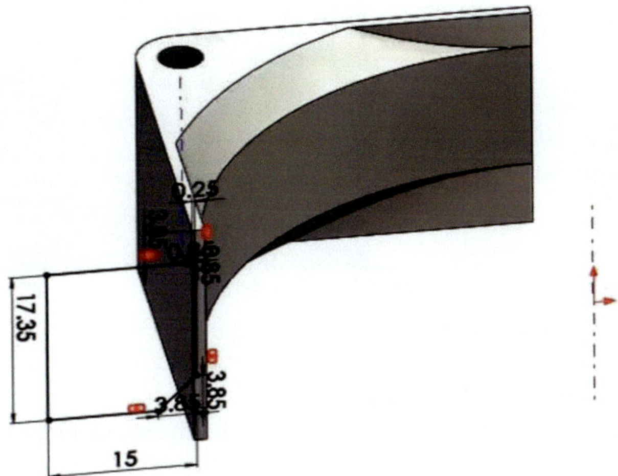

Fig. 3.56 Additional trimming of part

24.   Create a rectangle with trimmed corners.

24a.   Click and hold the scroll wheel to rotate the part into position. Select the thinner outermost edge of the part. Place the Sketch plane perpendicular to the screen.

24b.   On the Sketch plane draw a rectangle of dimensions 17.35 × 15 with 0.25 depth into the material and 3.85 mm from the outer edge. Draw with approximate dimensions and sharpen with Smart Dimension.

24c.   Cut the corners using the Chamfer tool with dimensions of 0.85 and 3.85 mm.

24d.   Visualize the temporary axes needed for rotation.

24e.   Trim the piece using the Revolved Cut tool.

Left-click, place it perpendicular to the screen.

24a.   Click and hold the scroll wheel to rotate the part into position. Select the thinner outermost edge of the part. Place the Sketch plane perpendicular to the screen.

Fig. 3.57   Select thinner edge to start sketch

Select sketch and draw the rectangle

24b.   On the Sketch plane, draw a rectangle of dimensions 17.35 × 15 with 0.25 depth into the material and 3.85 mm from the outer edge. Draw with approximate dimensions and sharpen with Smart Dimension.

24b

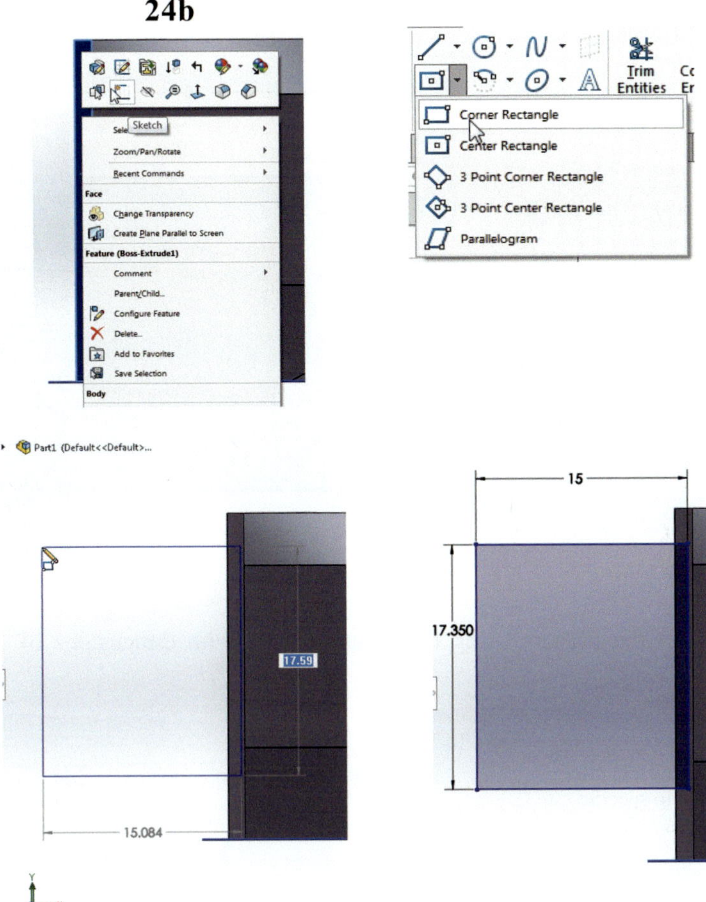

Fig. 3.58 Create a rectangle on the thin edge

Measure the distances from the side of the piece to the side of the rectangle using Smart Dimension. Click and Hold the Crtl key while choosing the sides with the mouse.

Fig. 3.59 Apply Smart Dimension to refine mandrel

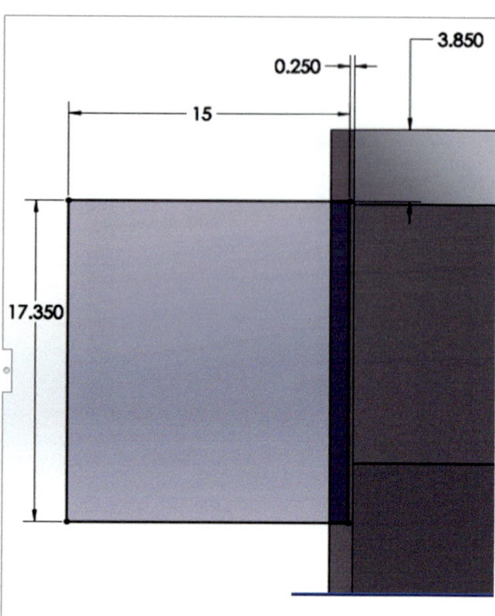

24c.   Cut the corners using the Chamfer tool with dimensions of 0.85 and 3.85 mm.

Fig. 3.60 Using Chamfer to cut corners to specific dimensions

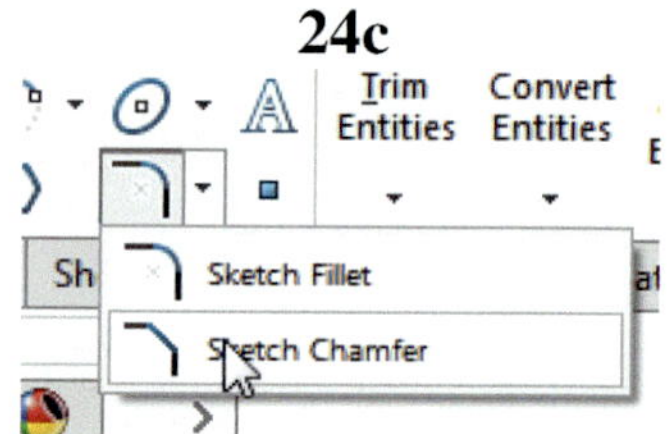

Fig. 3.61 Preparing first corner

Once the dimension of the Chamfer is annotated, the yellow-lined preview is generated by left-clicking inside the drawing.

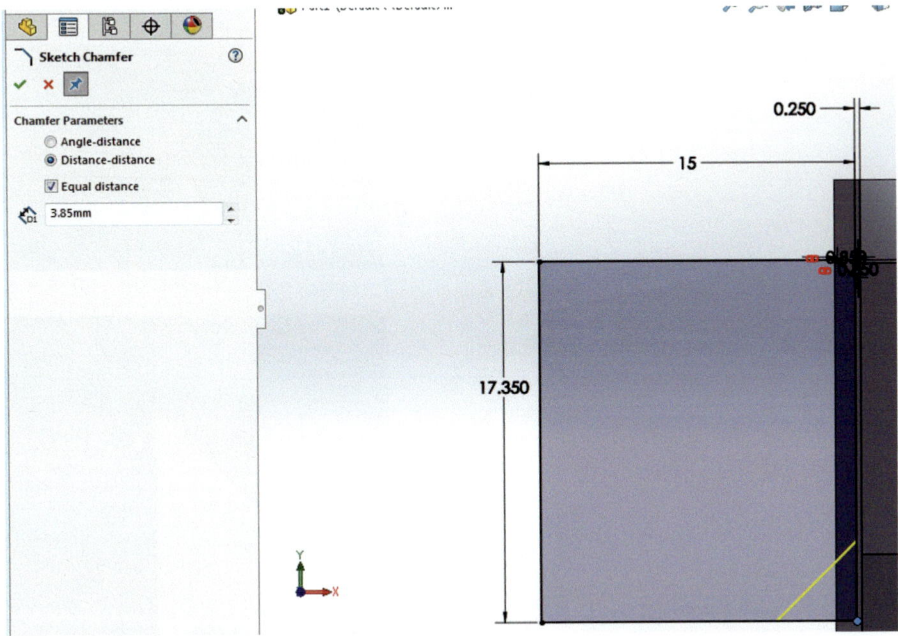

Fig. 3.62 Previewing Chamfer for second corner

Now, another design tool has used the Sketch Chamfer.

Fig. 3.63 Visualizing both
corners with proper
dimensions

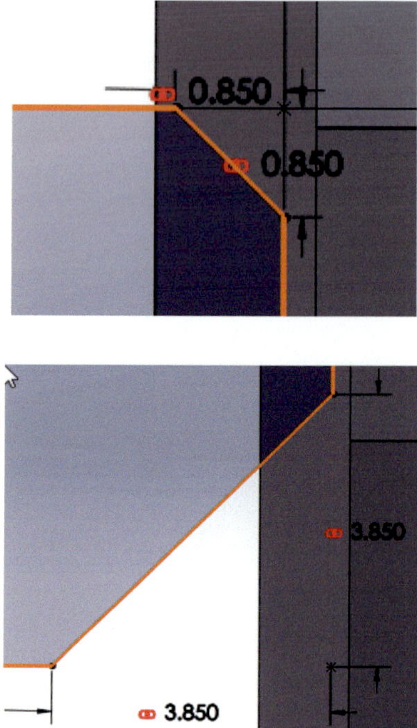

Rotation needs the visualization of Temporary Axes.

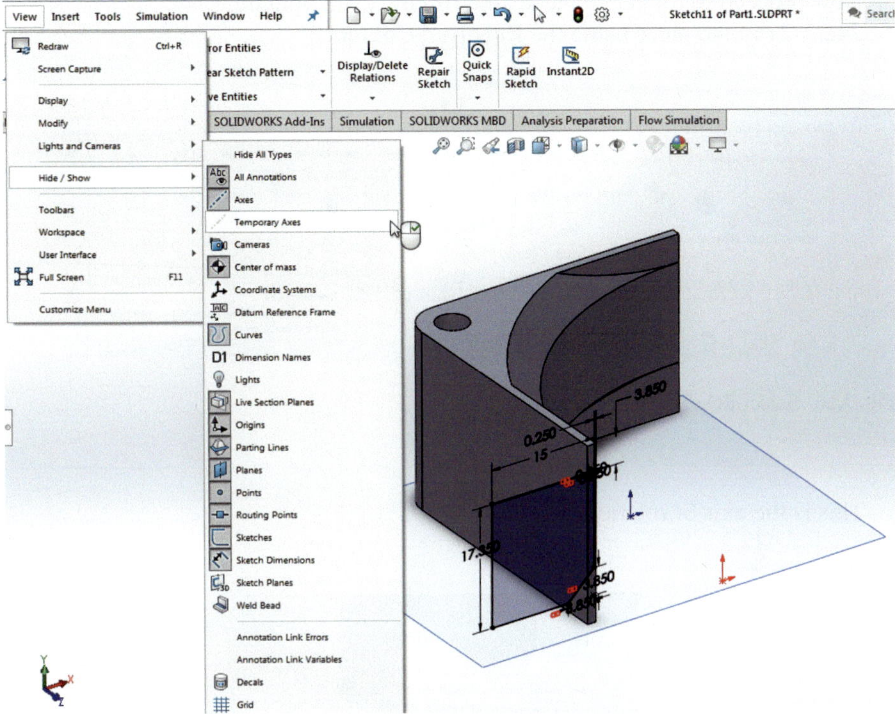

Fig. 3.64 Rectangular mandrel for rotation

24d. Visualize the temporary axes needed for rotation.

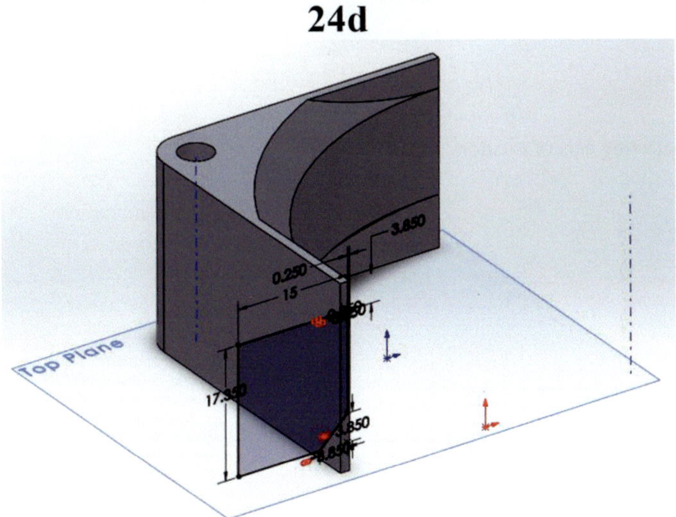

Fig. 3.65 Visualize axis of rotation

Now proceed to rotate the piece and trim the extra volume.

24e.   Trim the piece using the Revolved Cut tool.

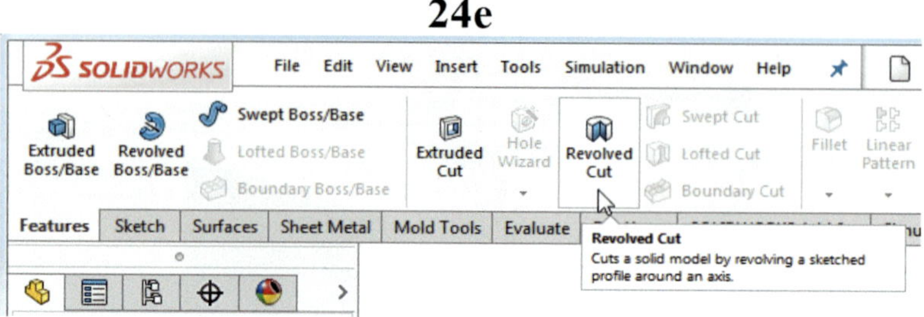

Fig. 3.66 Select Revolved Cut form menu

This is the axis of rotation.

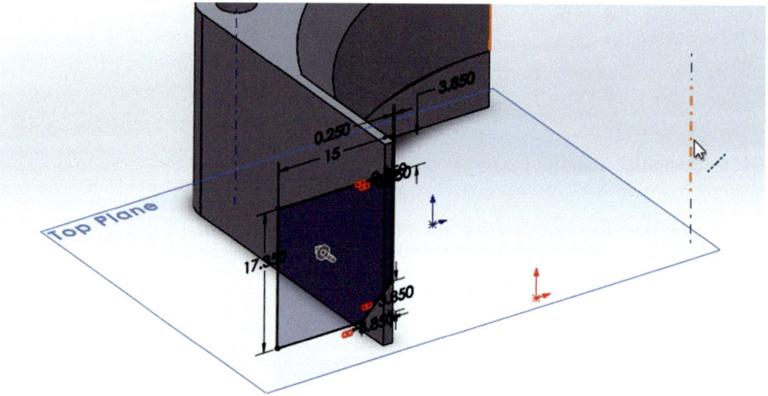

Fig. 3.67 Selecting axis of rotation

Cut revolve generates this figure. Finish by left-clicking on the green check mark.

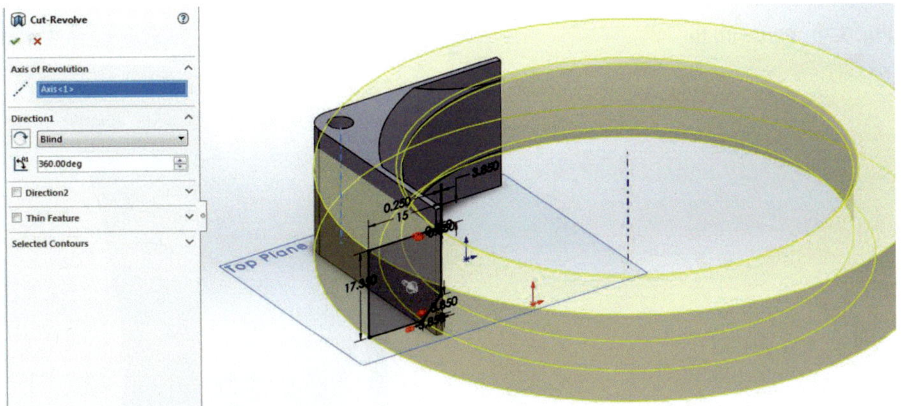

Fig. 3.68 Revolved Cut from menu

The design up to this point looks like the one below.

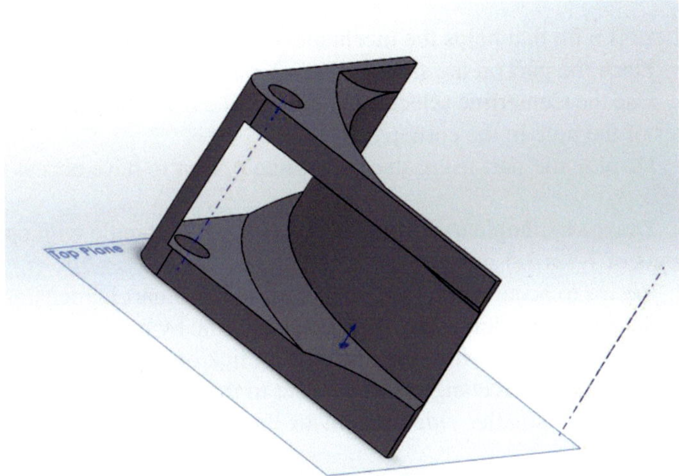

Fig. 3.69 Part design with additional detail

Add now a fin that requires the Convert Entities utility from the drawing menu.

Fig. 3.70 Part with fin

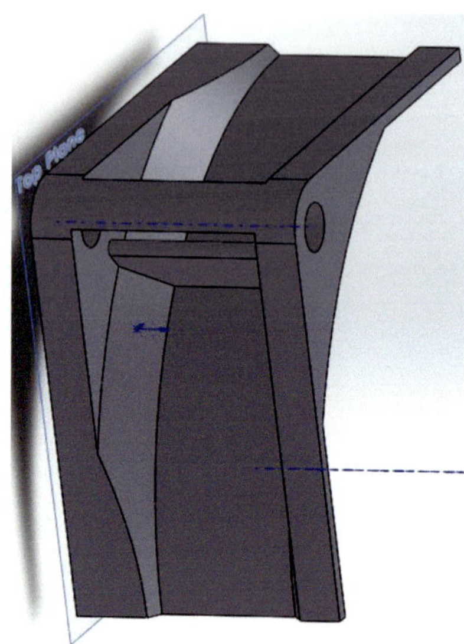

25.   There is a fin that helps the mechanical support of the part.
    25a.   Place the part on the side. Left-click on the mouse and select Sketch.
    25b.   Use the Centerline selection to draw a line from the center of symmetry to the center of the hole in the corner.
    25c.   Display the part using the Wireframe option to have access to the part contours.
    25d.   Draw a rectangle using the 3-Point Corner Rectangle with approximate measurements of 7.1 × 1.5 mm.
    25e.   Resize to accurate measurements using the Smart Dimension tool.
    25e.   Select the two long sides and the center line by holding pressed the Ctrl key and left-clicking over each of the lines. Left-click on the selection Symmetric; this action will center the rectangle with respect to the center line.
    25f.   Make the smaller side tangent to the uppermost line segment below the circle.
    25g.   Return to Shaded with Edges to visualize the current overall drawing.
    25h.   Extrude the profile using Extruded Boss/Base. Choose the option Up to Surface where the other surface is the opposite face of the part. The option to Merge is not checked.
    25i.   Trim the excess material from the part, use the Intersect option that is located under Features. Two objects are needed: one is the slab just created and the other is the rest of the part.
    The excess geometry is checked under Regions to Exclude, in this case Region 2. Merge the parts to get a single object.

25j.   Round the top of the fin. In the Features menu, left-click on Fillet and Choose Fillet. From the left-hand menu select Full Round Fillet, the three visible sides of the fin are the surfaces to use.

25k.   Use change of transparency on the side elements of the drawing in order to appreciate better the new geometry of the fin.

25l.   Make the entire part visible again. The parts with modified transparency are no longer selections that are available. Go to Options at the top of the screen, left-click on Selection and remove the check mark Enable selection through transparency. Now the hidden parts can be selected and the transparency modified back to its default value.

25a.   Place the part on the side. Left-click on the mouse and select Sketch

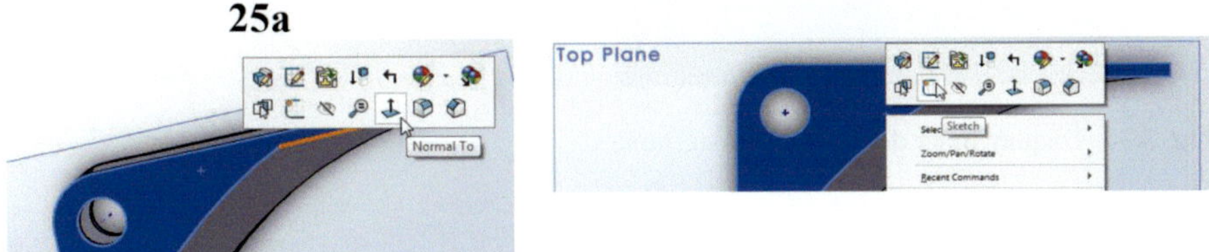

Fig. 3.71 Select plane to create sketch

25b.   Use the Centerline selection to draw a line from the center of symmetry to the center of the hole in the corner.

Fig. 3.72 Select Centerline

Fig. 3.73 Draw a construction centerline

25c.   Display the part using the Wireframe option to have access to the part contours.

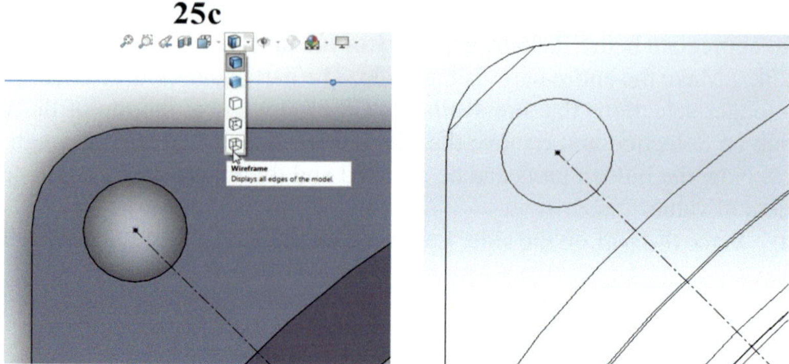

Fig. 3.74 Display inner details using Wireframe

25d.   Draw a rectangle using the 3-Point Corner Rectangle with approximate measurements of 7.1 × 1.5 mm.

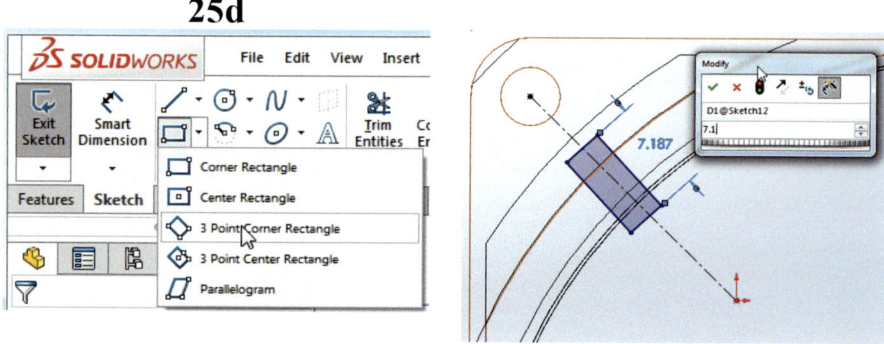

Fig. 3.75 Draw a rectangle in radial direction

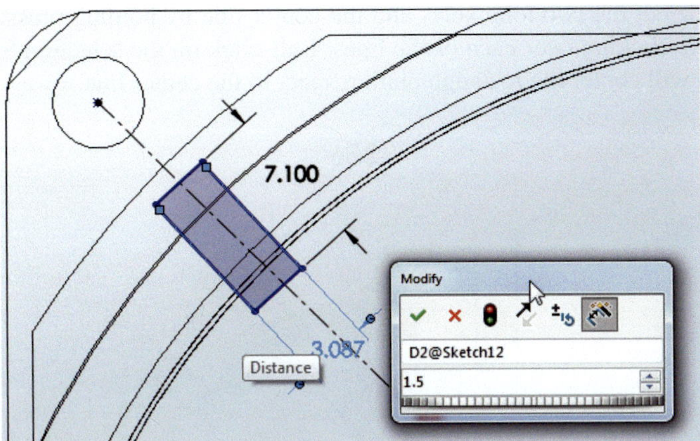

Fig. 3.76 Refine measurements using Smart Dimension

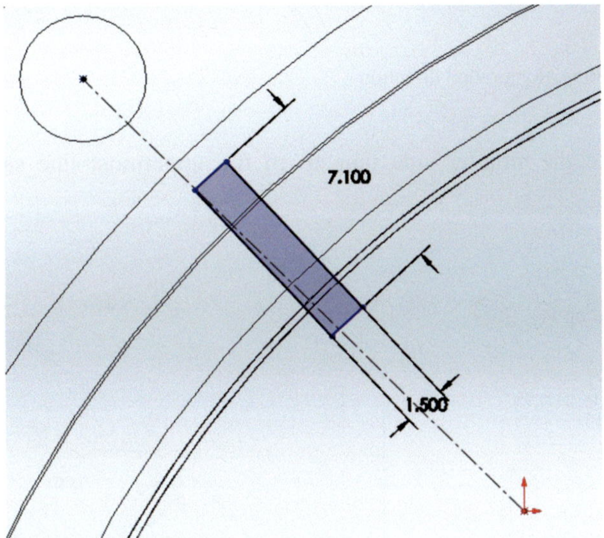

Fig. 3.77 Misaligned rectangular construction

25e.   Select the two long sides and the center line by holding pressed the Ctrl key and left-clicking over each of the lines. Left-click on the selection Symmetric; this action will center the rectangle with respect to the center line.

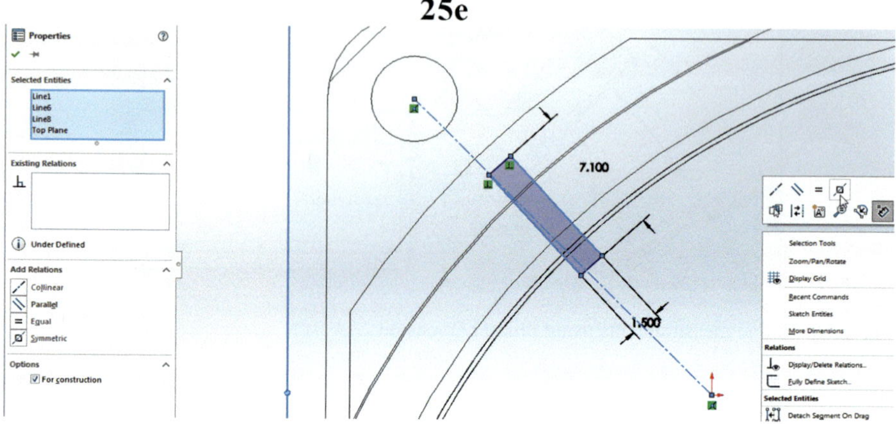

Fig. 3.78   Centering along radial direction

25f.   Make the smaller side tangent to the uppermost line segment below the circle.

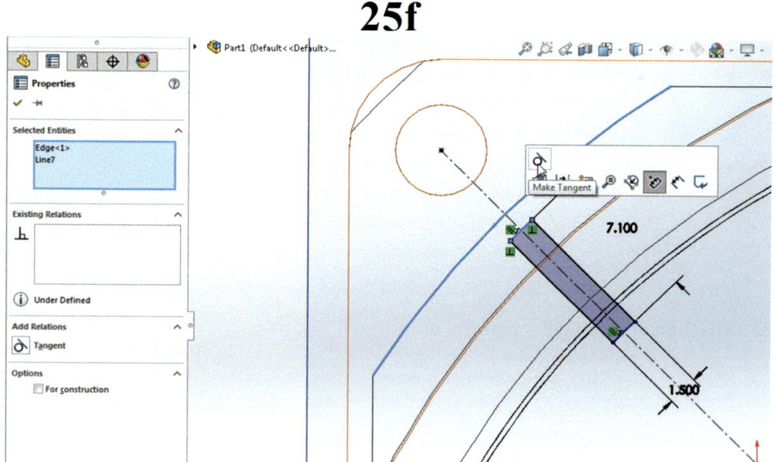

Fig. 3.79   Making tangential element

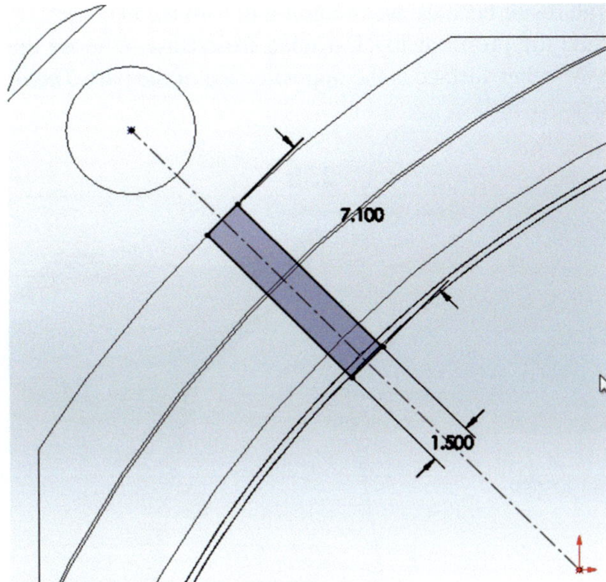

Fig. 3.80 Rectangular element properly centered

25g.   Return to Shaded with Edges to visualize the current overall drawing.

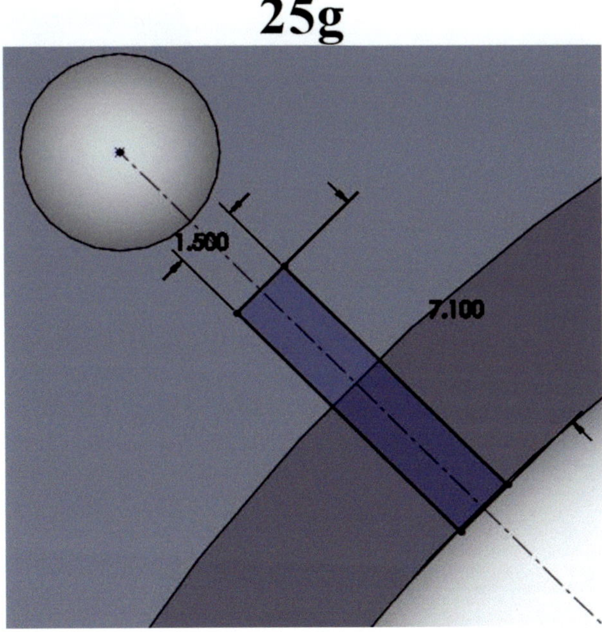

Fig. 3.81 Visualize using Shaded with Edges

Do not Merge result because we are going to trim the excesses.

25h.   Extrude the profile using Extruded Boss/Base. Choose the option Up to Surface where the other surface is the opposite face of the part. *The option to Merge is not checked.*

25h

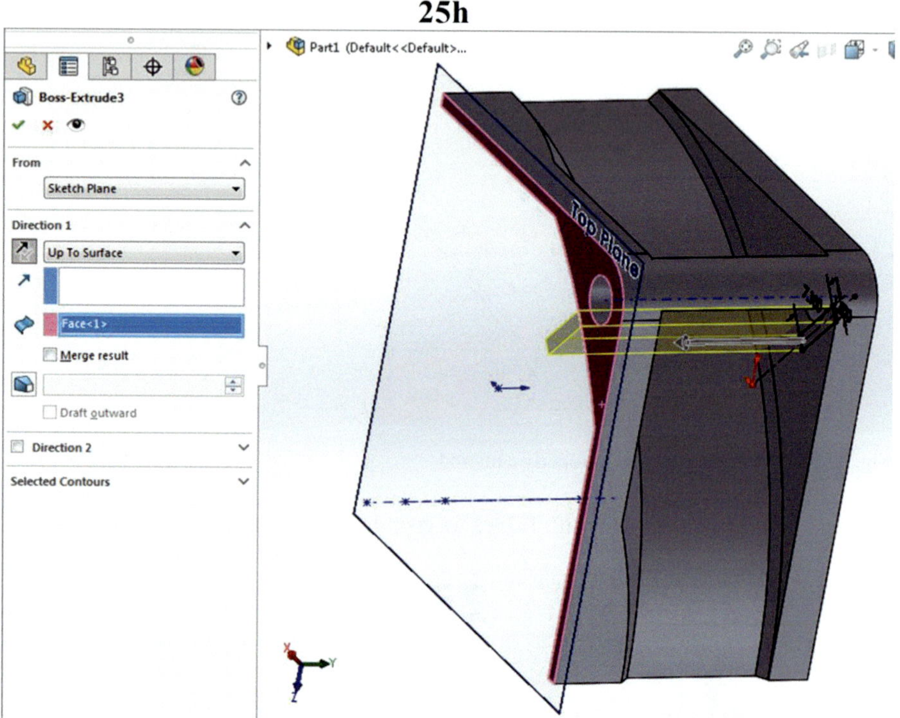

Fig. 3.82 Extrude up to Surface

25i.   Trim the excess material from the part, use the Intersect option that is located under Features. Two objects are needed: one is the slab just created and the other is the rest of the part. The excess geometry is checked under Regions to Exclude, in this case, Region 2. Merge the parts to get a single object.

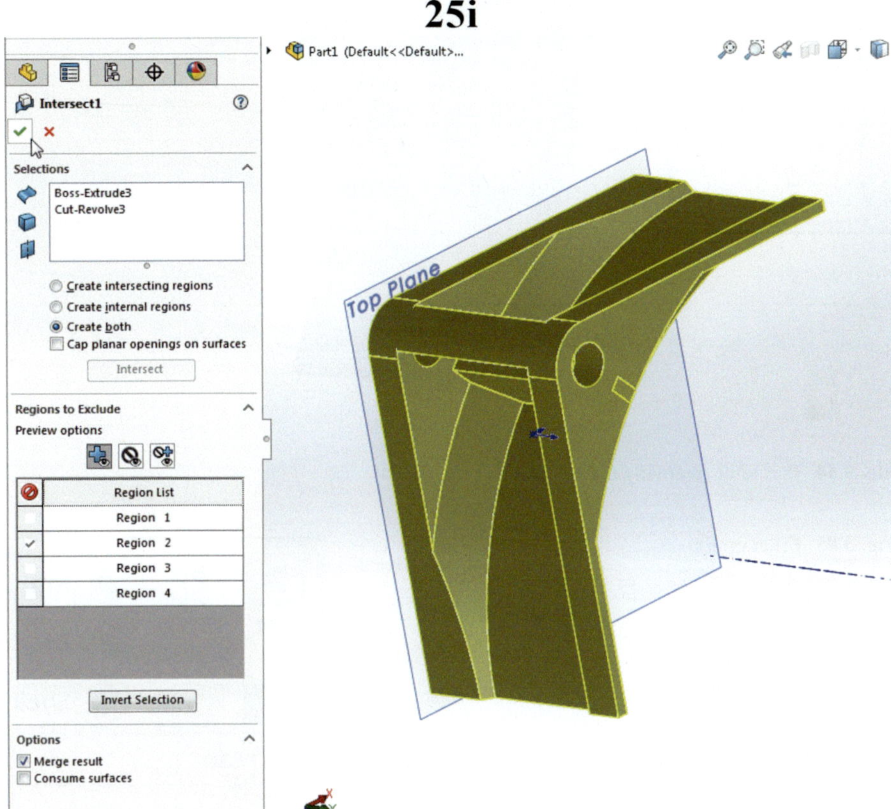

Fig. 3.83 Remove excess material

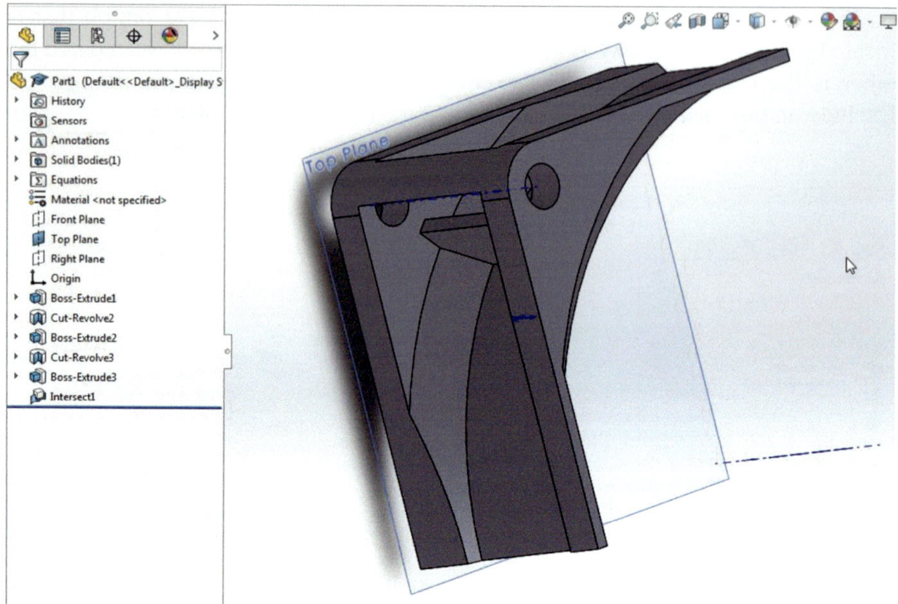

Fig. 3.84 Part after removing excess material

Fig. 3.85 Fillet option

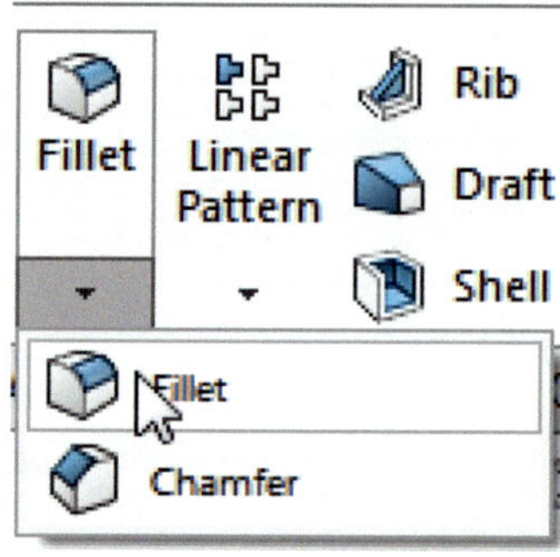

25j.   Round the top of the fin. In the Features menu, left-click on Fillet and Choose Fillet. From the left-hand menu select Full Round Fillet, the three visible sides of the fin are the surfaces to use.

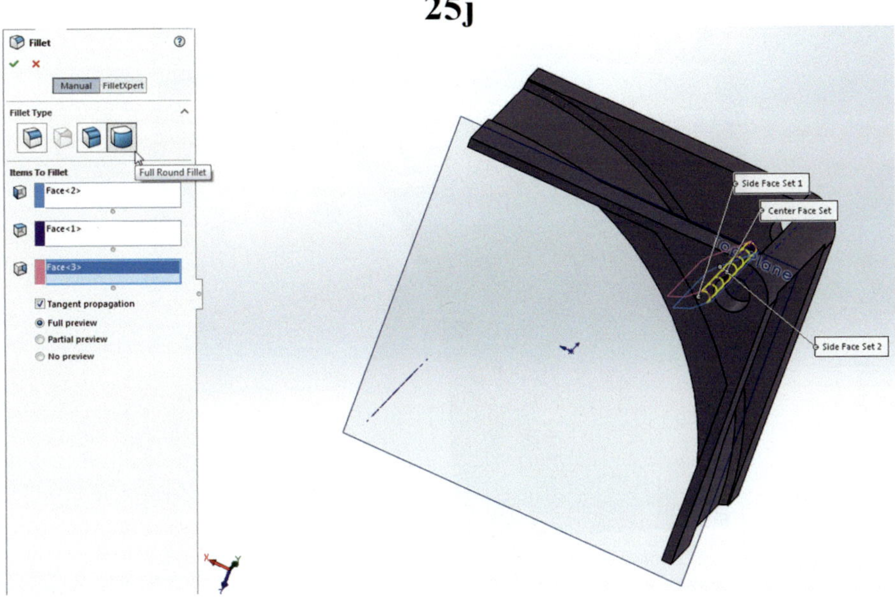

Fig. 3.86 Use Full Round Fillet to smooth fin

25k.   Use change of transparency on the side elements of the drawing in order to appreciate better the new geometry of the fin.

Fig. 3.87 Revise inner details of fin using Change Transparency

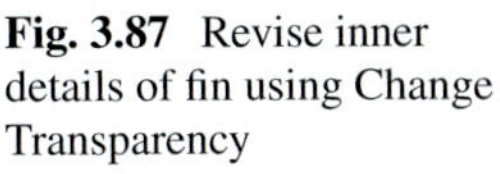

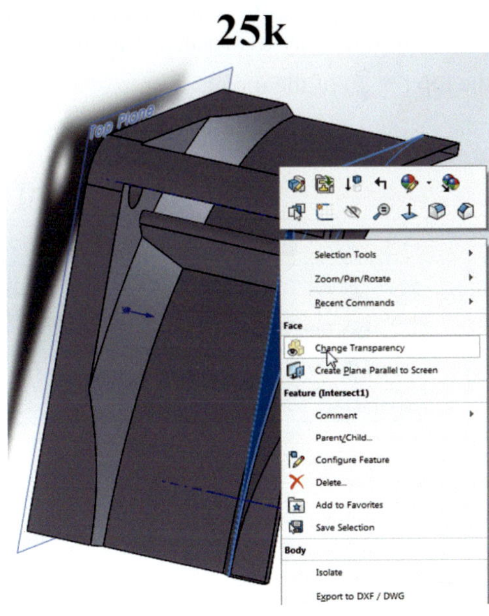

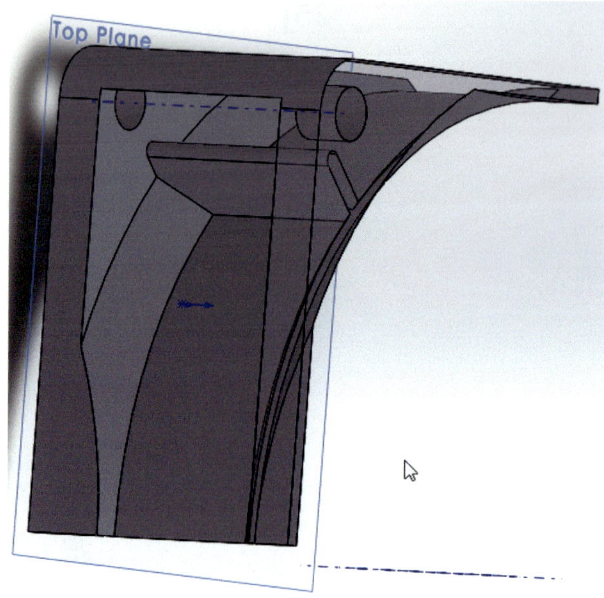

Fig. 3.88 Transparent wall created

25l.   Make the entire part visible again. The parts with modified transparency are not available for selection. Go to Options at the top of the screen, left-click on Selection and remove the check mark Enable selection through transparency. Now the hidden parts can be selected and the transparency modified back to its default value. Another way to proceed is to undo the last step using the backward arrow at the top ( ) of the menu.

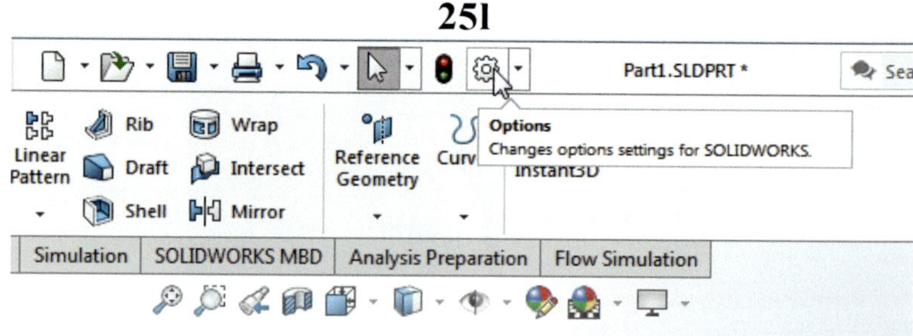

Fig. 3.89 Resetting view mode in Options

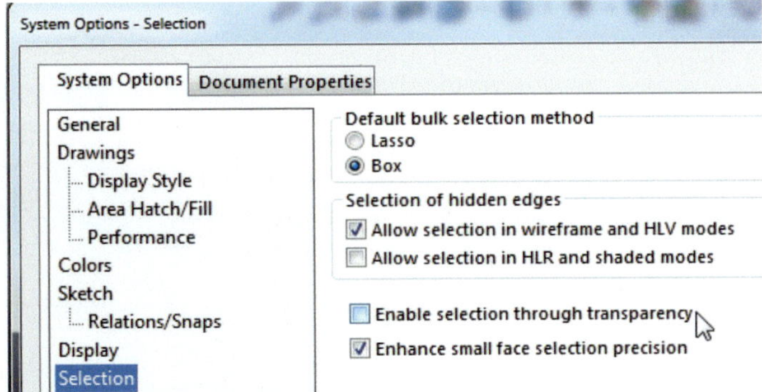

Fig. 3.90 Enable selection through transparency

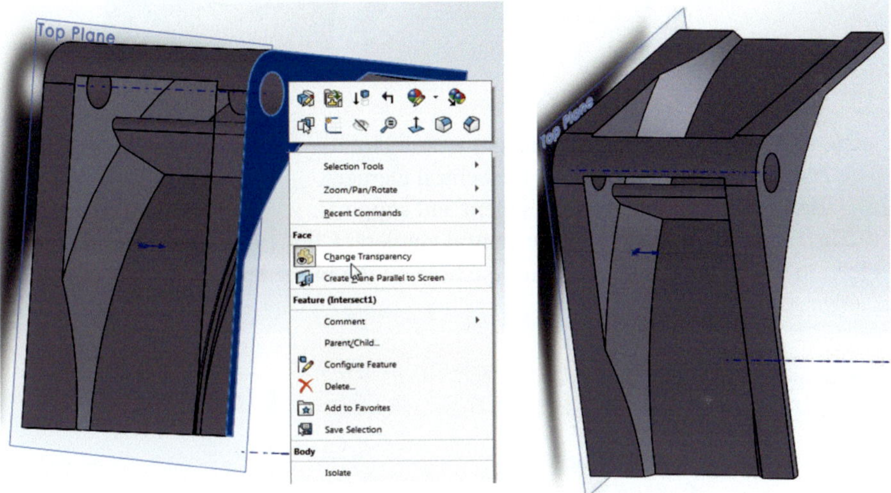

Fig. 3.91 Return to solid view

Fig. 3.92 Part with
additional details

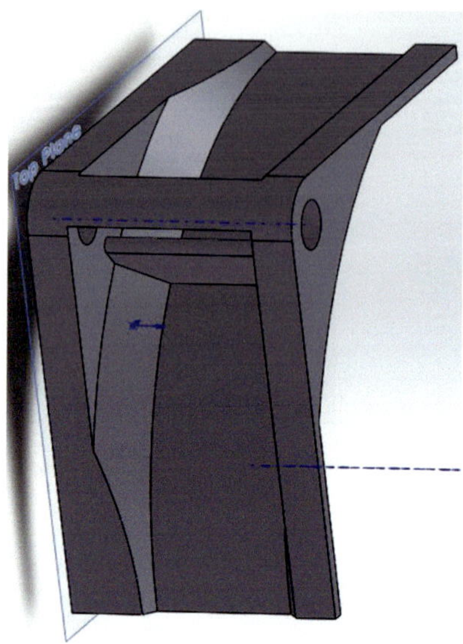

26.   Create the last central element with a circular segment and a linking arm.
   26a.   Create the sketch for the central element.
   26b.   Extrude the central element and supports.

26a.   Create the sketch for the central element. Click twice on the normal plane
selection to accommodate part with the hole on the lower left.

26a

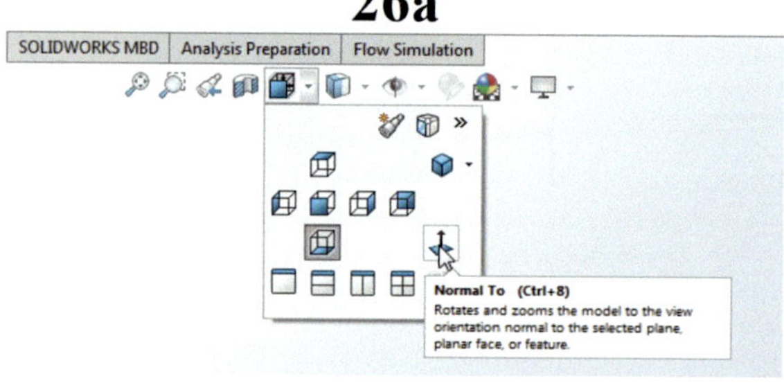

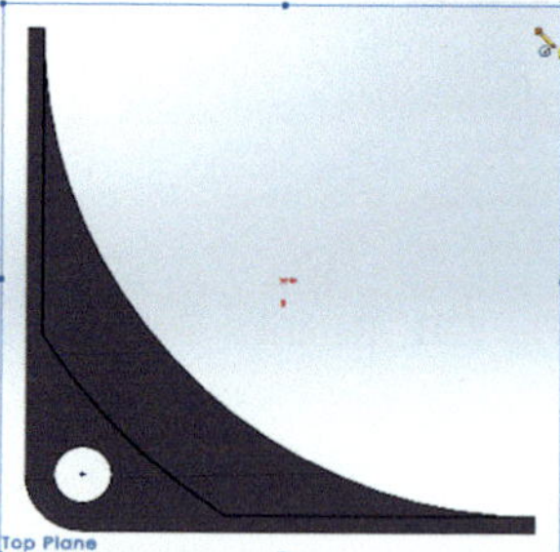

Fig. 3.93 Select sketch plane

Draw a circle with a diameter of 35 mm for construction. This means the circle is a guide structure.

Fig. 3.94 Select circle geometry

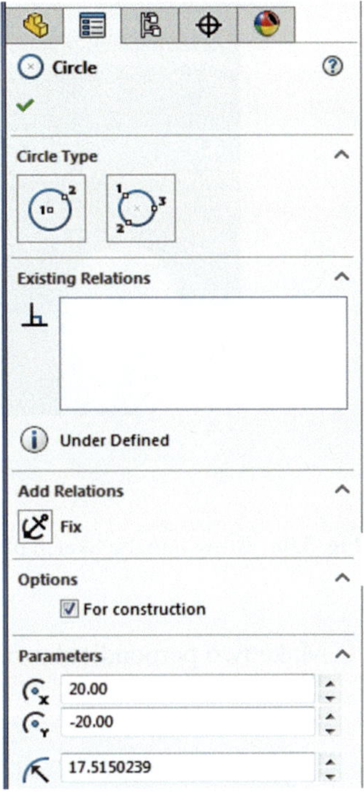

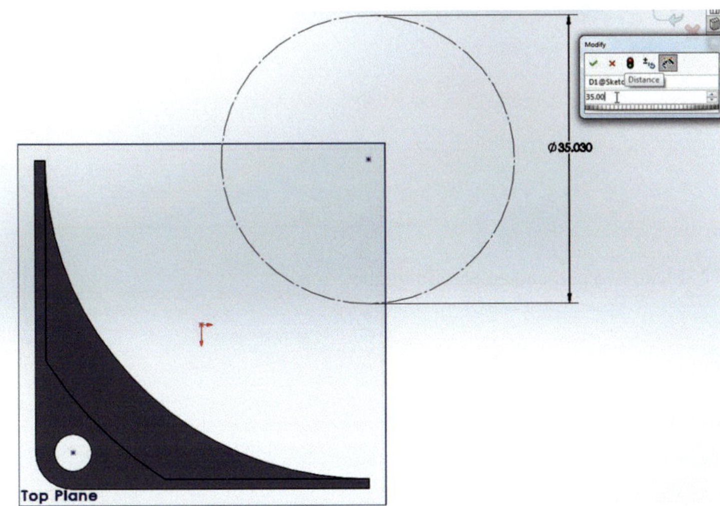

Fig. 3.95 Draw a circle with a 35-mm diameter

Select the middle section of the internal outline (second to the border line) and transfer the shape to the Sketch plane using Convert Entities.

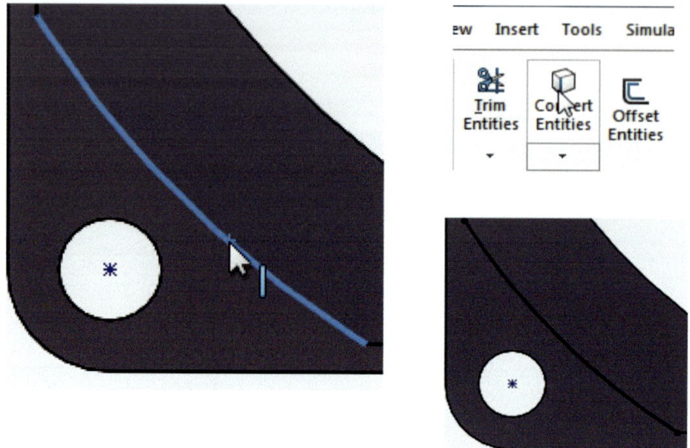

Fig. 3.96 Bring into the Sketch plane a chord

Make two perpendicular radii in the circle.

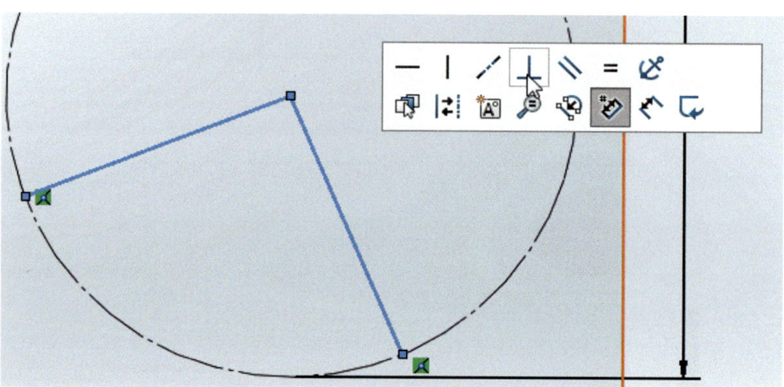

Fig. 3.97 Make two perpendicular radii on the circle

Construct four parallel lines with a distance of 1.75 mm between long and short lines. The lines angles are at 18.25° with respect to the horizontal line. A radius segment line is made parallel to the first short line.

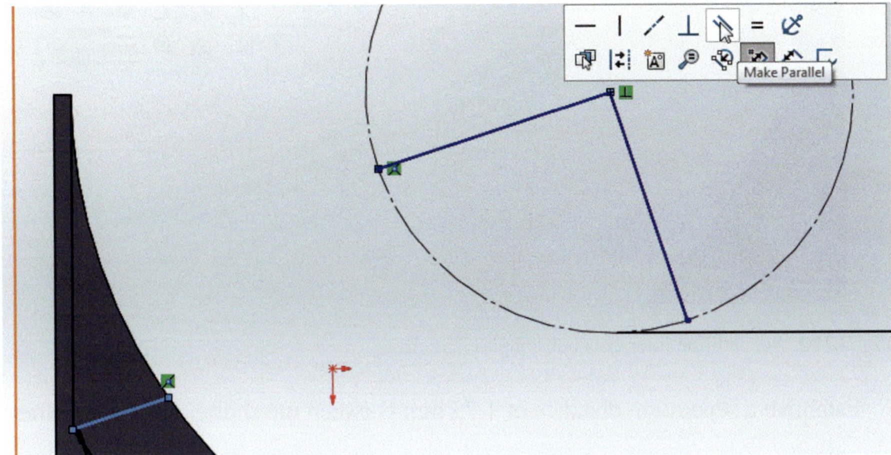

Fig. 3.98 First line at 18.25° inclination

Draw a new line that intersects the circle and is parallel to the previous line.

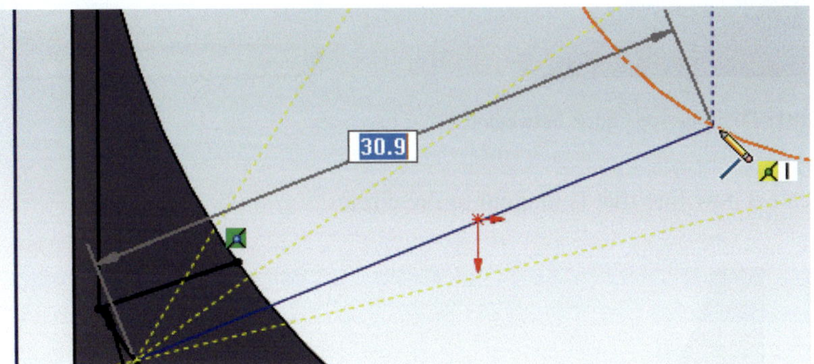

Fig. 3.99 Second line at 18.25° inclination

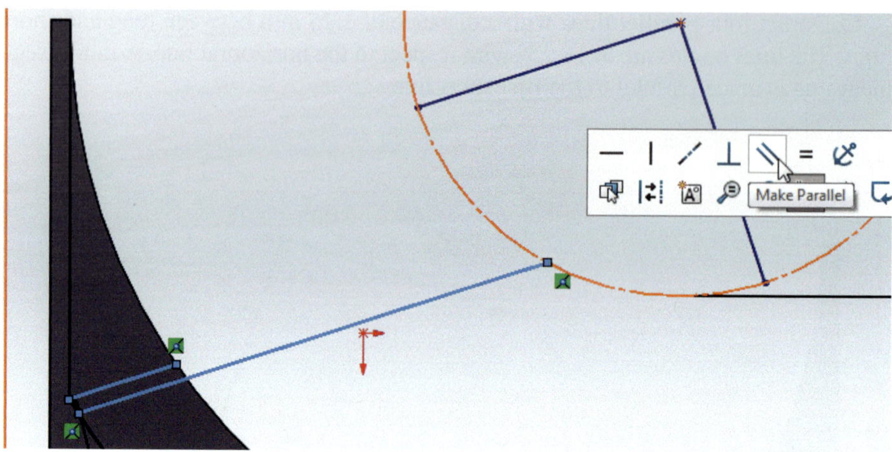

Fig. 3.100 Second line intersects current part and circle

Establish a separation distance of 1.75 mm between the shorter and longer lines.

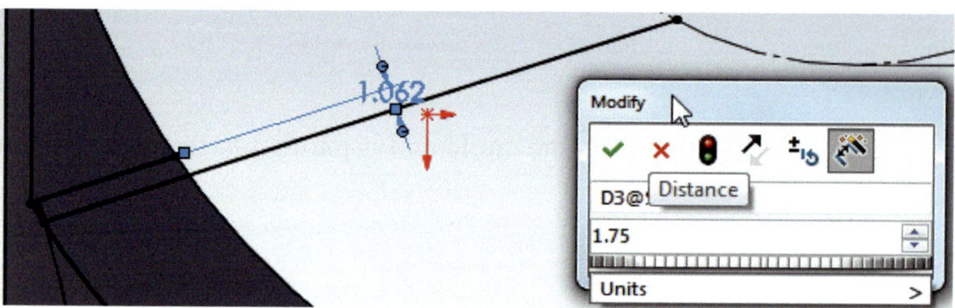

Fig. 3.101 Defining separation between the first two lines

Create a new line that is tangent to the circle.

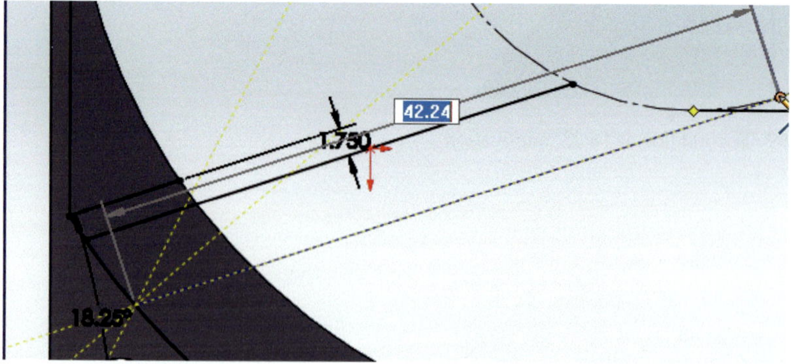

Fig. 3.102 Adding a third tangential line

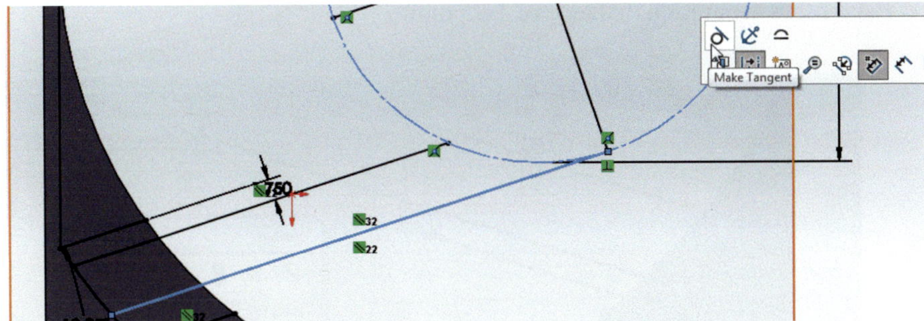

Fig. 3.103 Third tangential segment

Draw a new line at a different direction than the previous ones that ends on the open canvas.

This is because there are a number of restrictions added that work better with an unconstrained line.

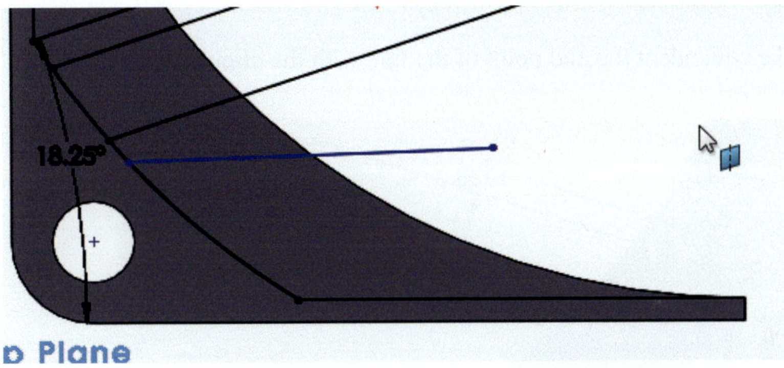

Fig. 3.104 Add a fourth line segment

Make the tangent line and the new line parallel.

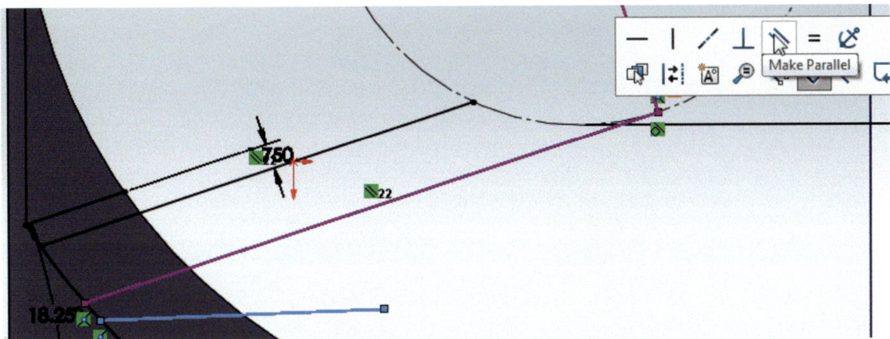

Fig. 3.105 Make fourth line parallel to the previous ones

Establish a separation distance of 1.75 mm.

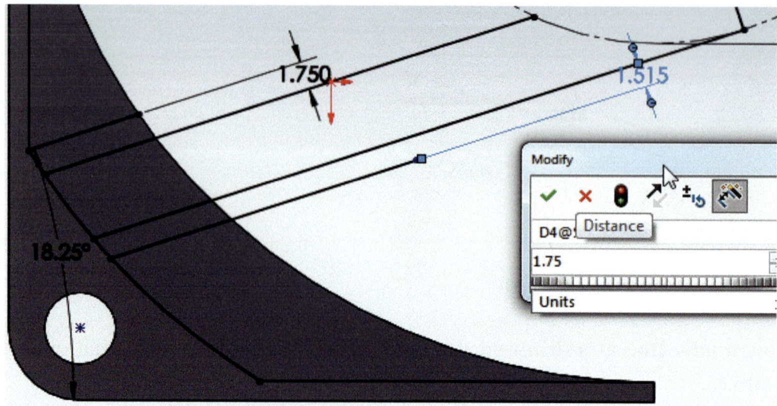

Fig. 3.106 Establish separation for the fourth line segment

Make coincident the end point of the line with the circumference.

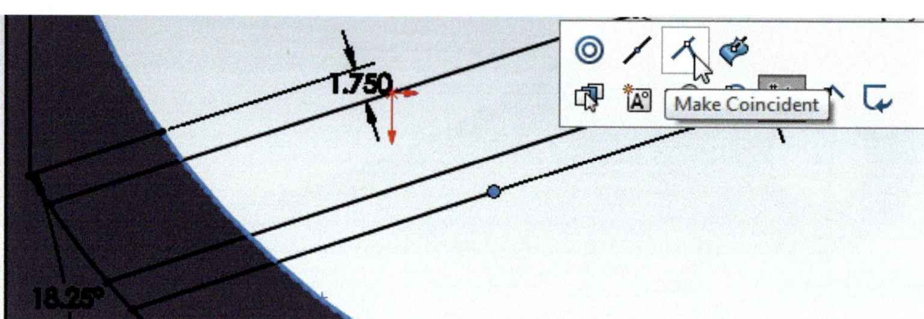

Fig. 3.107 Making line segment end coincident with circle

At this point, the part looks as indicated below.

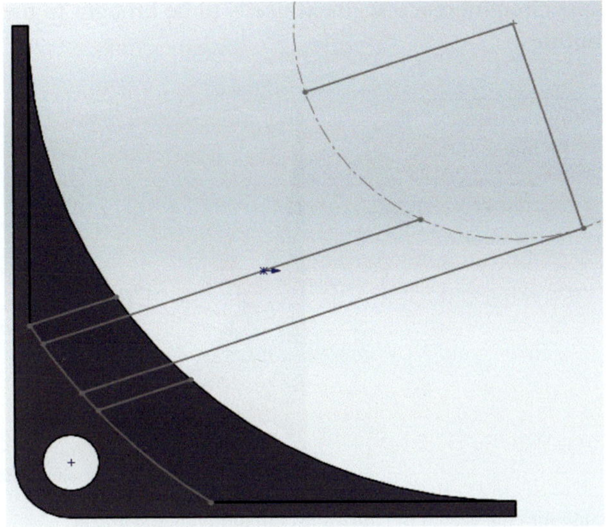

Fig. 3.108 Current structure of line segments

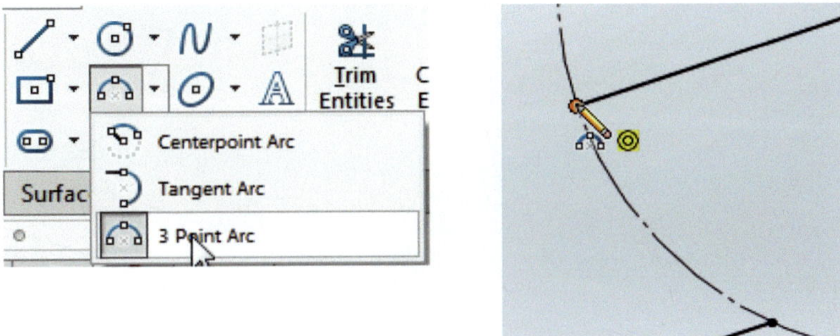

Fig. 3.109 Drawing an arc between lines

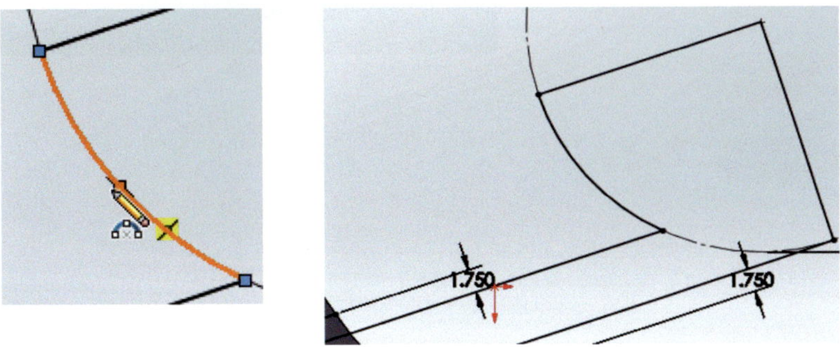

Fig. 3.110 Start and end of arc in sketch

The innermost circumference segment needs to be brought to the Sketch plane with Convert Entities.

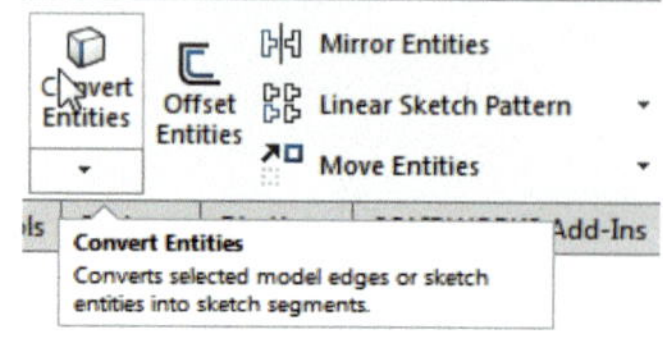

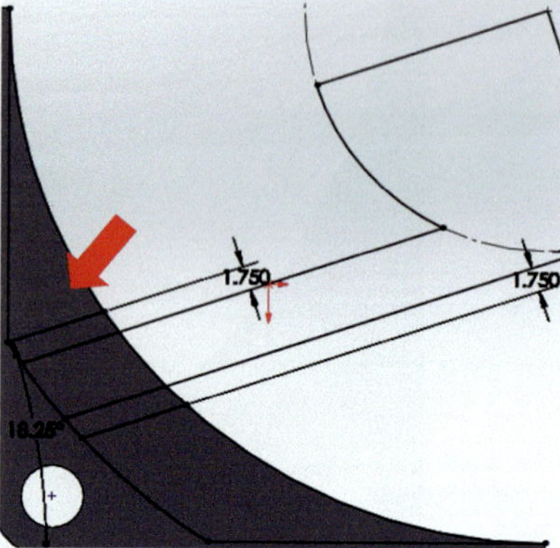

Fig. 3.111 Bringing into the sketch the innermost circumference segment

Fig. 3.112 Elements for profile construction

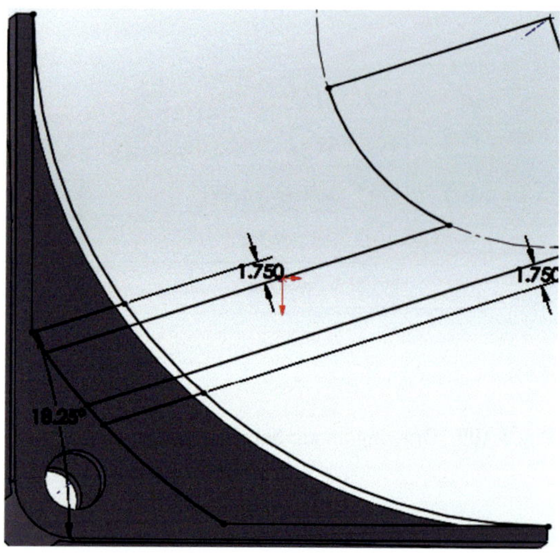

There are three additional arcs that need to be finished using the three-point arc. In order to draw these right, for each arc, *select the initial point, the final point, and the middle point in that order*.

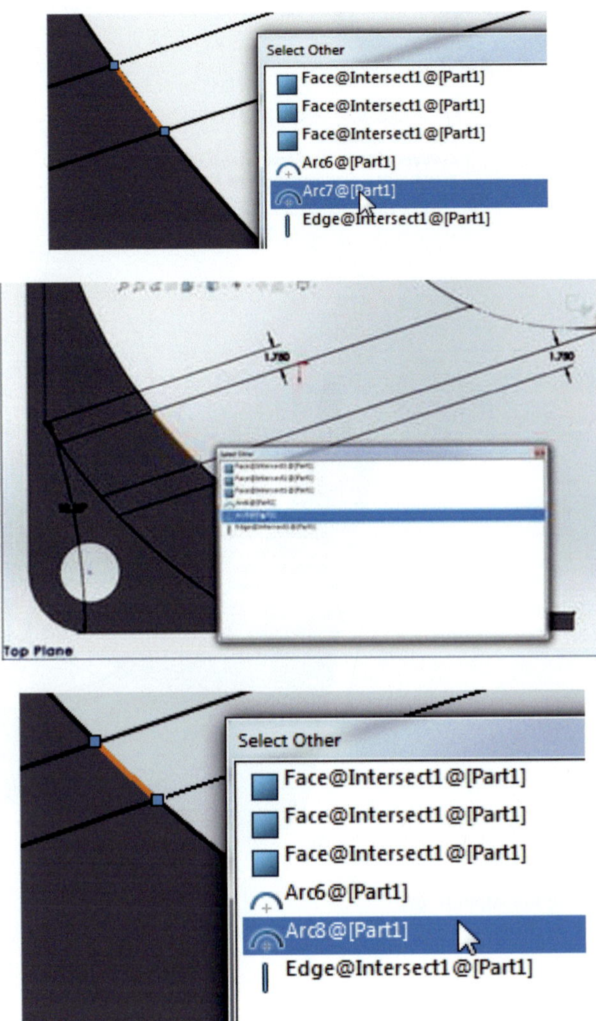

Fig. 3.113 Creating three Arcs over the circle circumference

26b.   Extrude the central element and supports with a thickness of 3 mm.

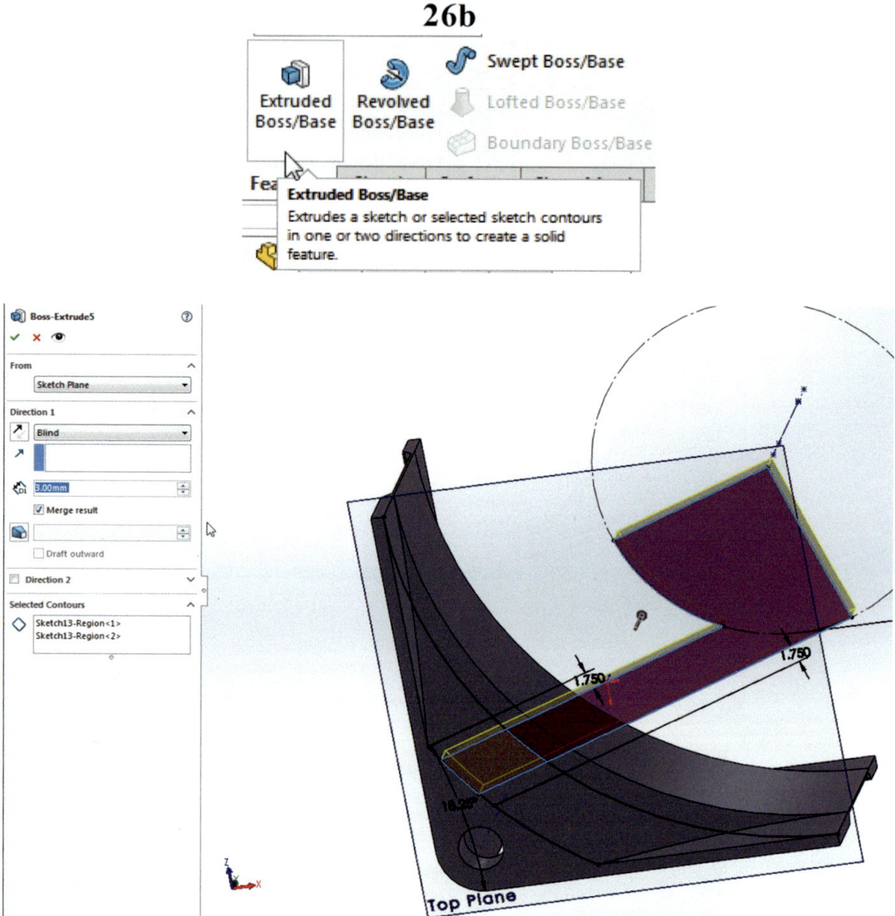

Fig. 3.114 Extruding fan support arm

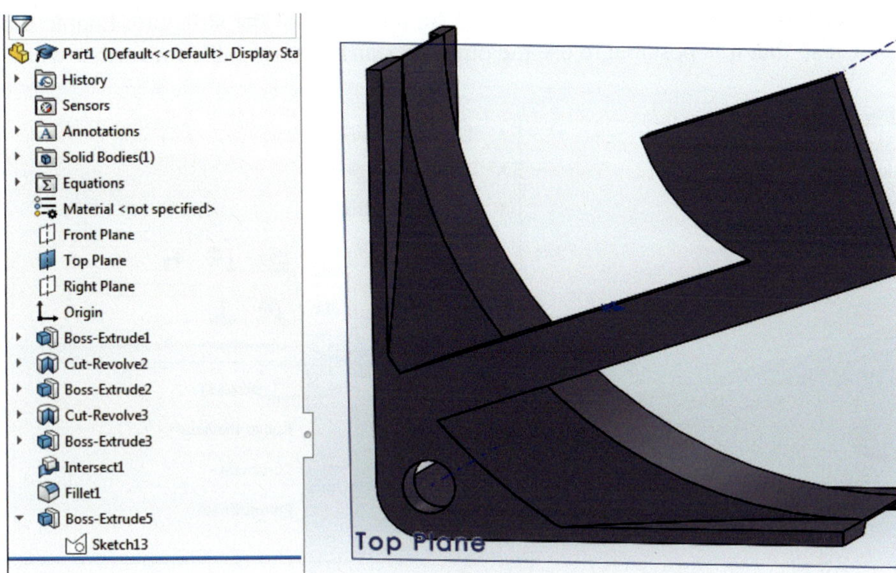

Fig. 3.115 Extruded support arm

Need to complete other details on the support basis of the structure. Enable the sketch so that it is possible to use the other two surfaces available for further work.

Fig. 3.116 Visualizing the Sketch to create additional segments

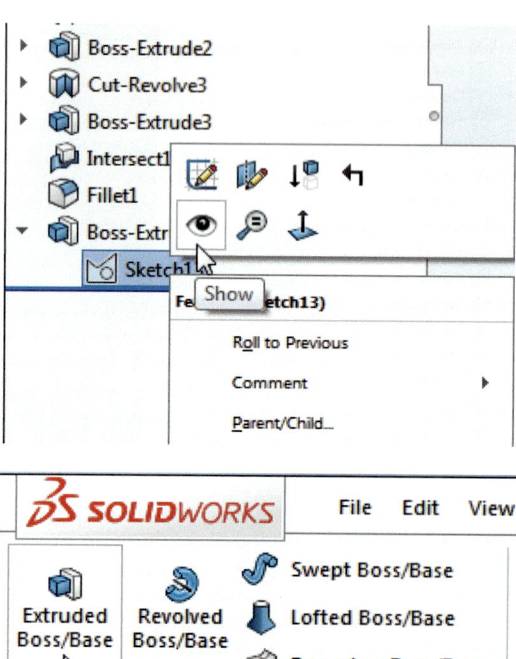

Look carefully that if we left-click on the *outer edge* of the area we would like to extend we get an area with an undesirable *tail*. Consequently, we need erase this automatic selection area and select the three segments of interest explicitly.

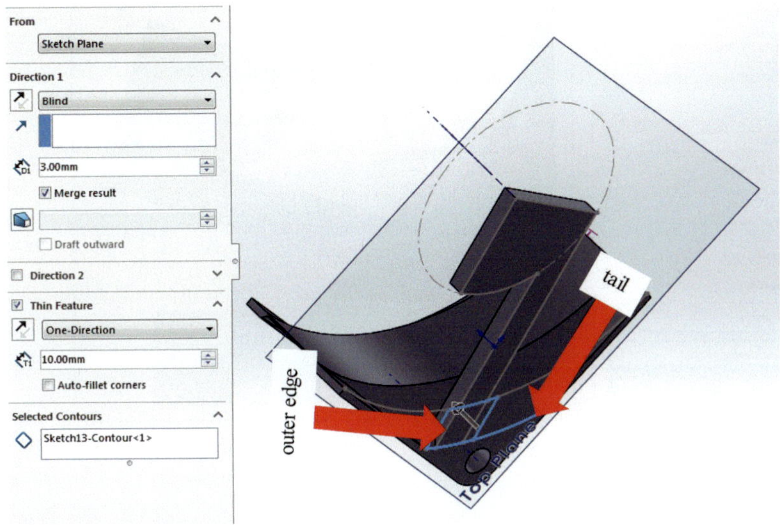

Fig. 3.117 Select contour Sketch13

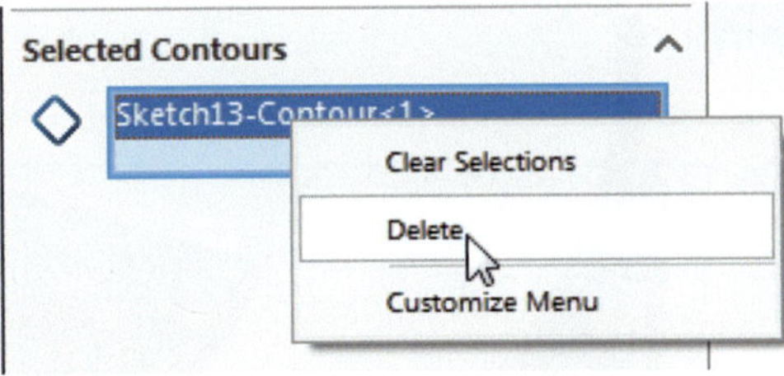

Fig. 3.118 Delete element from Sketch13 and select new areas

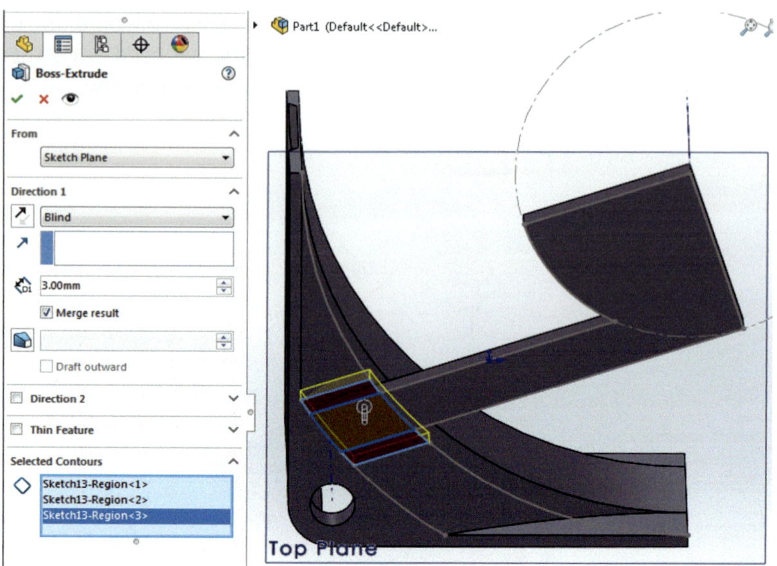

Fig. 3.118 (continued)

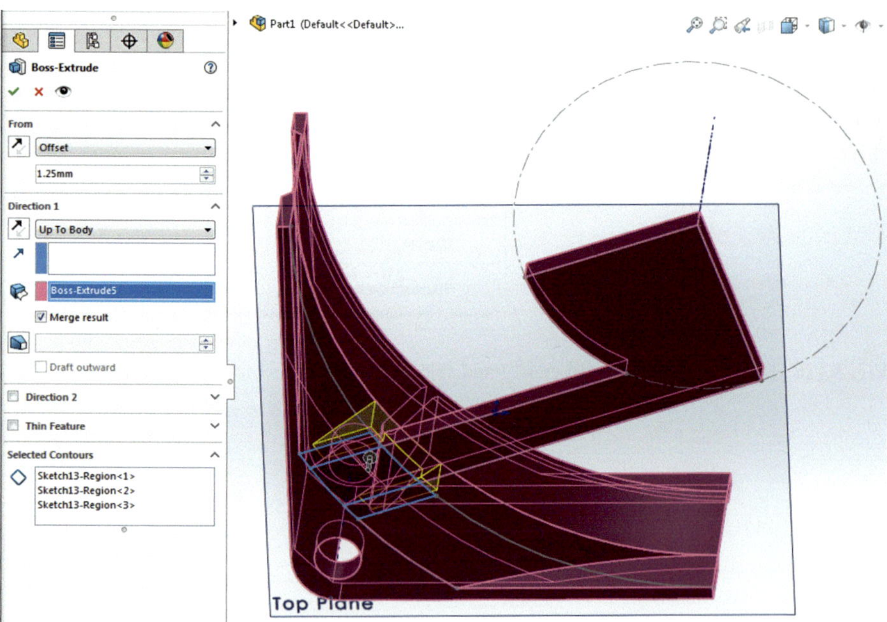

Fig. 3.119 Extrude up to Body and merge result

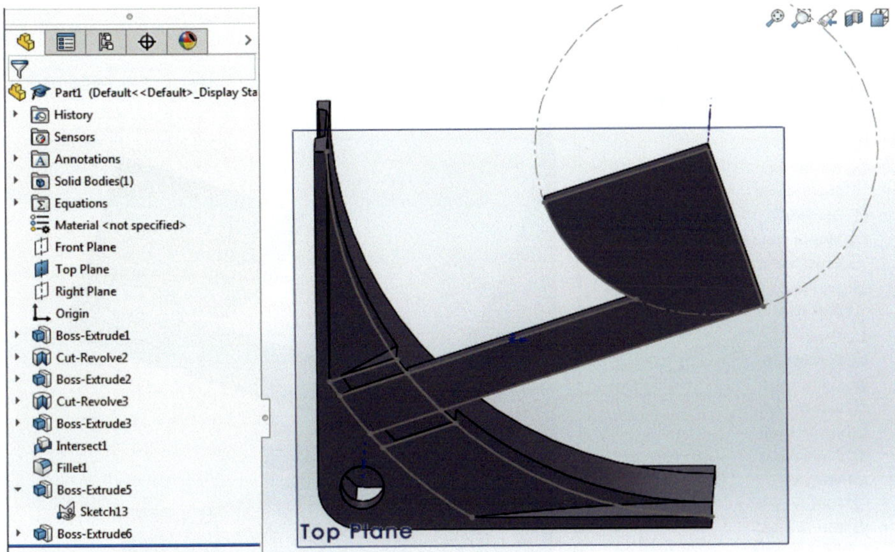

Fig. 3.120 Part with arm and support

Hide the Sketch plane in order to see the quarter of piece construction that has been achieved.

Fig. 3.121 Hide working Sketch

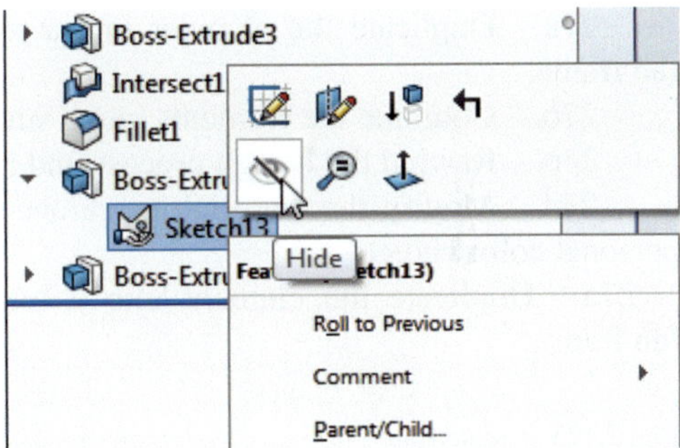

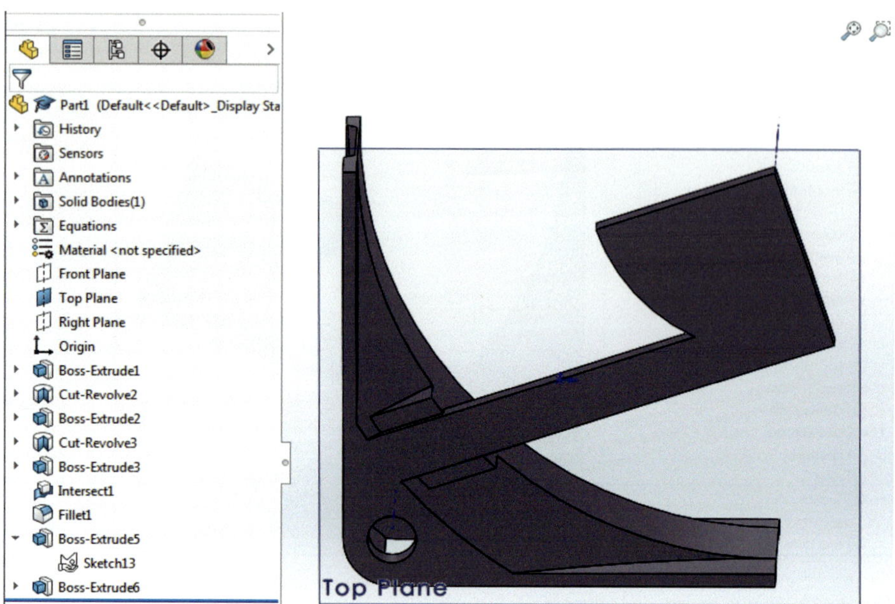

Fig. 3.122 Current part with arm and support

27.   This is a quarter of the total frame. A Linear Pattern is used to integrate the other three equal parts that are needed.

27a.   Duplicate the element into a total of four elements to complete the fan frame.

27b.   Combine the elements into a single piece.

27c.   Rewind the design process and add details to the part.

27d.   Modify the material appearance and express a creative streak with a personal color choice.

27a.   Duplicate the element into a total of four elements to complete the fan frame.

Fig. 3.123 Create four
identical elements using
Circular Pattern

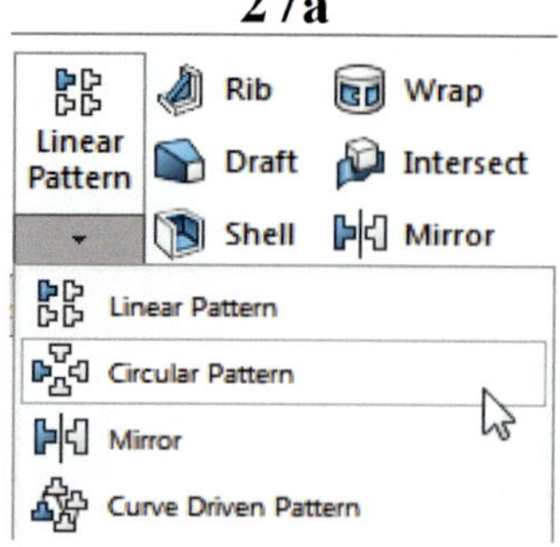

The first choice is the axis of rotation on the edge of the wing of the part. Make sure to *type in the appropriate number of replicas* and that the selection is made via a *body selection*.

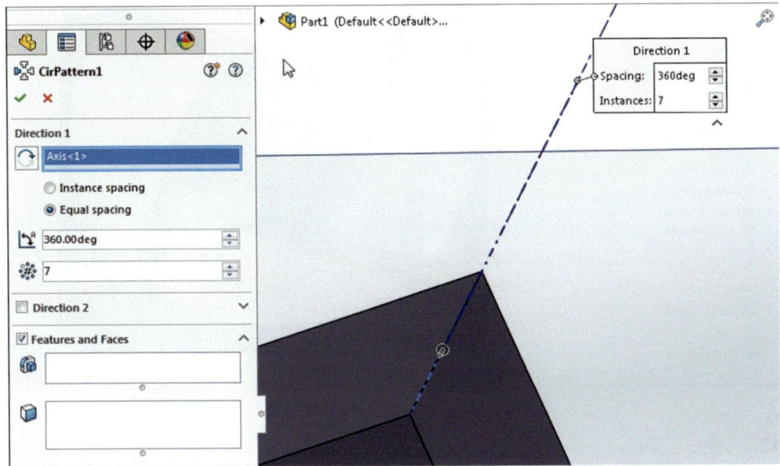

Fig. 3.124 Choose axis for circular pattern

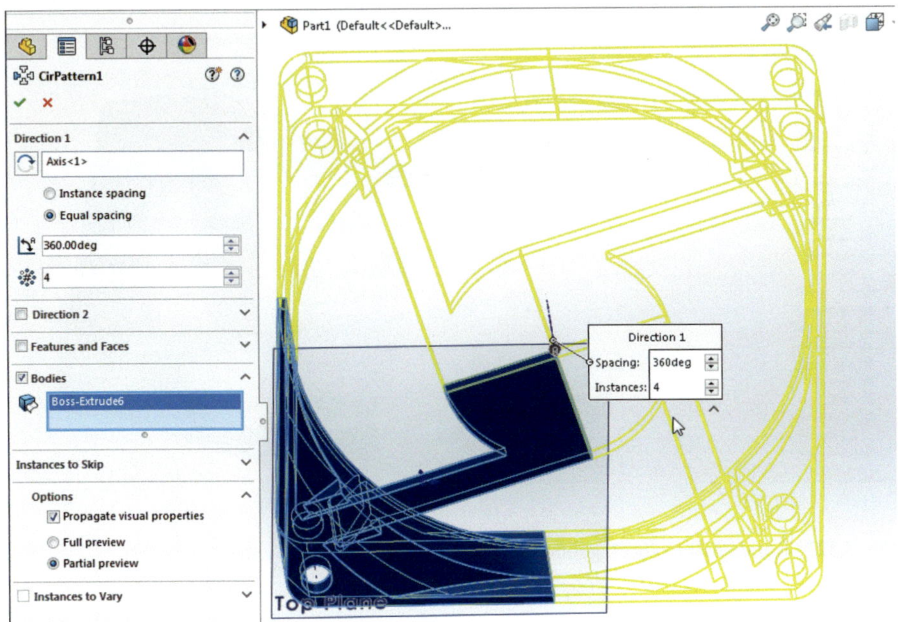

Fig. 3.125 Preliminary view of circular pattern

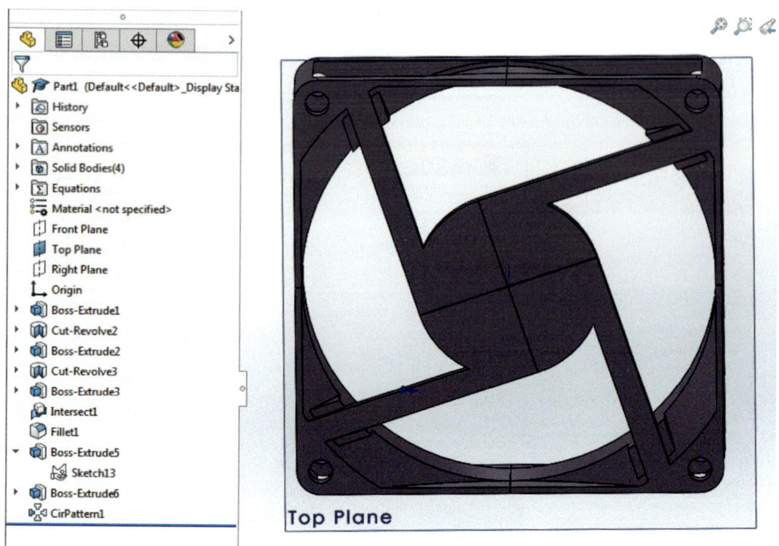

Fig. 3.126 Pattern with four identical parts

Turn the piece around to see the other side and now it is evident that this part has a shape similar to that of a desk computer fan.

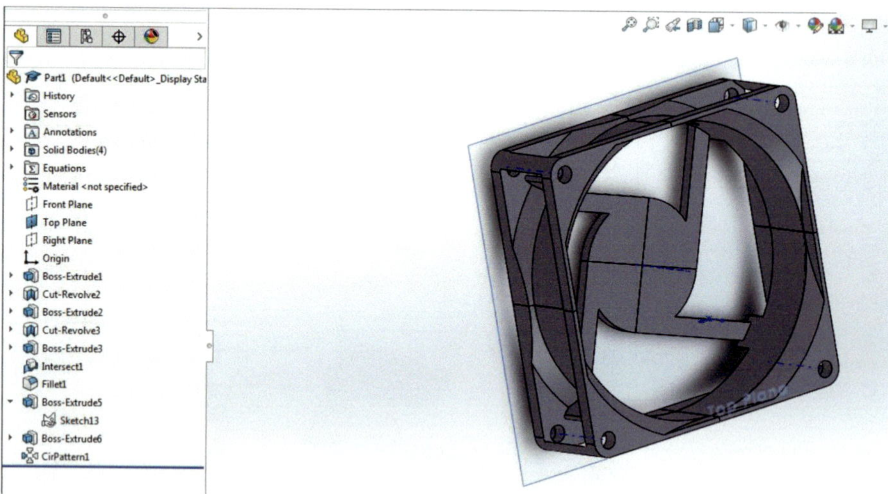

Fig. 3.127 Rotate part to check geometry

There are a few steps to improve on this part. The first steps combine the parts, the second rolls back part construction for the addition of details and the third adds details to the part surface.

27b.   Combine the elements into a single piece.

27b

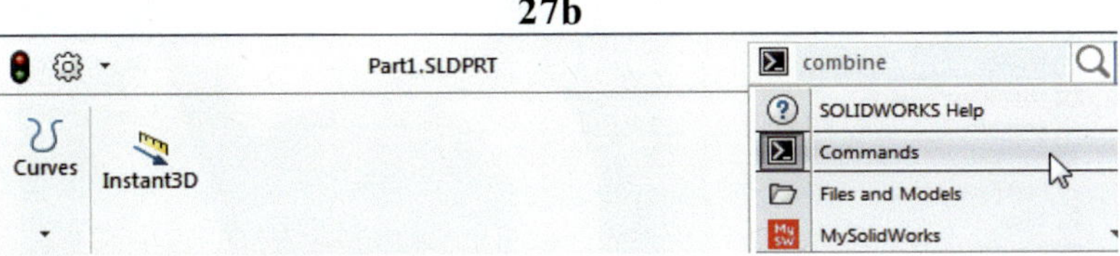

Fig. 3.128   Search for the Combine command

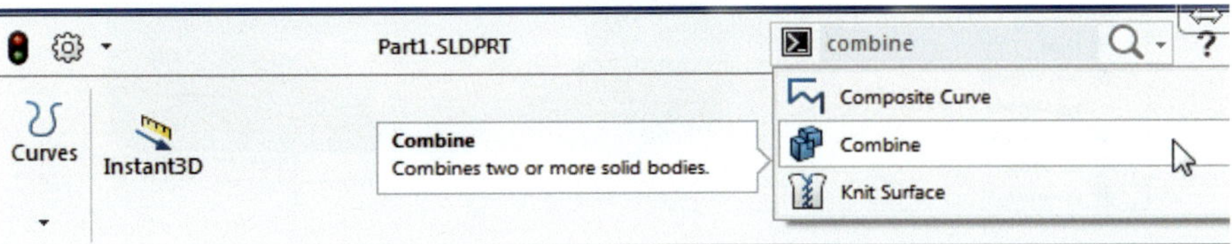

Fig. 3.128   (continued)

Select the four bodies to fuse into a single object.

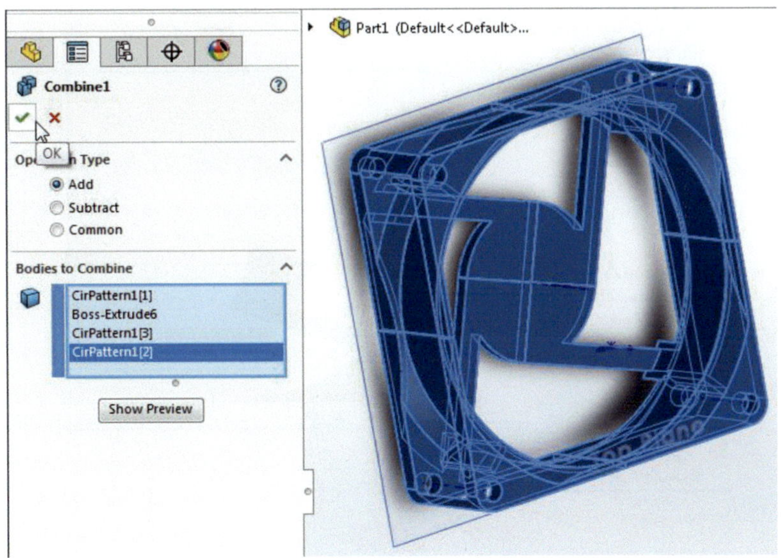

Fig. 3.129   Combine the four parts into a single unit

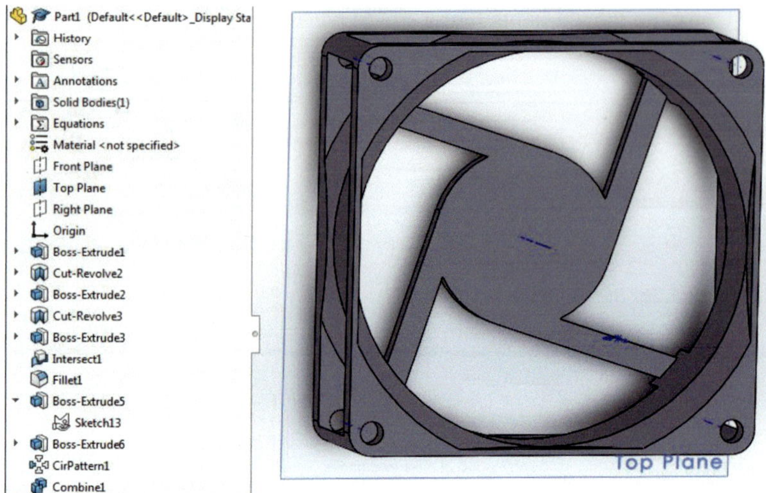

Fig. 3.130 Resulting single unit

Retract the design to a third to last step to add some additional detail to the part.
27c.   Rewind the design process and add details to the part.

27c

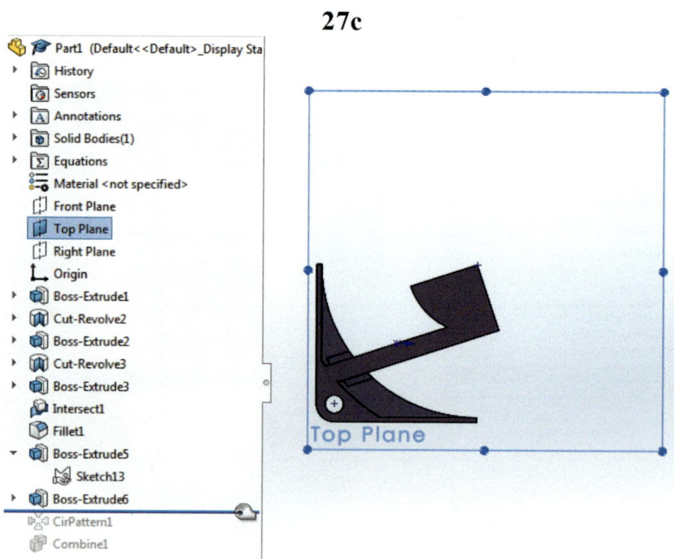

Fig. 3.131 Rewind design process

Relocate the part as shown and select the corner surface as your Sketch plane.

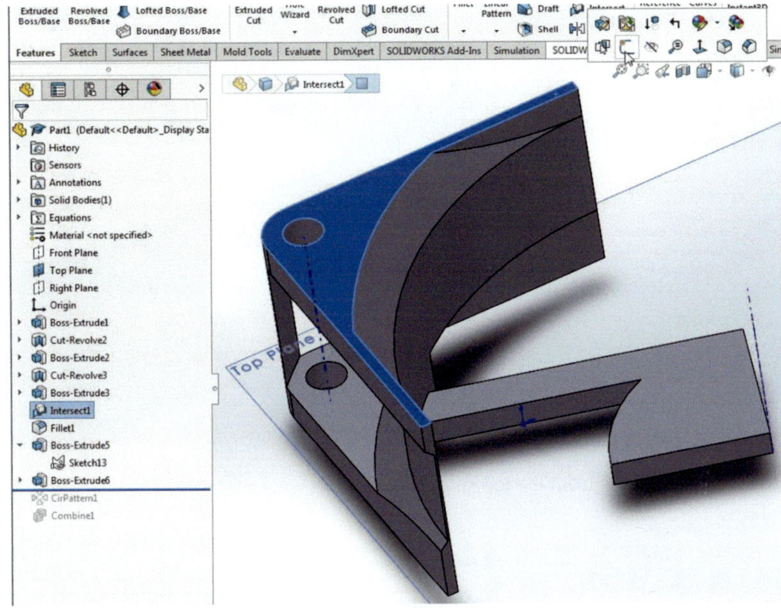

Fig. 3.132 Select the Sketch plane

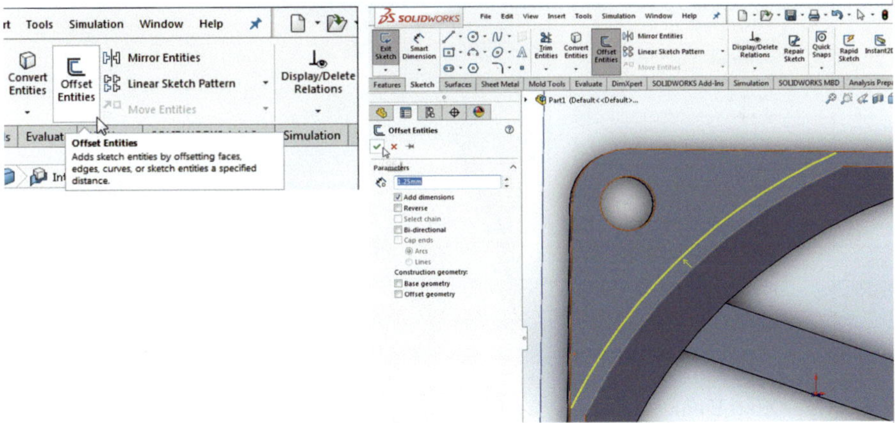

Fig. 3.133 Offset Entities from the circular edge

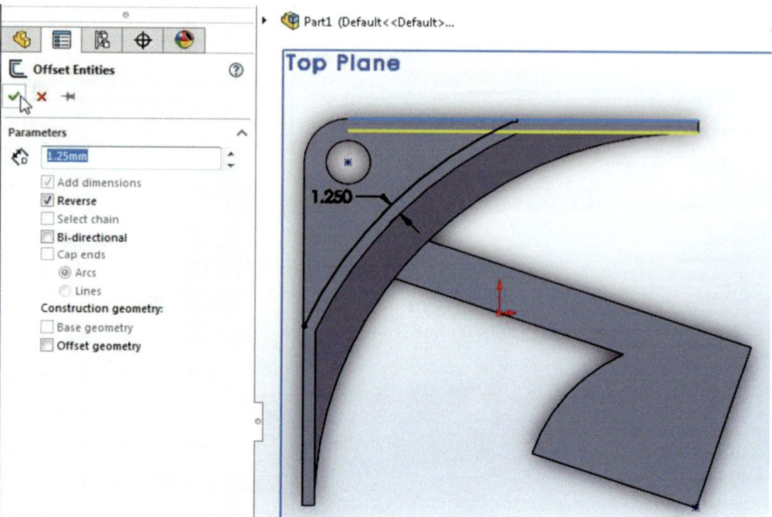

Fig. 3.134 Offset Entities from the top straight edge

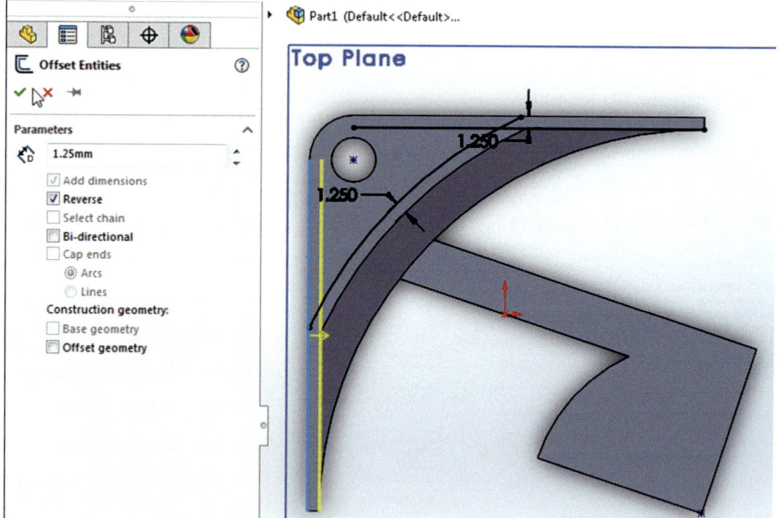

Fig. 3.135 Offset Entities from the side edge

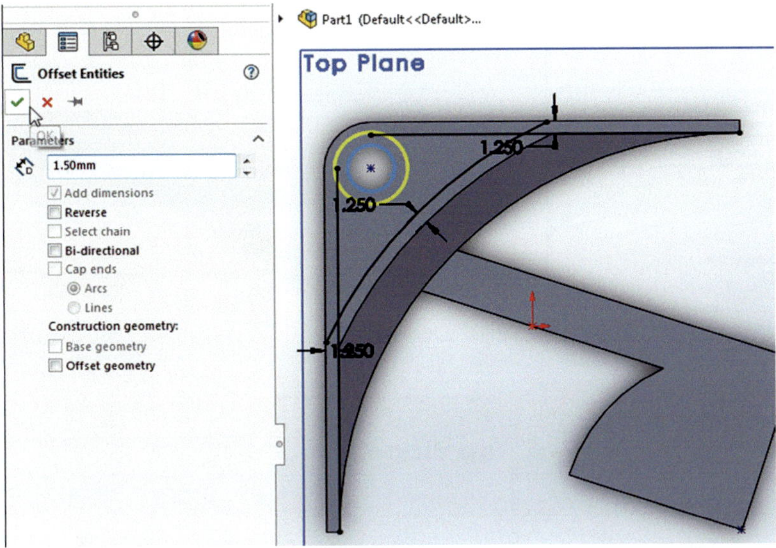

Fig. 3.136 Offset Entities on the circle on the corner

This is the preliminary profile to imprint. The trim tool is used to eliminate unnecessary segments.

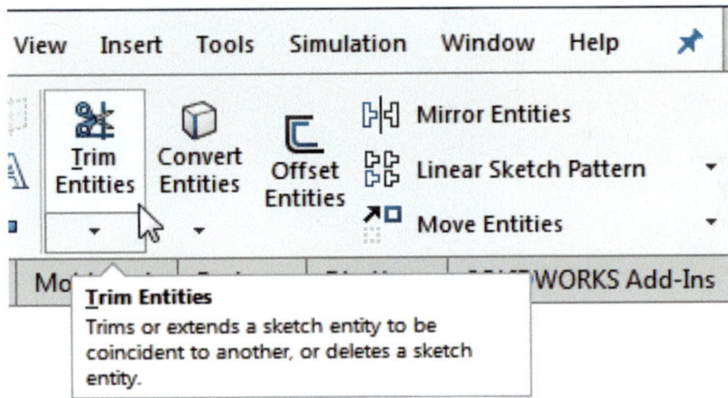

Fig. 3.137 Trim longer edges of the construct

Drag the mouse over the extra whisker lines on the top-right and bottom-left to trim them. Adjust the center similarly.

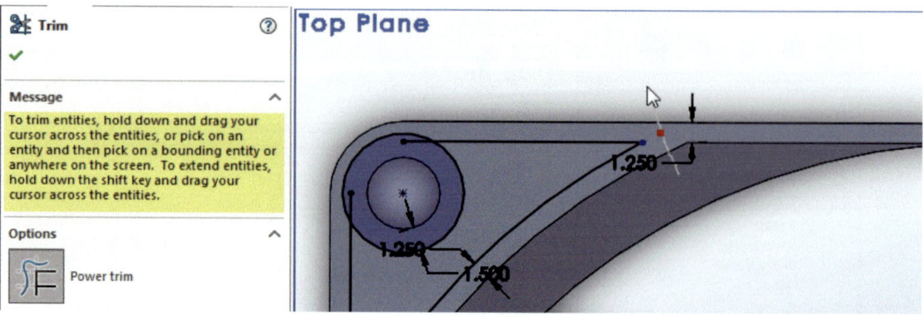

Fig. 3.138 Use Power trim to cut longer whiskers

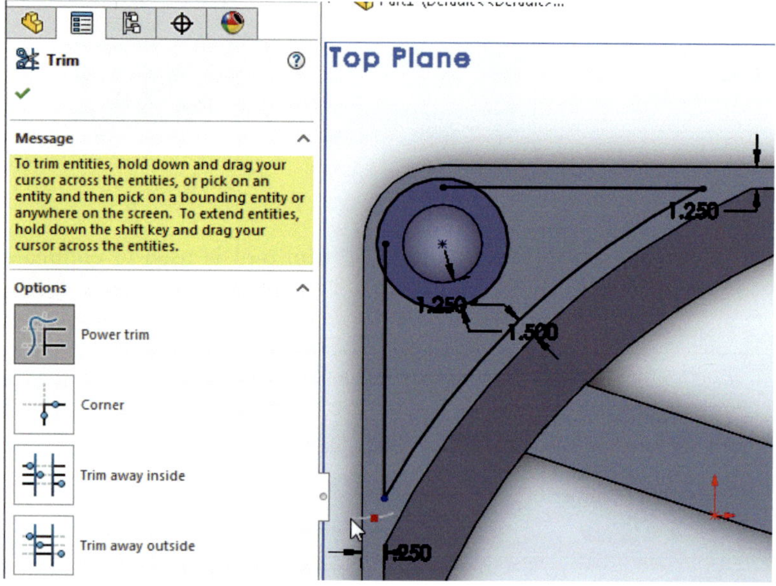

Fig. 3.138 (continued)

Fig. 3.139 Extruded Cut the composed figure

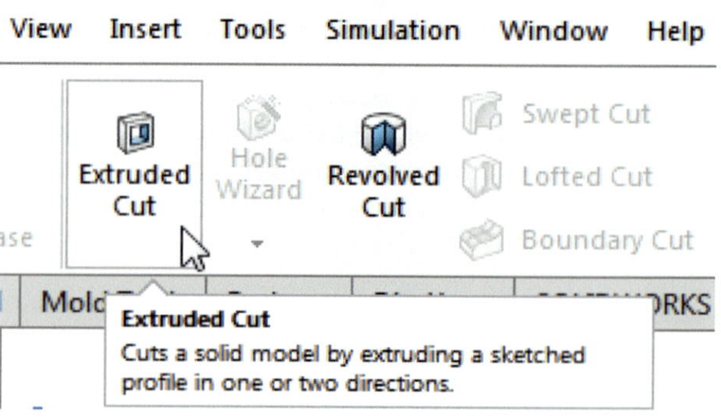

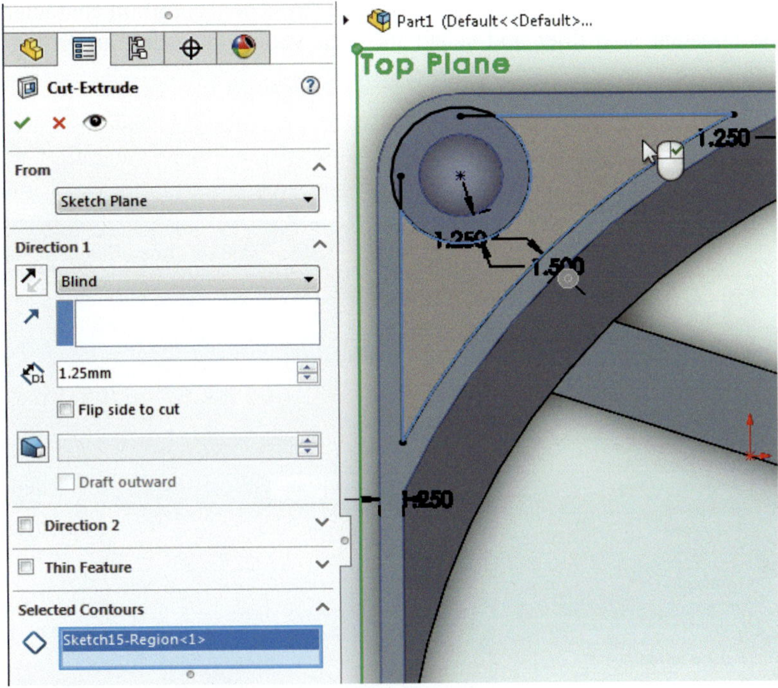

Fig. 3.140 Enlarged view of delineated profile to cut extrude

A portion of the material has been removed leaving the shape to the specified depth.

Fig. 3.141 Enlarged view of cut-extrude profile

Eliminate sharp edges that are also assigned stress concentration characteristics. Use the constant size fillet and select the four vertex lines that are going to be rounded.

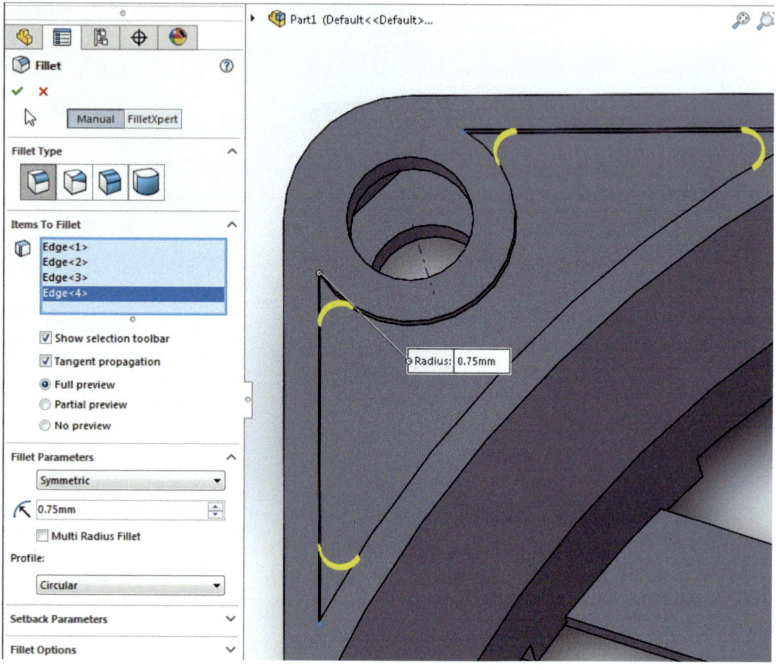

Fig. 3.142 Use the Fillet menu option to round corners

Fig. 3.143 Section with rounded contours

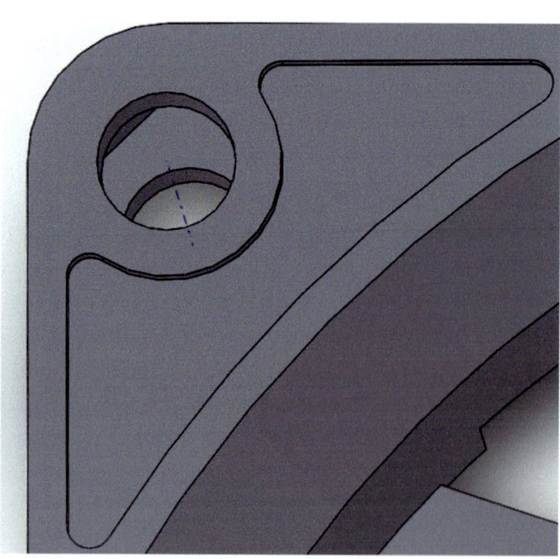

The actions are rolled forward to redraw the entire part with the new details included.

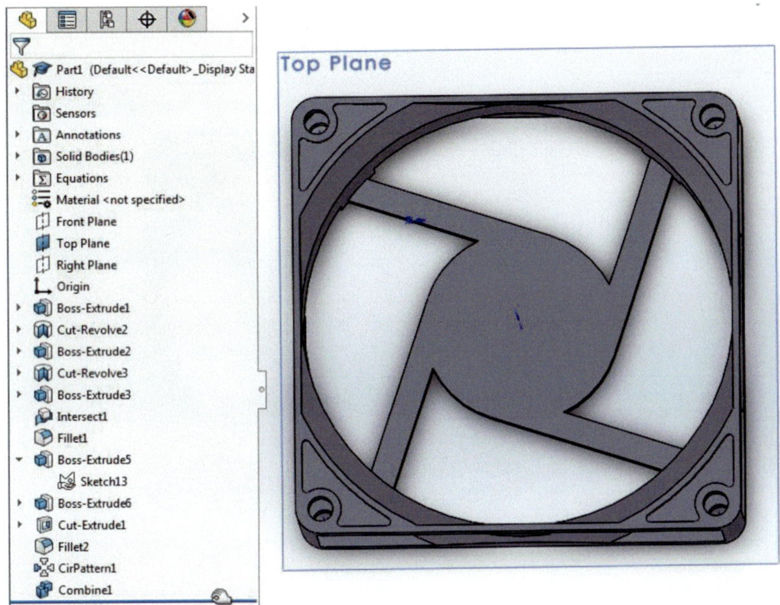

Fig. 3.144 Re-execute the drawing commands

The edges are sharp on the part's contour. Use the Fillet menu selection to smooth out all exposed edges.

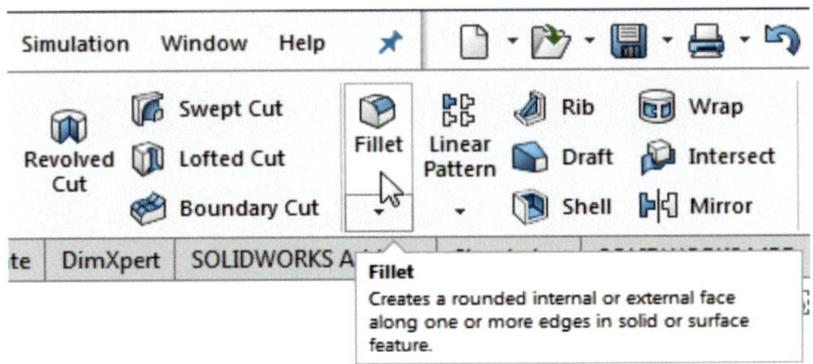

Fig. 3.145 Fillet command to round additional corners

In order to select, enclose with a selection box the entire part. This is done by left-clicking outside the figure from the top left corner and keeping the mouse button pressed and drag the mouse pointer beyond the lower right-hand corner.

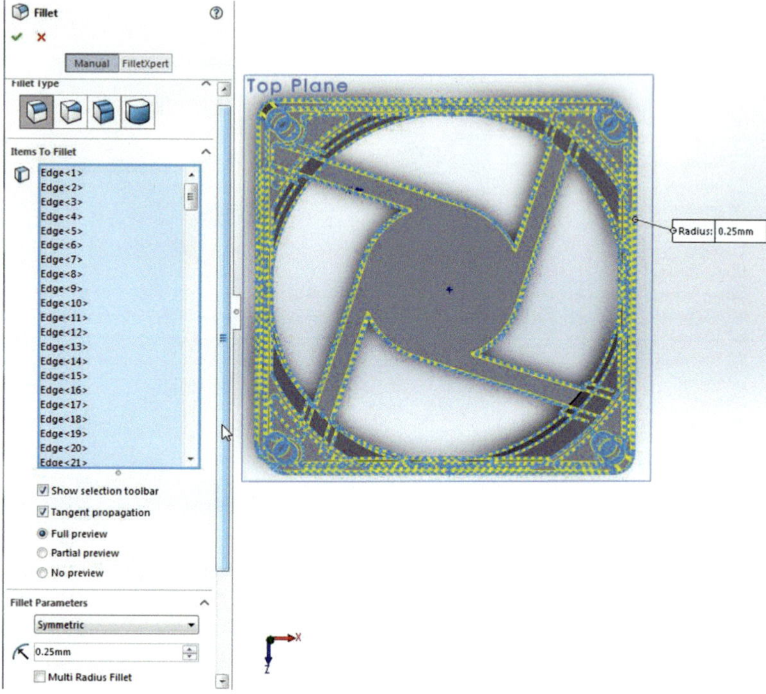

Fig. 3.146 Round corners of the figure (a larger radius of curvature may be needed for fewer finite elements)

At this step between 2017 and 2018, some warnings seemed to have crept in from the twilight zone, let the FeatureXpert fix them for you and continue. It is likely that some features are too thin to allow for a radius of curvature of 0.25 mm (try 0.0025 mm curvature radius and no warning messages appear).

27d.   Modify the material appearance and express a creative streak with a personal color choice.

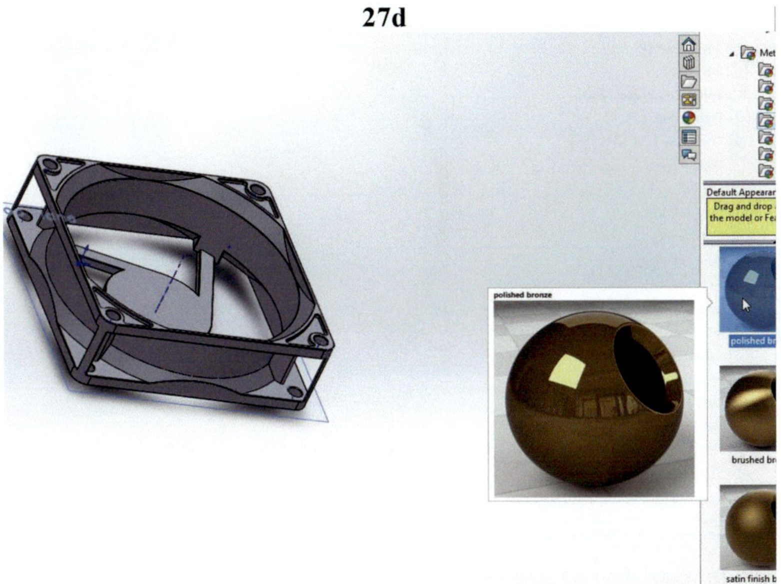

Fig. 3.147   Selecting material and surface finish

28.   To add a polished bronze finished, select on the right the colored icon: Appearances, Scenes, and Decals, to the Metals submenu and select Polished bronze. Click and Hold on Polished bronze and drag over the part to an area on the surface.

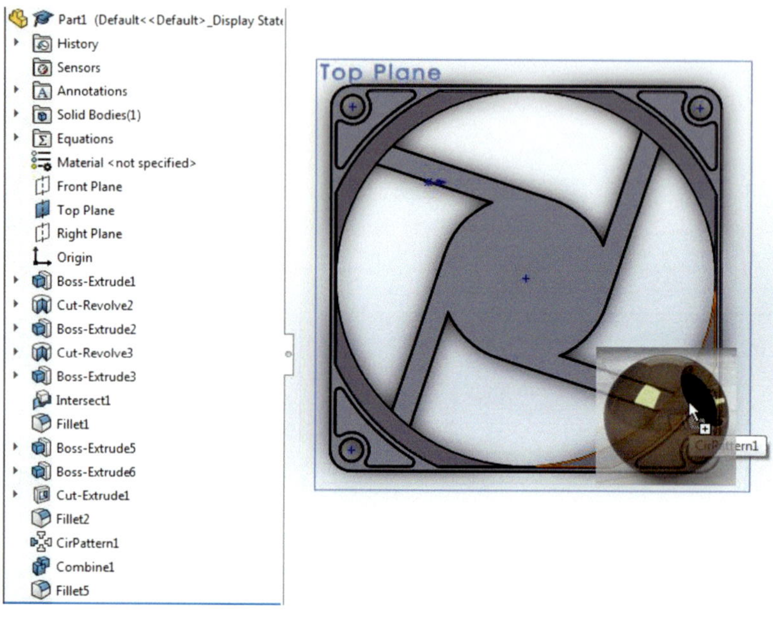

Fig. 3.148 Drag and drop material appearance on frame

Fig. 3.149 Select body to assign new material to part

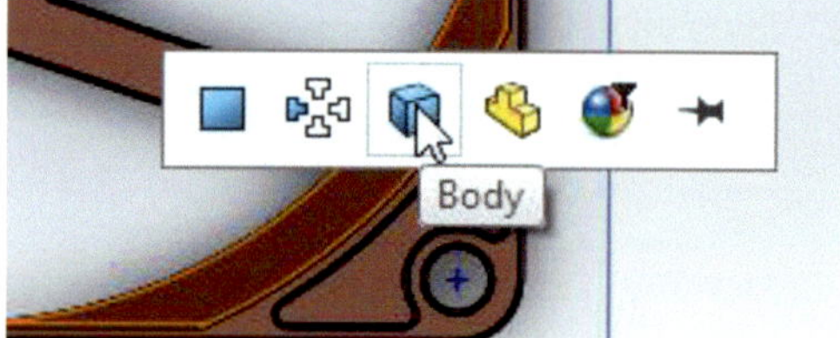

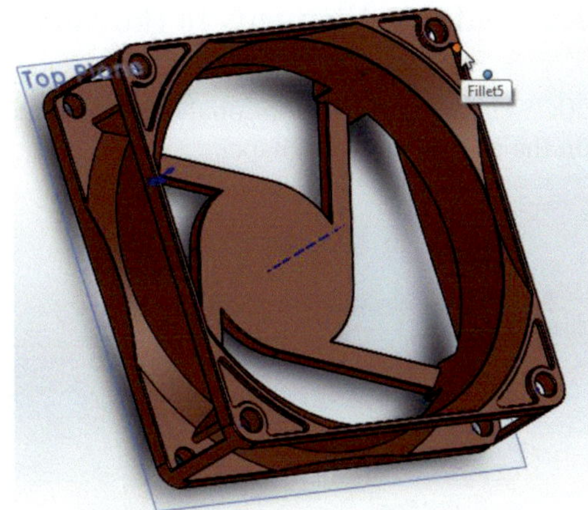

Fig. 3.150 Appearance of frame with new material assignment

29.   Feel free to experiment and color this accomplishment to your heart's desires. There are no instructions for this extra step, rock on.

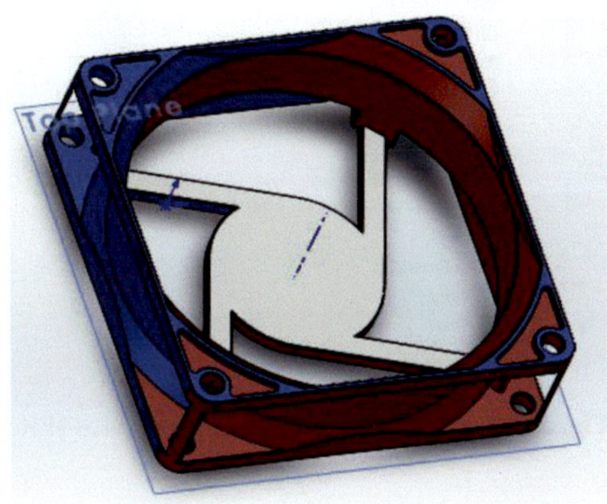

Fig. 3.151 Assigning specialty colors to your frame

3.2   Adding Elements to the Solid Frame

30.   There are some additional details to the center piece. Make a new Sketch plane on the central circular shape.

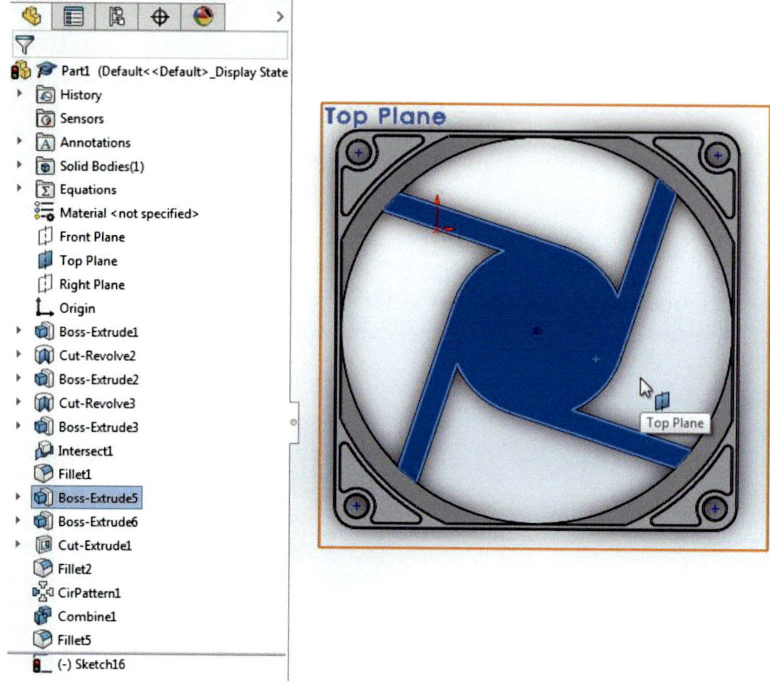

Fig. 3.152   Select the central circular region as sketch plane

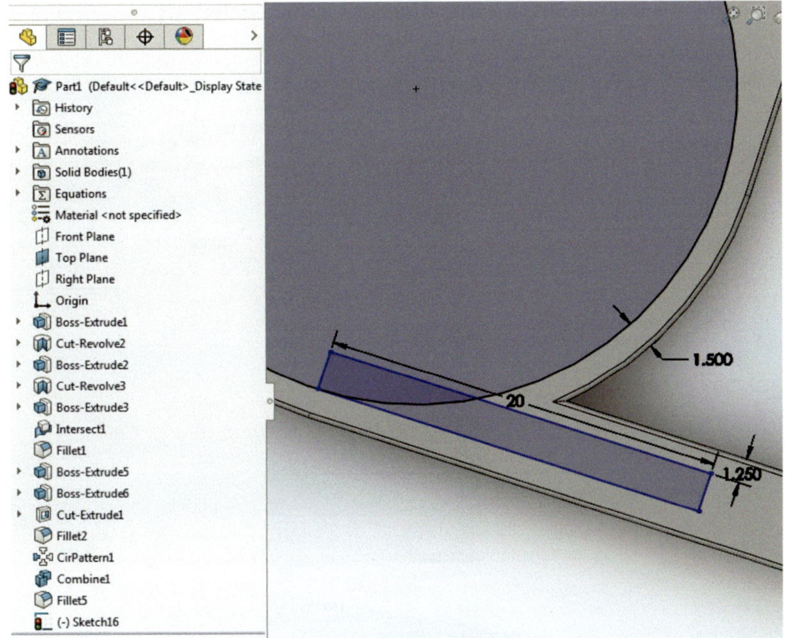

Fig. 3.153 Draw a circle and a rectangle

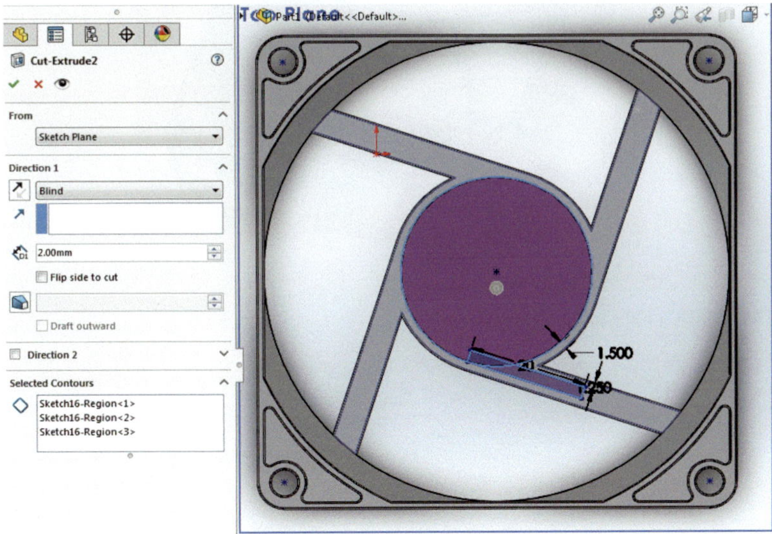

Fig. 3.154 Select Cut-Extrude for rectangle and circle

31.   Now make a hole of 2.1-mm diameter at the center.

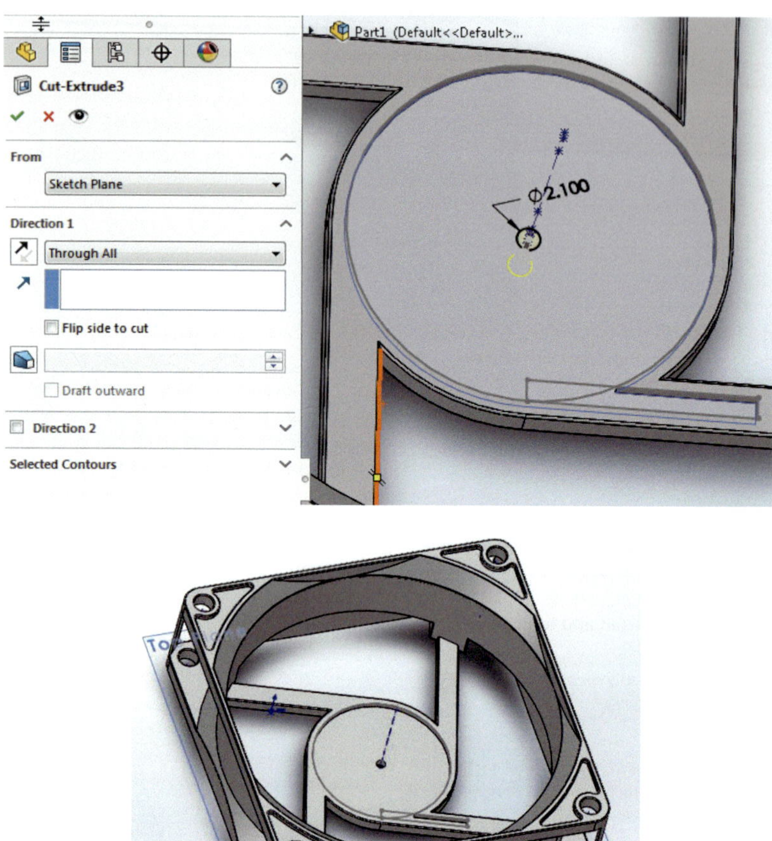

Fig. 3.155 Make a hole in the center of the frame

32. Make a new Sketch on the central circular section. Apply convert entities to the circular hole. The circular geometry of the hole is now on the Sketch plane. Make an annular region and Boss-Extrude with the option of Thin Feature.

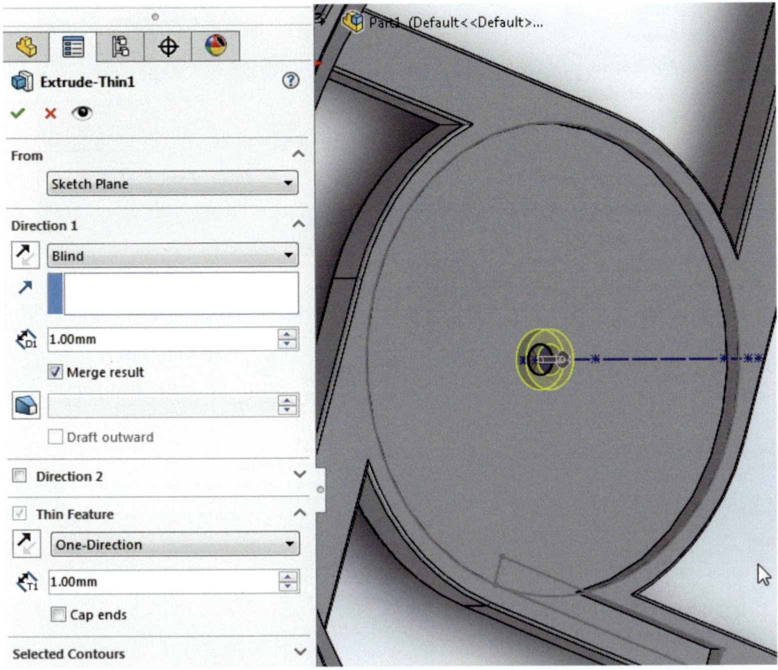

Fig. 3.156 Add annular regions using Thin Feature

33.°Now draw another annular region creating a Sketch plane on the top surface of the annular region just created and projecting the outer circumference using the Convert Entities tool.

Fig. 3.157 Convert Entities of the external larger circumference

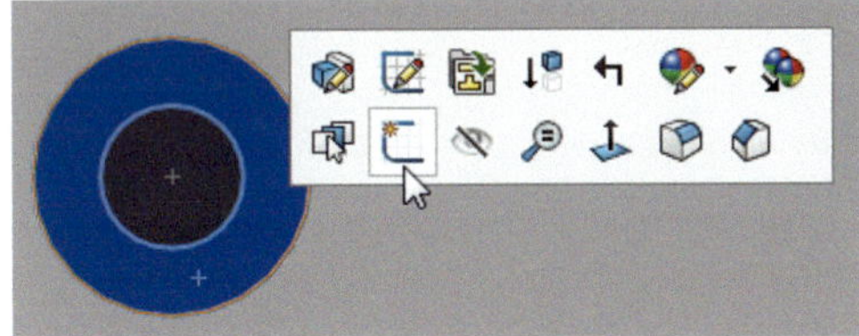

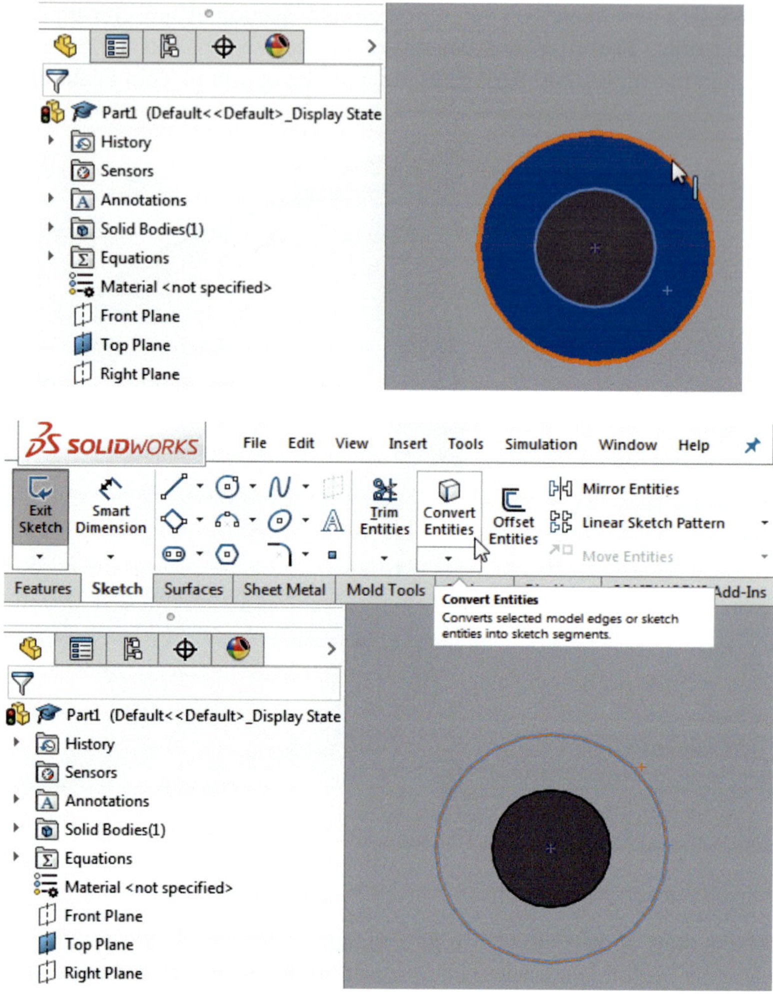

Fig. 3.158 Larger circle circumference on the sketch plane

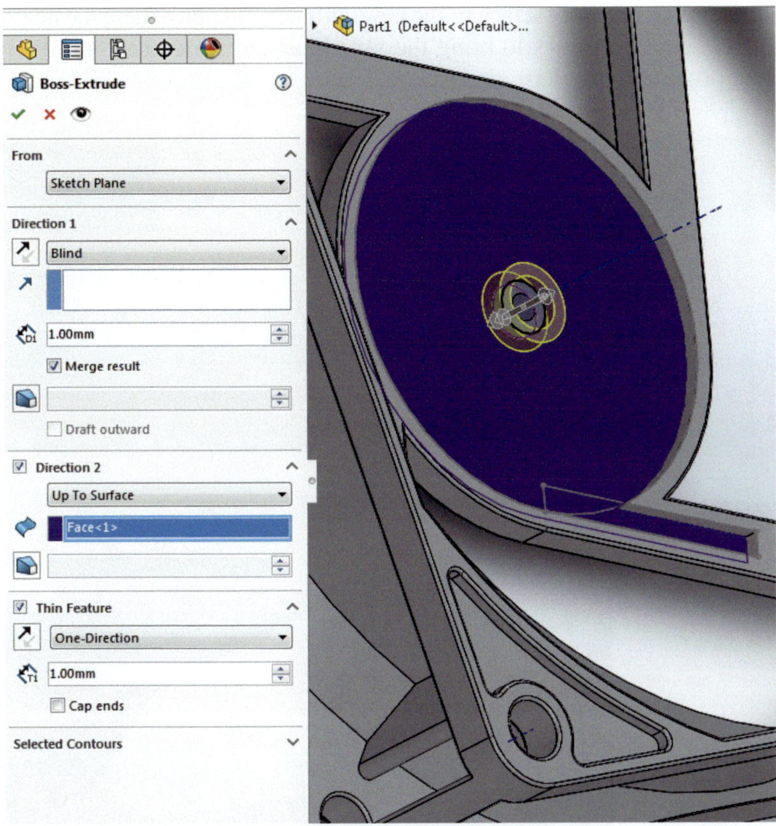

Fig. 3.159 Extrude the ring to create a protruded shape

This is the casing part up to this point.

Fig. 3.160 Front ring
section to accommodate
additional parts

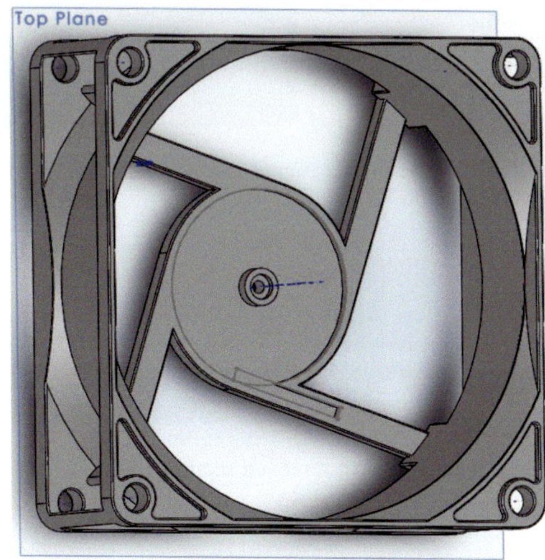

Take the upper annular surface of the new cylinder and place there the new sketch plane. Use Convert Entities to bring the innermost circumference onto the plane. After that proceed to use Extruded Cut.

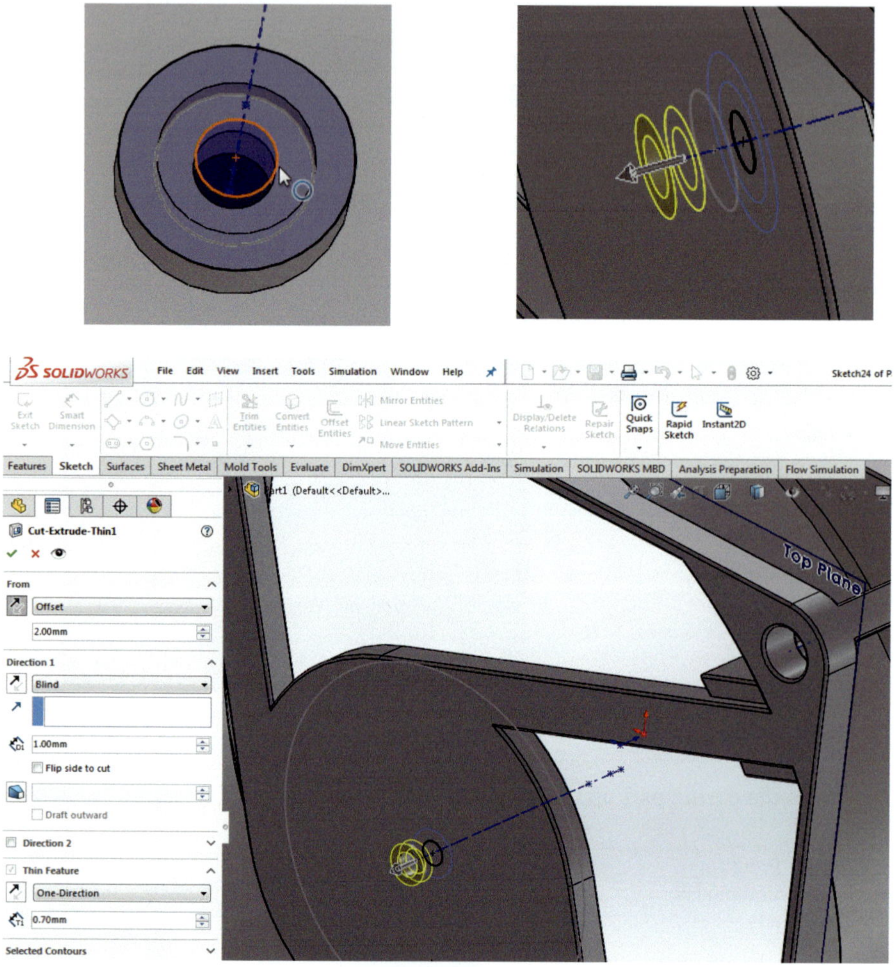

Fig. 3.161 Remove a ring section from the center of the frame

Fig. 3.162 Detail of removed section

3.3   Creating Propeller for Fan

In this section the propeller for the fan is created.

Fig. 3.163 Propeller for the fan

34.   Build the cylinder upon which to place the blades for the computer fan on the Top Plane.

34a.   Remove sharp edges from the cylinder at the current top face.

Fig. 3.164 Select units to use in new canvas

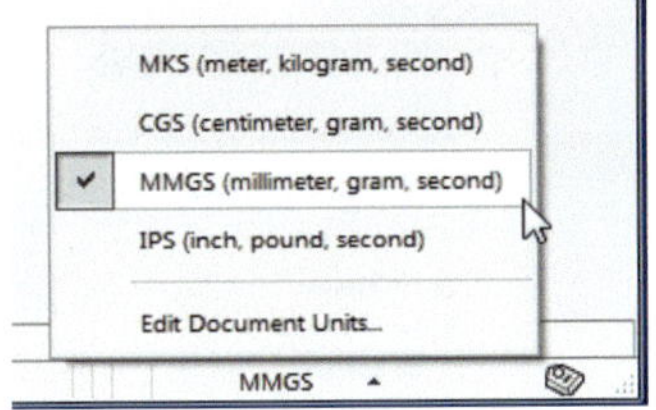

The cylinder is 35 mm in diameter and 18.75 mm in height.

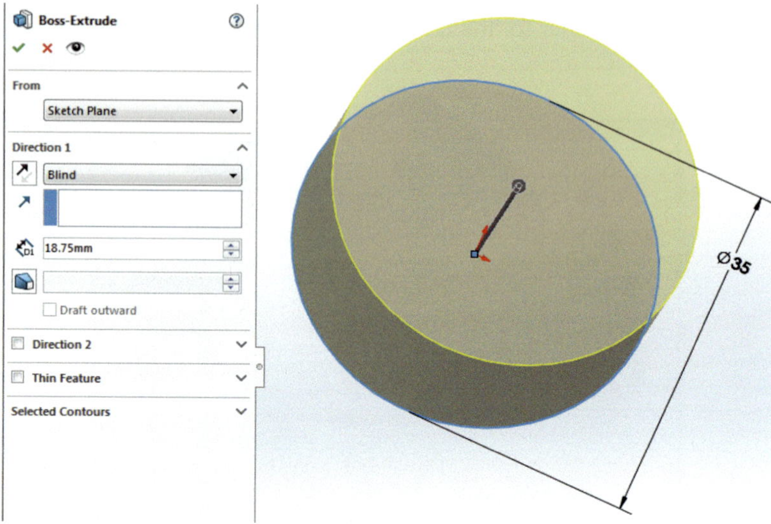

Fig. 3.165 Extrude a cylinder

Remove sharp edges from the cylinder at the current top face.

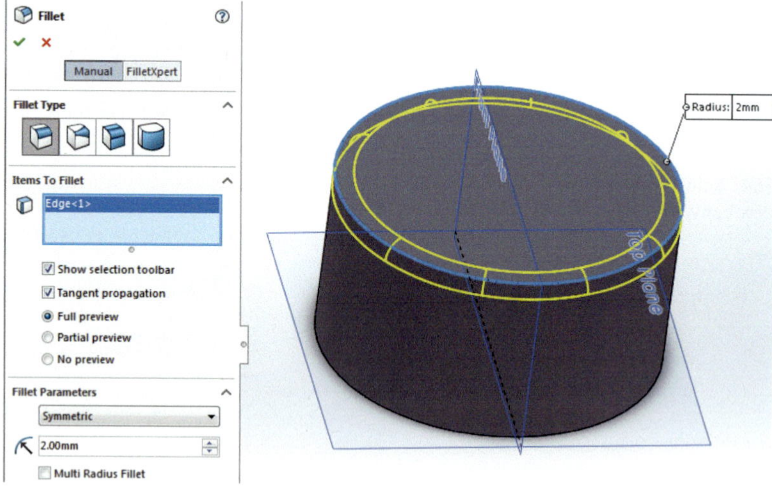

Fig. 3.166 Smooth the sharp edges on one side

35.   Create two planes upon which draw the profiles for the fan's blade.

35a.   Activate the view of the Front Plane and create a new Reference Geometry Plane parallel to the Front Plane at a distance of 40 mm.

35b.   Choose the Front Plane and select to Sketch on that Plane. This is a bit different because the drawing is being done in the plane that is in the back.

35c.   On the Sketch plane activate the temporary axis to obtain a middle line of reference.

35d.   The task is to generate two profiles one on the Front plane and another on Plane 1. These two profiles when joined generate a fin for the fan. Another drawing element is used: Point.

35e.   Draw a point at the base and a line segment at the top with the indicated measurements. The segment at the top is vertical with a size of 0.25 mm and at a distance of 2.75 mm of the centerline.

35f.   Using the three-point arc from the drawing menu, draw two arcs that join the top two points with the bottom point. For now the curvature of the segments needs to be smooth, something easy to draw.

35g.   The radii of curvature are 21 and 24.35 mm for the right-most and left-most line segments, respectively. These are three-point arcs, *draw starting and finishing points of the arc first*, and the middle point of the arc is the one to draw last.

35h.   Now select Plane 1 as the Sketch plane to draw another two line profiles.

35i.   Now two additional profiles are drawn on Plane 1. Place a Line segment at the top of the cylinder and a Point at the Bottom. Two three-point arcs are added with radii of 46.5 and 36.5 mm.

35j.   Create a new Sketch plane on rounded end of the cylinder.

35k.   Wire view of line segment to be drawn.

35l.   The radius of curvature for the arc is 33.25 mm.

35m.   Now enable a Sketch plane at the bottom of the cylinder.

35n.   The light blue circle is converted to the current Sketch plane (this is the inner circle on the top face).

35o.   The new circular element on the plane is better as a construction curve.

35p.   A straight line segment is drawn at an angle of 112° and three-point arc is drawn with a curvature radius of 24.50 mm. The starting point of the linear segment is at the Point element on the front plane and its end point is on the construction circle brought into the plane via Convert Entities at an angle of 112°.

35a.   Activate the view of the Front Plane and create a new Reference Geometry Plane parallel to the Front Plane at a distance of 40 mm.

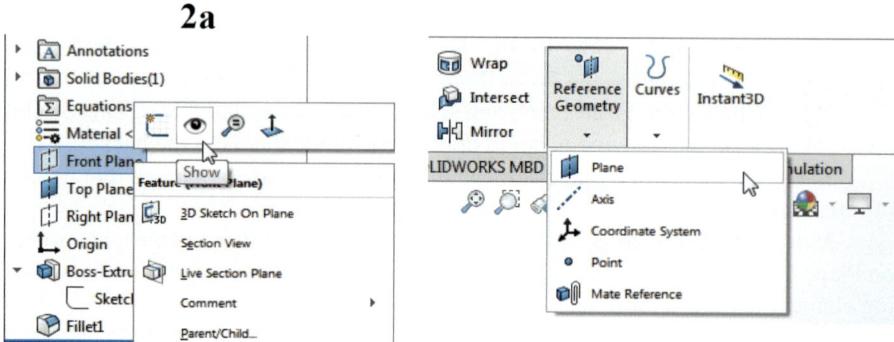

Fig. 3.167 Create a new Reference Geometry Plane

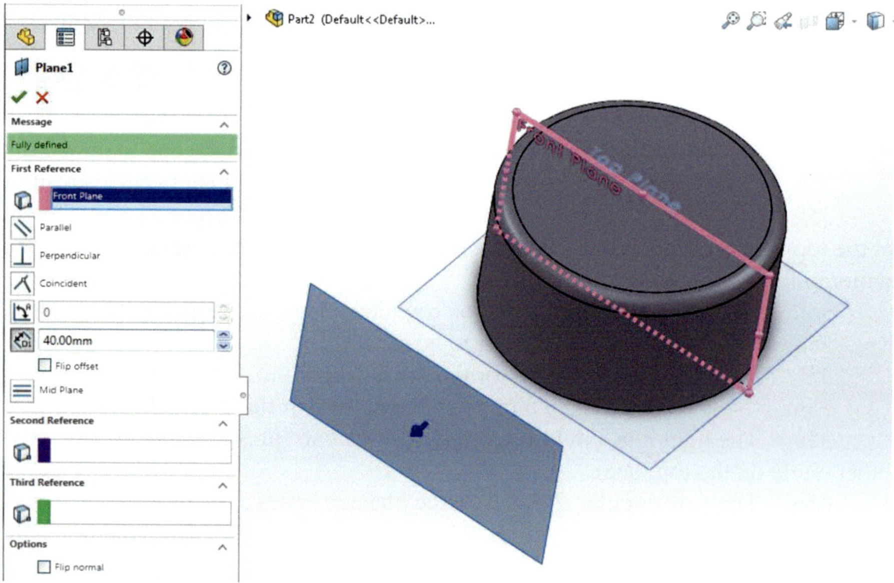

Fig. 3.168 Create an additional plane offset from Top Plane

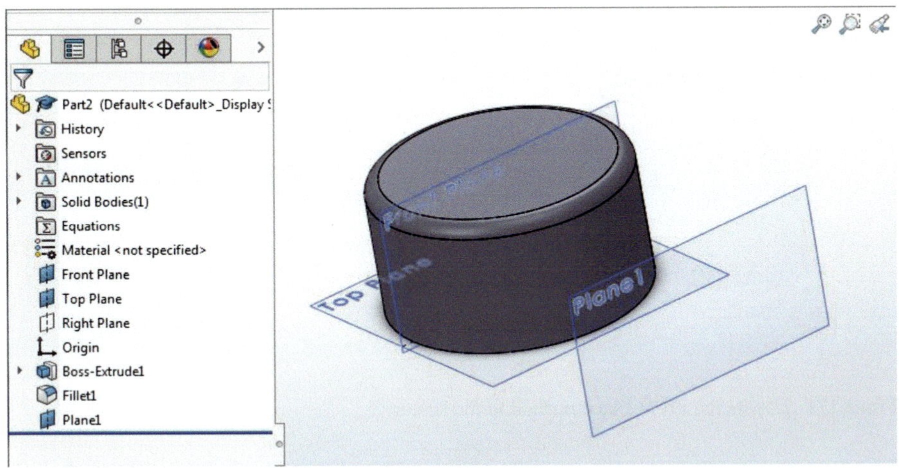

Fig. 3.169 Planes for sketching propeller components

35b. Choose the Front Plane and select to Sketch on that Plane. This is a bit different because the drawing is being done in the plane that is in the back.

35b

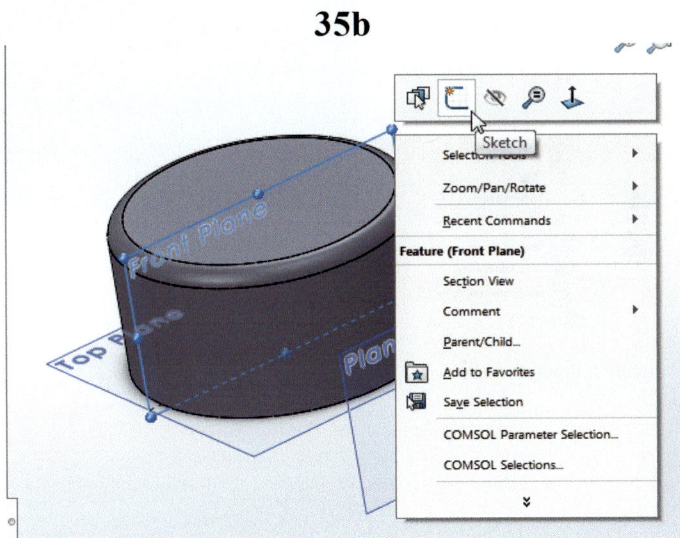

Fig. 3.170 Create a sketch plane on the Front Plane

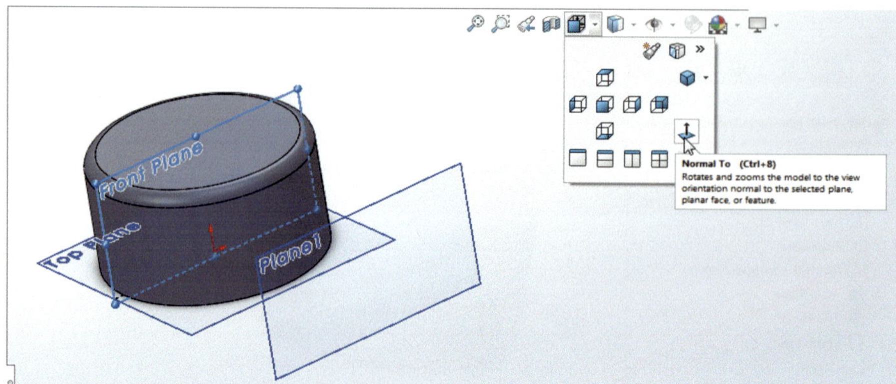

Fig. 3.171 Locate the Front Plane normal to the screen

35c.   On the Sketch plane activate the temporary axis to obtain a middle line of reference.

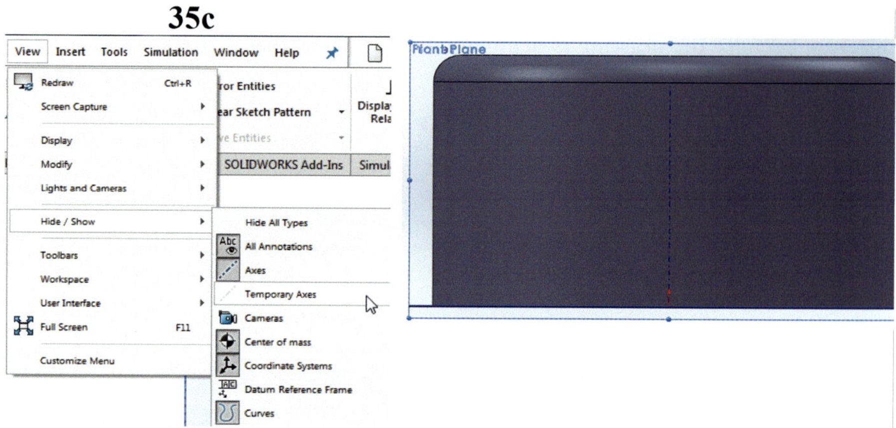

Fig. 3.172 Activate temporary axis

35d.   The task is to generate two profiles one on the Front plane and another on Plane 1. These two profiles when joined generate a fin for the fan. Another drawing element is used: Point.

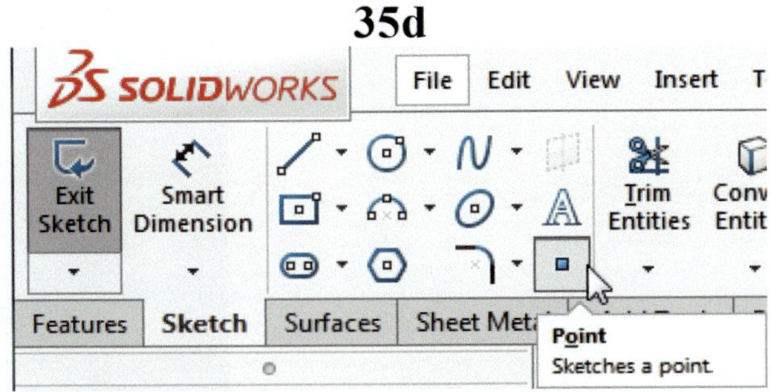

Fig. 3.173 Find on the menu the option to draw a single point

35e.   Draw a point at the base and a line segment at the top with the indicated measurements. The segment at the top is *vertical* with a size of 0.25 mm and at a distance of 2.75 mm of the centerline.

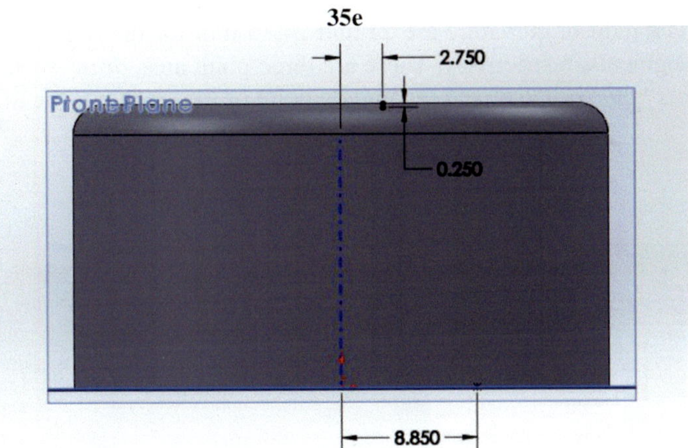

Fig. 3.174 Draw a line segment and a point

35f.   Using the three-point arc, draw two arcs that join the top two points with the bottom point. For now, the curvature of the segments needs to be smooth, something easy to draw.

Fig. 3.175 Draw two arcs for the surfaces of the fan blade

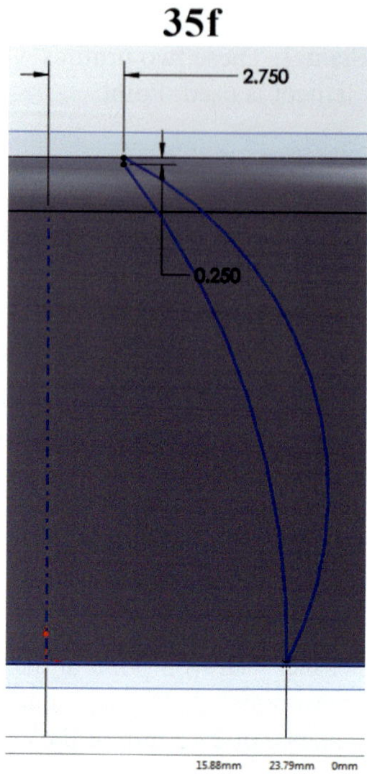

35g. The radii of curvature are 21 and 24.35 mm for the right-most and left-most line segments, respectively. These are three-point arcs, *draw starting and finishing points of the arc first*, and the middle point of the arc is the one to draw last.

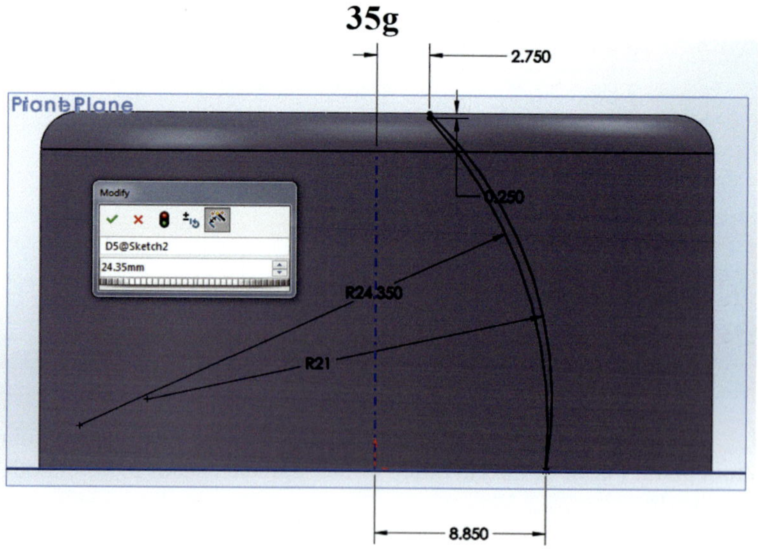

Fig. 3.176 Apply specific curvature radii

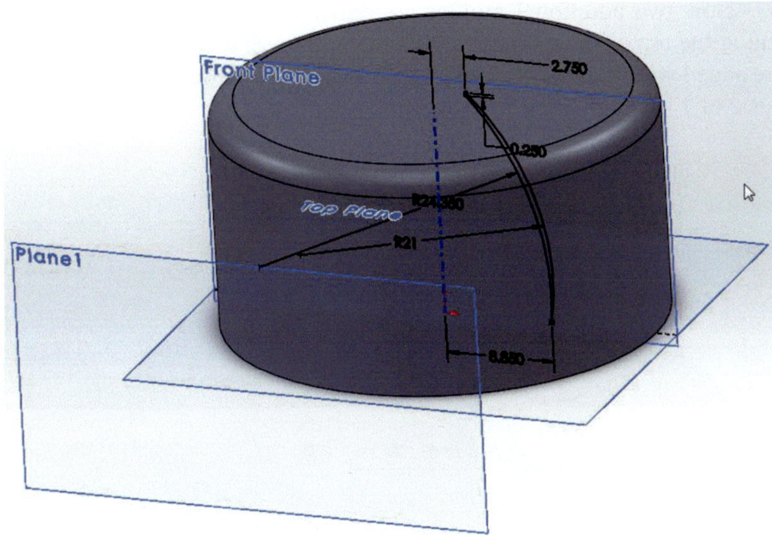

Fig. 3.177 View of current sketch plane

35h.   Now select Plane 1 as the Sketch plane to draw another two line profiles.

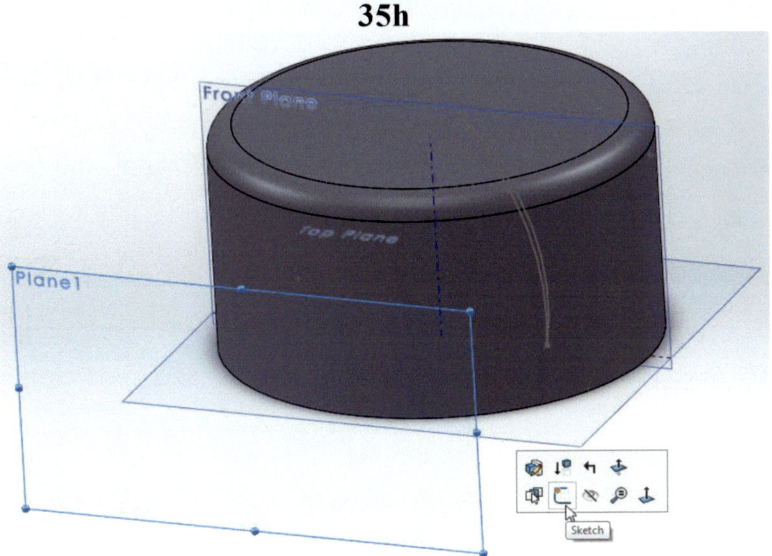

Fig. 3.178 Select Plane 1 to create another sketch

35i.   Now two additional profiles are drawn on Plane 1. Place a *vertical* line segment at the top of the cylinder and at the bottom draw a Point. Two three-point arcs are added with radii of 46.5 and 36.5 mm. At the bottom, an additional point at a distance of 13.25 mm from the centerline is present.

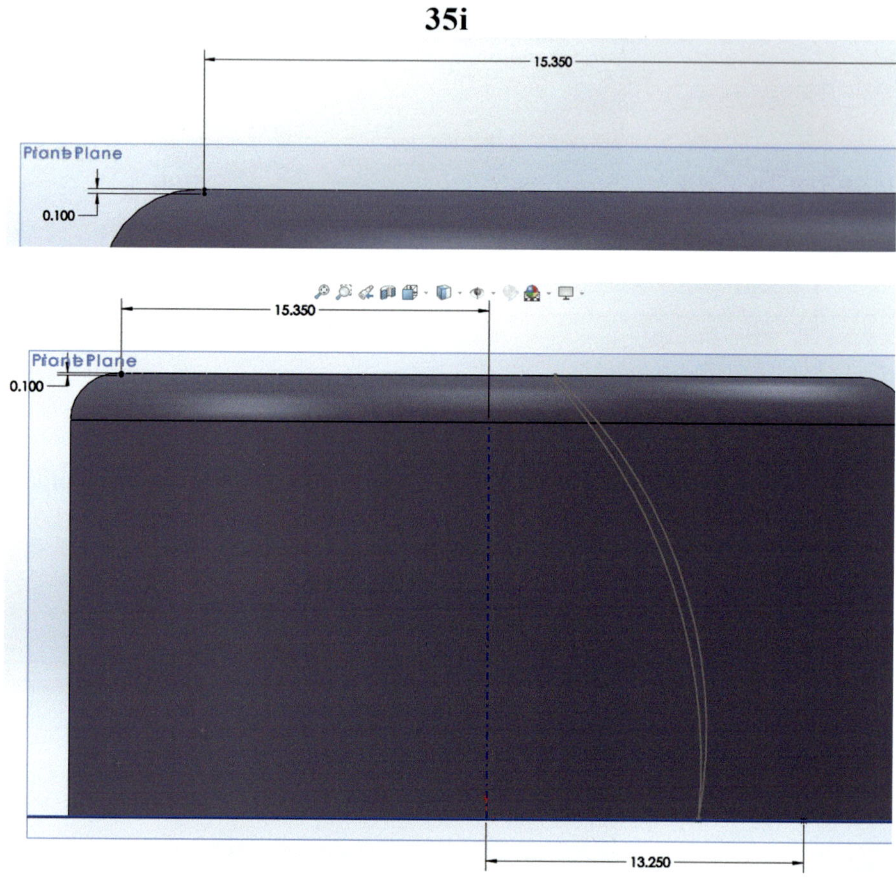

Fig. 3.179 Select Plane 1 to create another sketch

The two arcs do not cross each other.

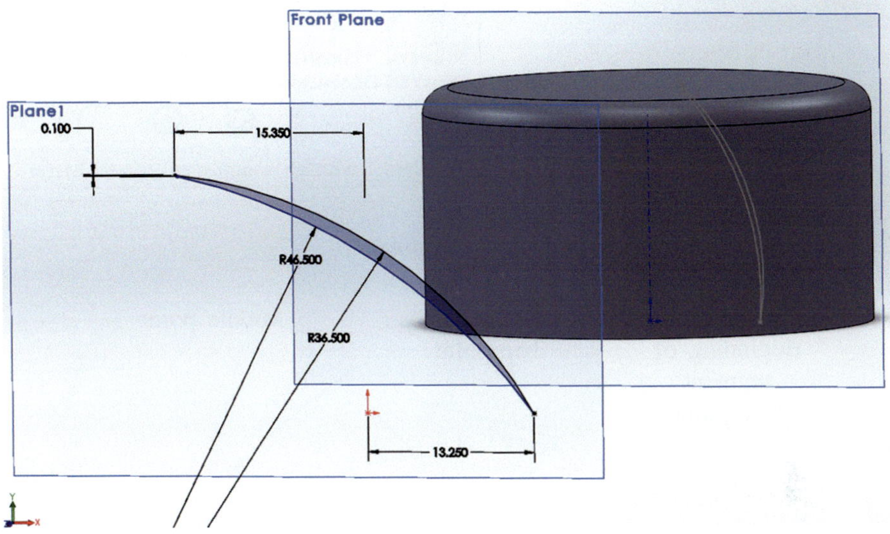

Fig. 3.180 Apply specific curvature radii

35j. Create a new Sketch plane on rounded end of the cylinder.

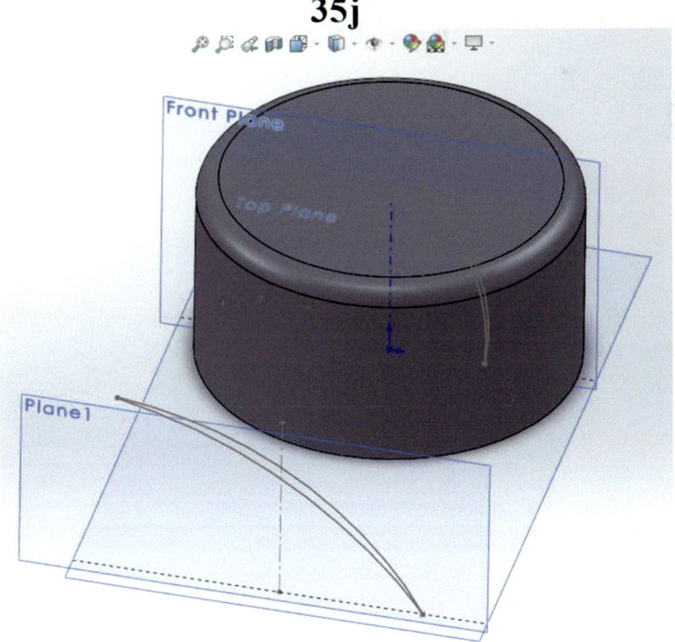

Fig. 3.181 Create a Sketch plane on the surface with rounded edges

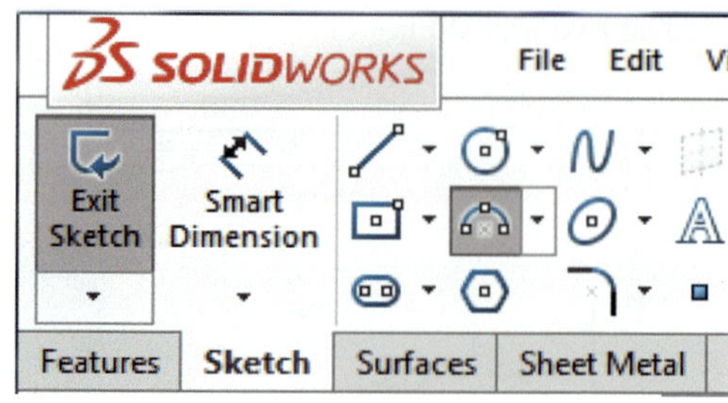

Fig. 3.182 Select the Three Point arc to make a frame for fan blade

35k.   Wire view of line segment to be drawn.

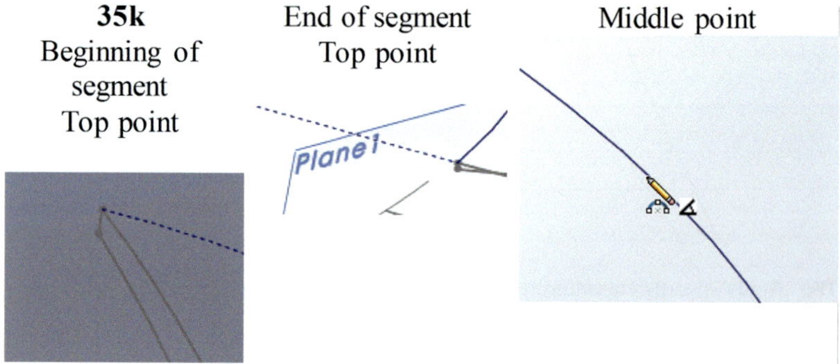

Fig. 3.183 Detail of arc profile between planes

35l.   The radius of curvature for the arc is 33.25 mm.

35l

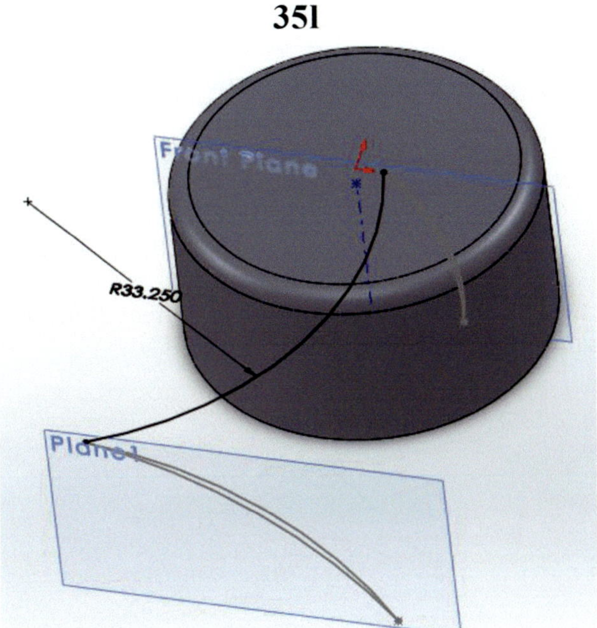

Fig. 3.184 Specifying the radius of curvature for arc between planes

35m.   Now enable a Sketch plane at the bottom of the cylinder (make the top plane invisible).

35m

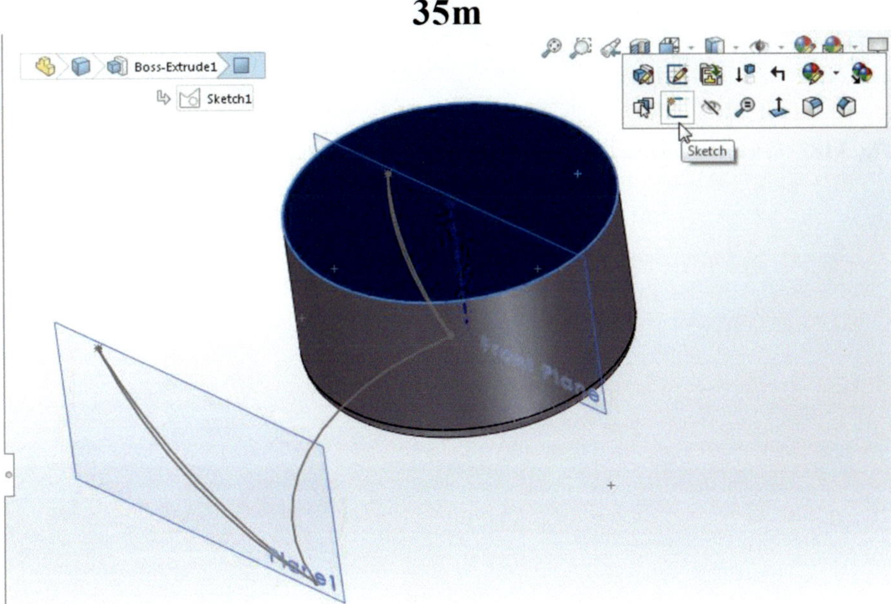

Fig. 3.185 The bottom surface turns of a solid. Create Sketch plane on that surface

Fig. 3.186 Select Convert
Entities from the drawing
menu

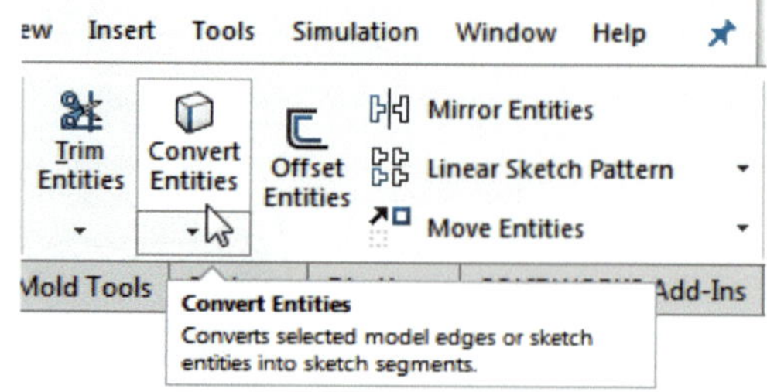

35n.   The light blue circle is converted to the current Sketch plane (this is the
inner circle on the top face).

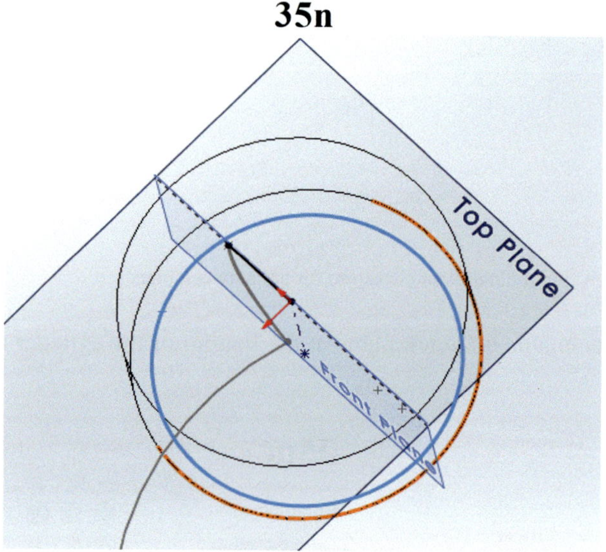

Fig. 3.187 Select the circular edge and bring into the plane with Convert Entities

35o.   The new circular element on the plane is better as a construction curve.

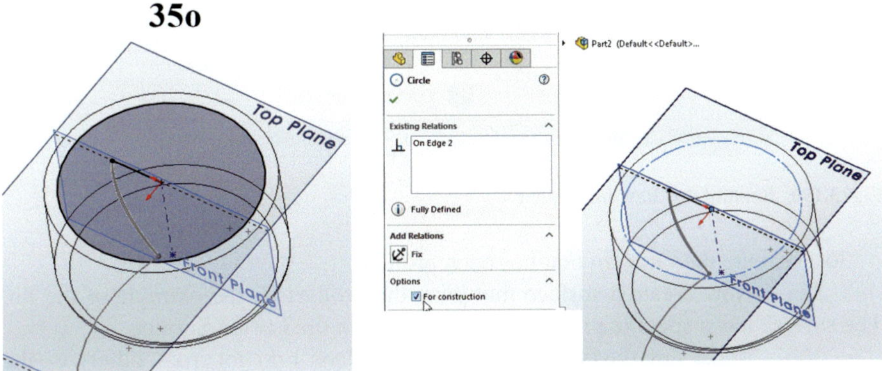

Fig. 3.188 The converted circular is brought in as a Construction curve

35p.   A straight line segment is drawn at an angle of 112° and three-point arc is drawn with a curvature radius of 24.50 mm. The starting point of the linear segment is at the *Point element* on the front plane and its end point is on the construction circle brought into the plane via Convert Entities at an angle of 112°.

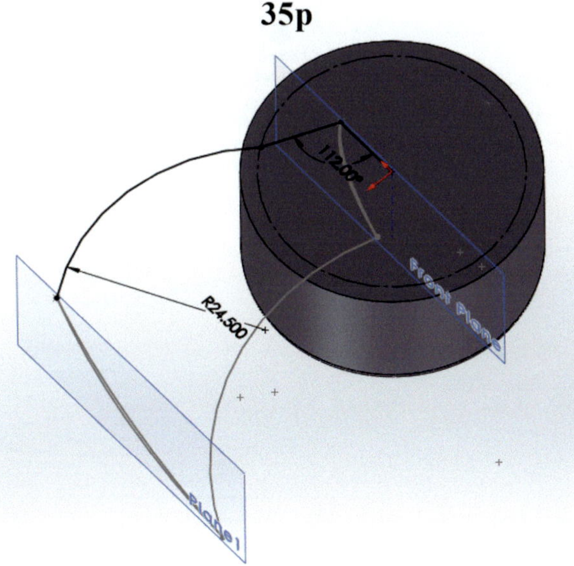

Fig. 3.189 Create a two segment construct for the fan blade

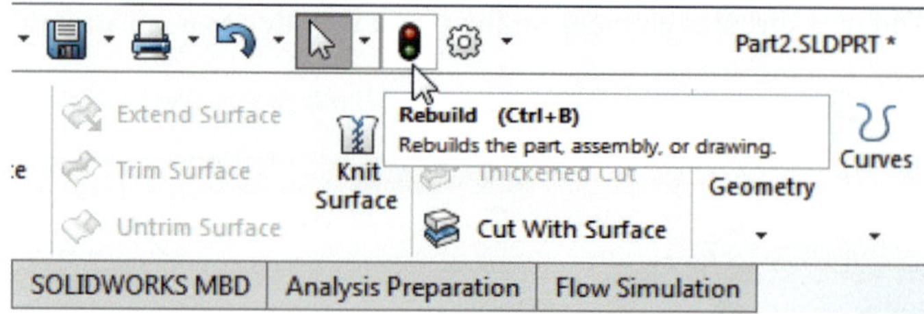

Fig. 3.190 Rebuild the drawing created

36.   Create the fan's fin that has been profiled.

36a.   Now create a surface that joins the profiles on the extrema of the fin. The surface has a specific profile that is specified in the Lofted Surface.

36b.   Make sure the two line segments on Plane 1 do not cross, otherwise the Lofted Boss/Base will not work.

36c.   The fin angle is modified using the Flex feature for a twisting angle of $-12°$. This allows for modifying the angle of the fin (blade).

36d.   Proceed to Combine the two elements of the part.

36e.   The first design of the fan is completed.

36a.   Now create a surface that joins the profiles on the extrema of the fin. The surface has a specific profile that is specified in the Lofted Surface.

Fig. 3.191 Find the Lofted Surface selection to join fan blade profiles

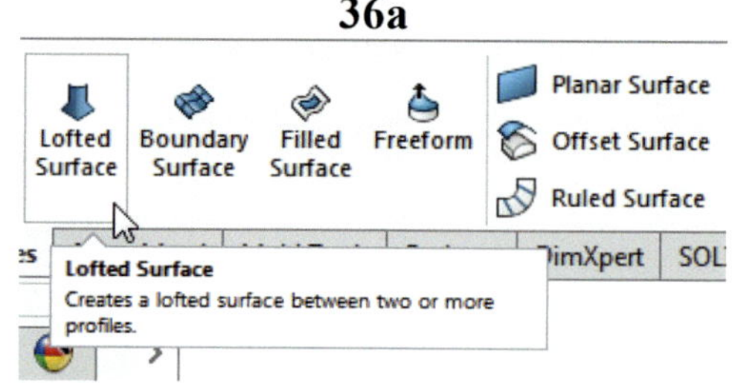

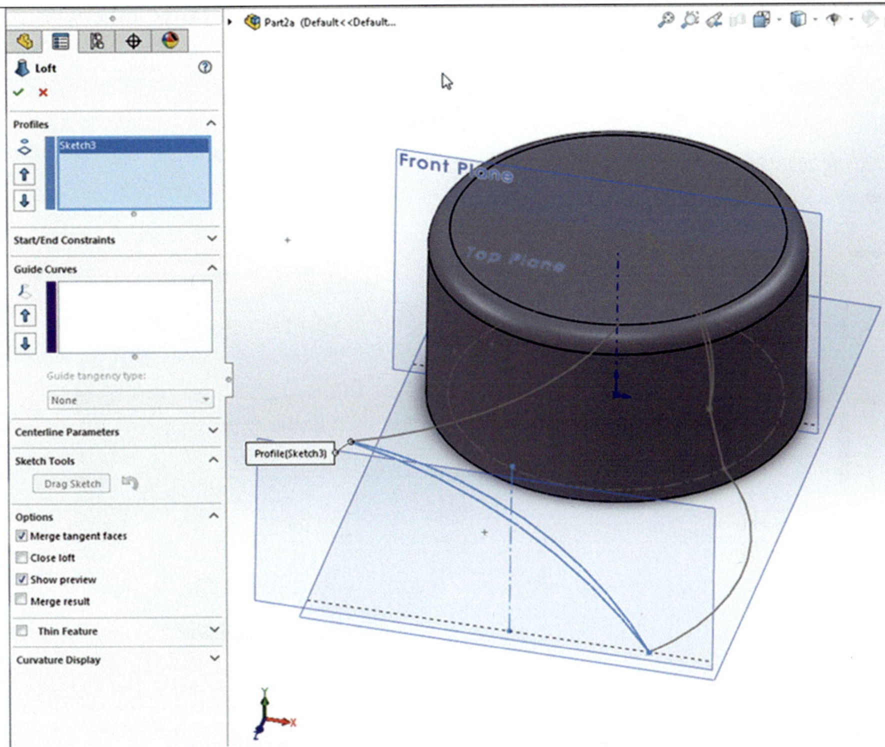

Fig. 3.192   Select the Profiles to guide the Lofted Surface

36b.   Make sure the two line segments on Plane 1 do not cross, otherwise the Lofted Boss/Base will not work.

36b

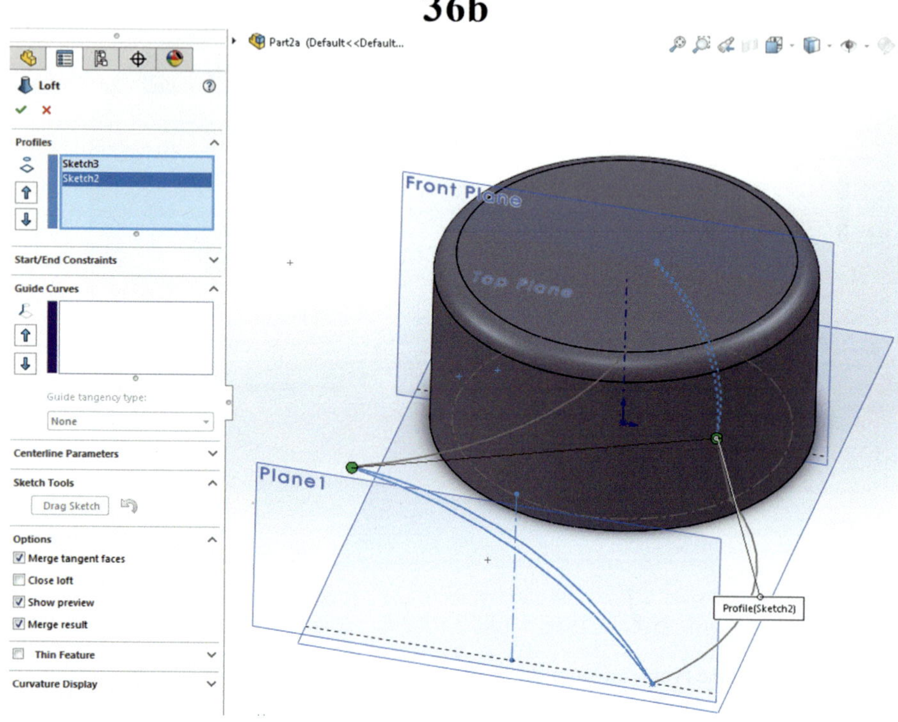

Fig. 3.193 View of the two profiles to guide Lofted Surface

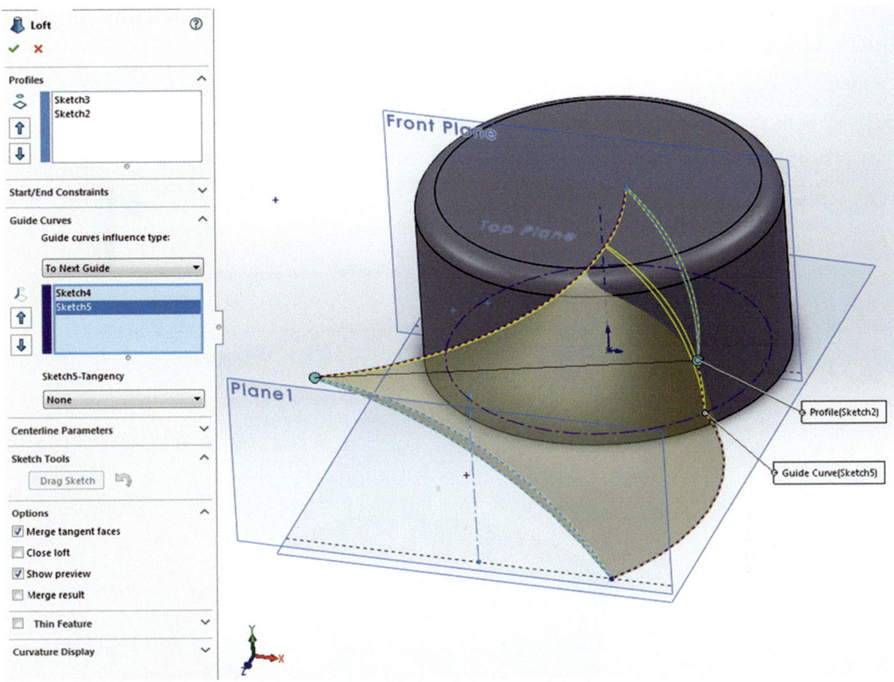

Fig. 3.194 Guide Curves to delimit the Lofted Surface

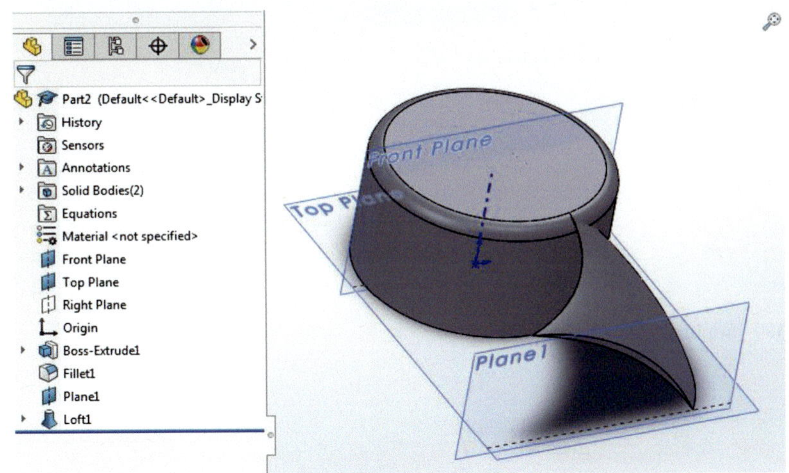

Fig. 3.195 Initial structure of a blade for the computer fan

36c.   The fin angle is modified using the Flex feature for a twisting angle of $-12°$. This allows for modifying the angle of the fin (blade).

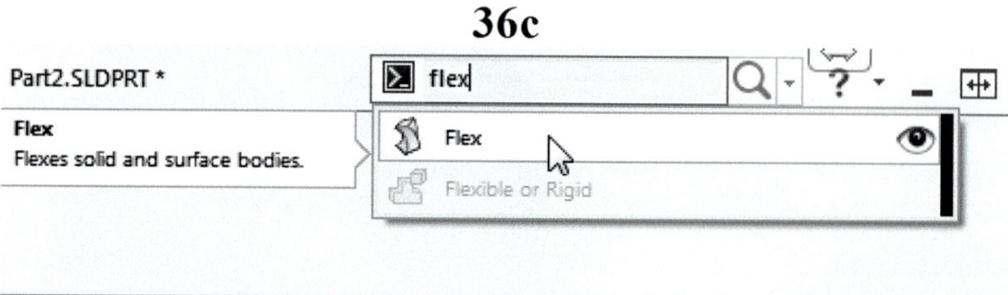

Fig. 3.196 Locate the Flex feature to set the blade angle

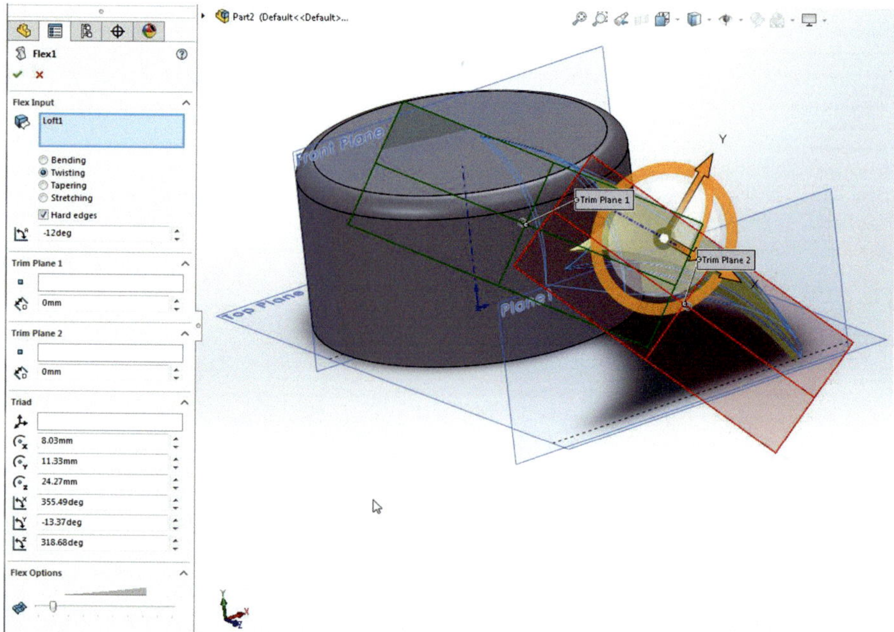

Fig. 3.197 Set the angle for the fan blade

36d.   Proceed to Combine the two elements of the part.

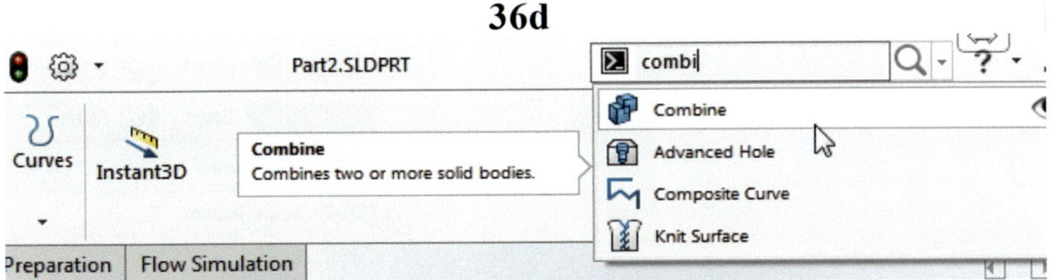

Fig. 3.198   Use the search tool to find the Combine option

If the Combine option is grayed out, this means that the part is already a single solid by the effect of the Merge in the Loft section.

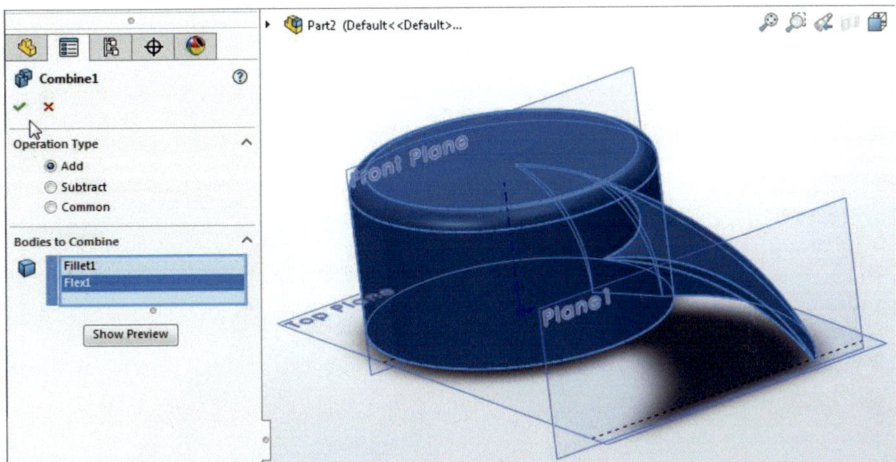

Fig. 3.199   Combine the elements created

Fig. 3.200 Select Circular
Pattern from the drawing
menu

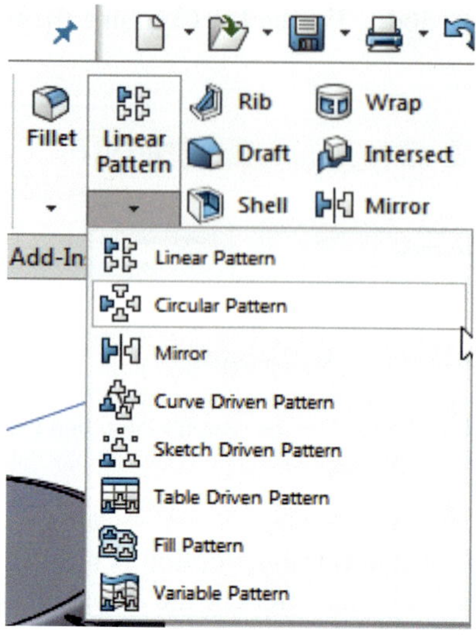

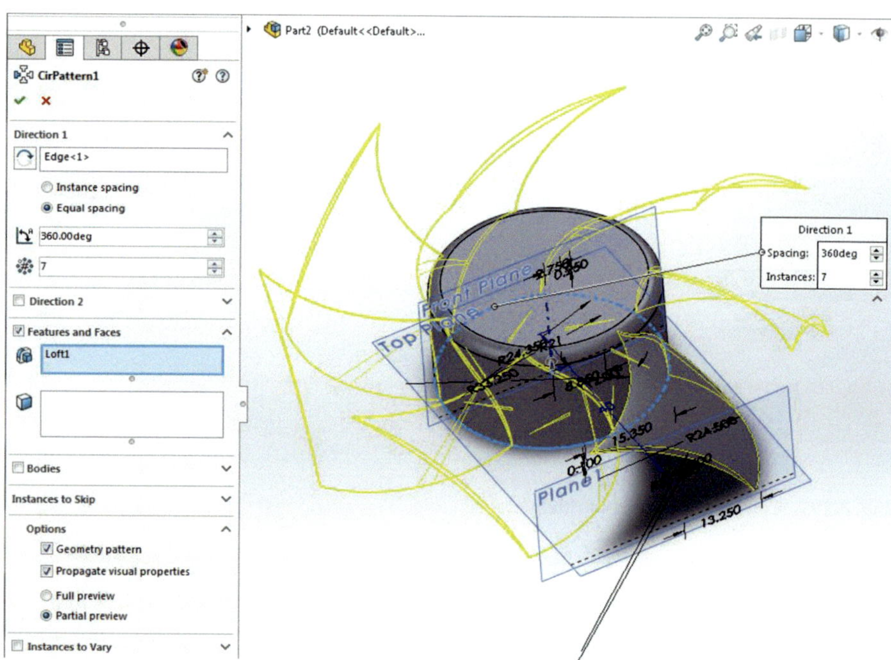

Fig. 3.201 Using CirPattern, add to the fan six additional blades

If there is a problem at this stage such that the Direction 1 cannot be selected as
the lower edge of the cylindrical element, then review the Flex1 step and make sure
that the Triad numbers are populated with values close to the ones in this document.
If needed enter -1 -2 -3 -4 … -12 in sequence for the Flex Input angle value and
verify the variation in Triad number values.

36e.   The first design of the fan is complete.

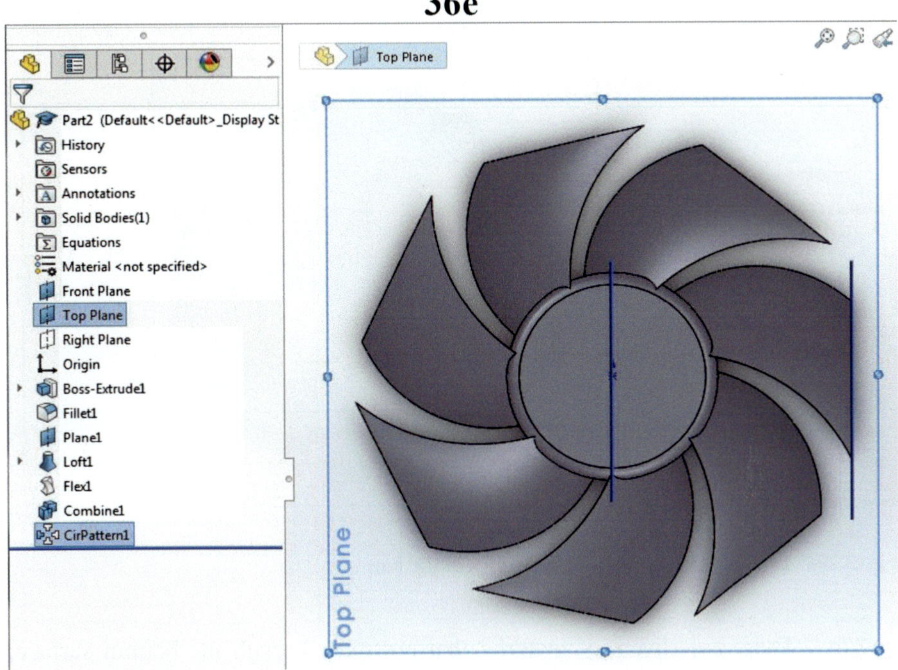

Fig. 3.202 Initial design of the fan is complete

37.   Trim the fan's propeller.

37a.   There are slivers of material that protrude through the bottom surface; these will be eliminated in a later stage when the bottom surface is configured to accommodate a rotation piece.

37b.   The sharp edge at the far end of each fin of the propeller is smoothed with the Fillet choice from the menu.

37c.   The Front Plane and Plane 1 are hidden in order to clear the part drawing.

37d.   Erode away part of the cylinder for the assembly.

37e.   Hide the top plane to have a less cluttered view.

37f.   Fillet the internal border to 1 mm.

Fig. 3.203 Select
Extruded Cut to eliminate
excesses

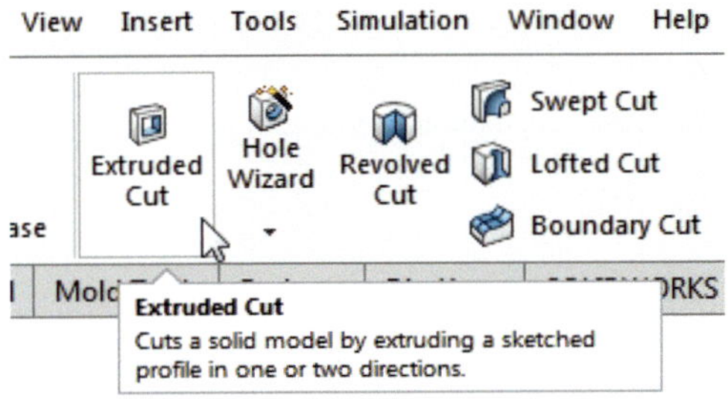

The option Flip side to cut makes possible to trim in the outward dimension.

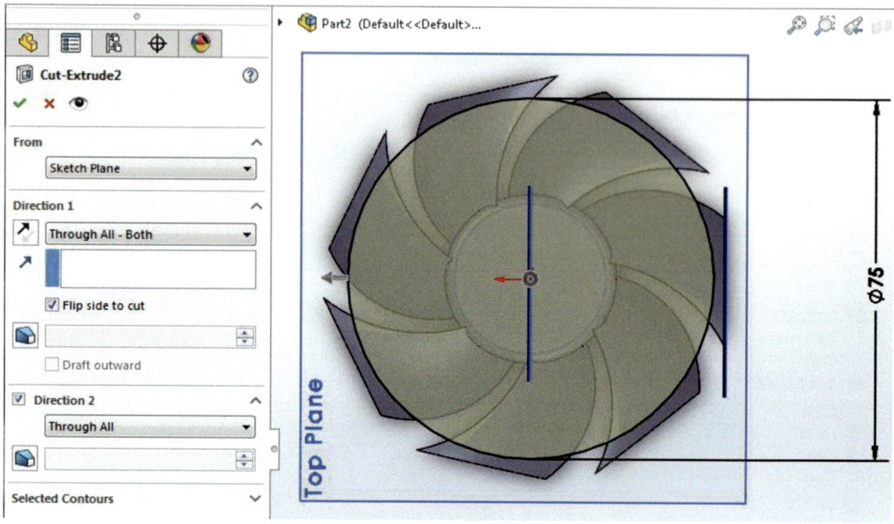

Fig. 3.204 Eliminate excesses from the blades of the fan

37a.   There are slivers of material that protrude through the bottom surface; these will be eliminated in a later stage when the bottom surface is configured to accommodate a rotation piece.

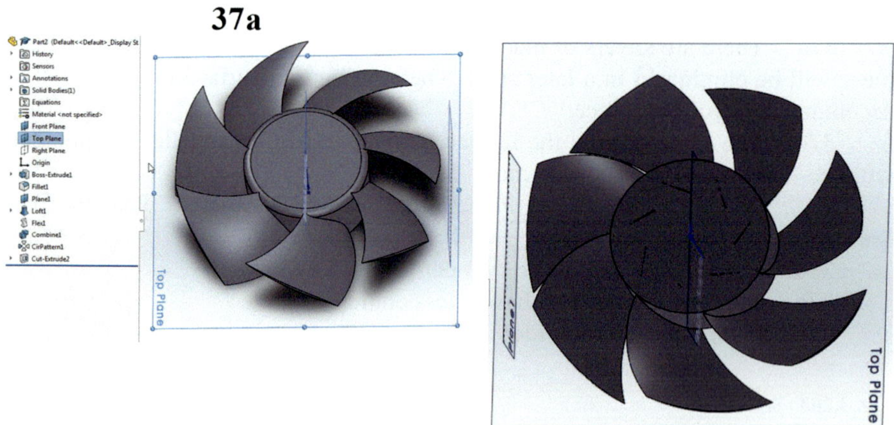

Fig. 3.205 Top/bottom views from part

37b.   The sharp edge at the far end of each fin of the propeller is smoothed with the Fillet choice from the menu.

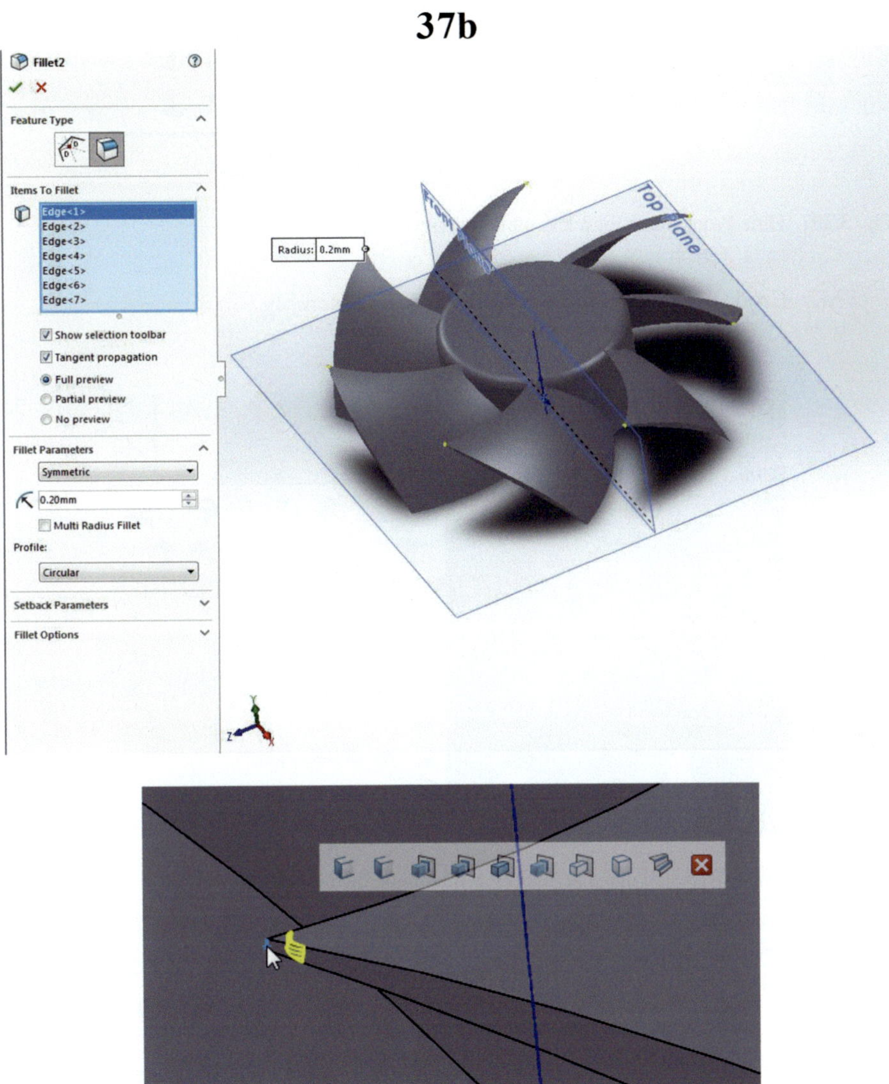

Fig. 3.206 Smoothing edges on fan blades

37c.   The Front Plane and Plane 1 are hidden in order to clear the part drawing.

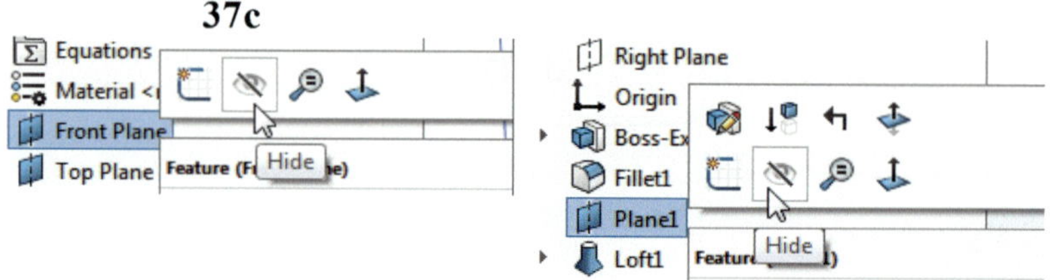

Fig. 3.207   Hide Front Plane for a less cluttered view

37d.   Erode away part of the cylinder for the assembly.

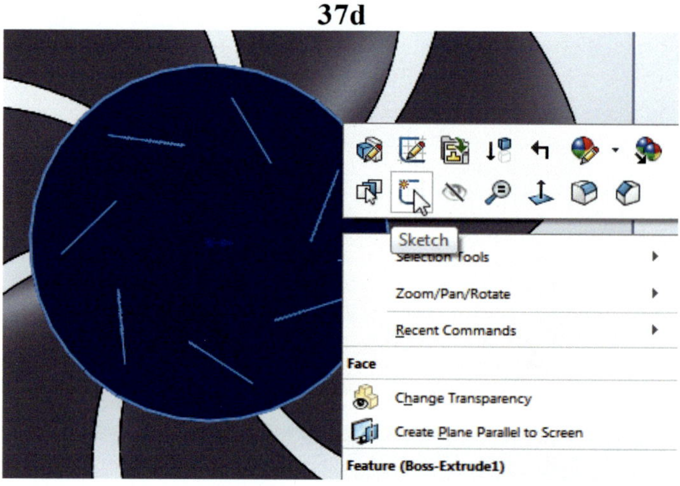

Fig. 3.208   Create a Sketch plane at bottom of the fan blade

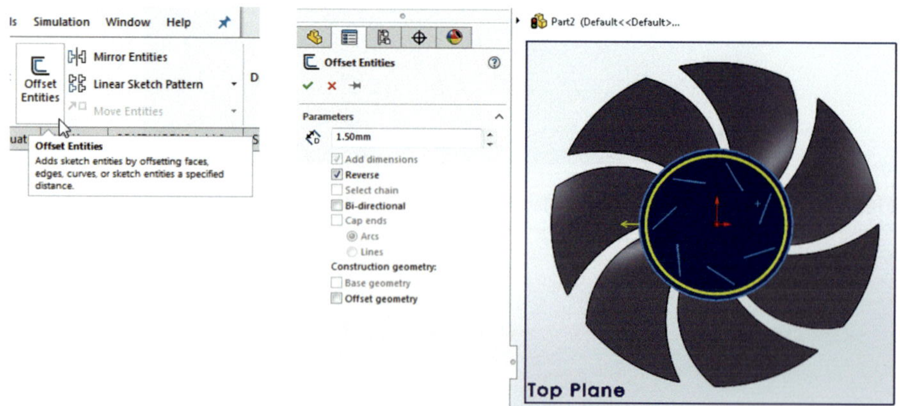

Fig. 3.209   Make a smaller circumference using Offset Entities

37e.   Hide the top plane to have a less cluttered view.

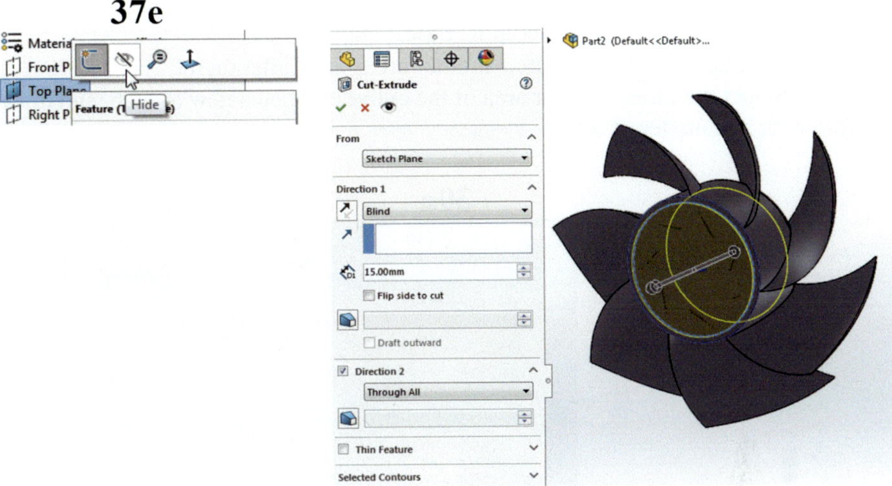

Fig. 3.210 Hide Top Plane and Cut Extrude a cylinder from the center

38f.   Fillet the internal border to 1 mm.

Fig. 3.211 Fillet the internal border to 1 mm

39.   There are additional features to the fan structure to be added.

39a.   Select the inner circular area of the cylinder, open a new Sketch, and draw a circle with a diameter of 3 mm.

39b.   Fillet the new cylinder at its base.

39c.   Select the two indicated annular areas and Fillet them.

39a.   Select the inner circular area of the cylinder, open a new Sketch, and draw a circle with a diameter of 3 mm.

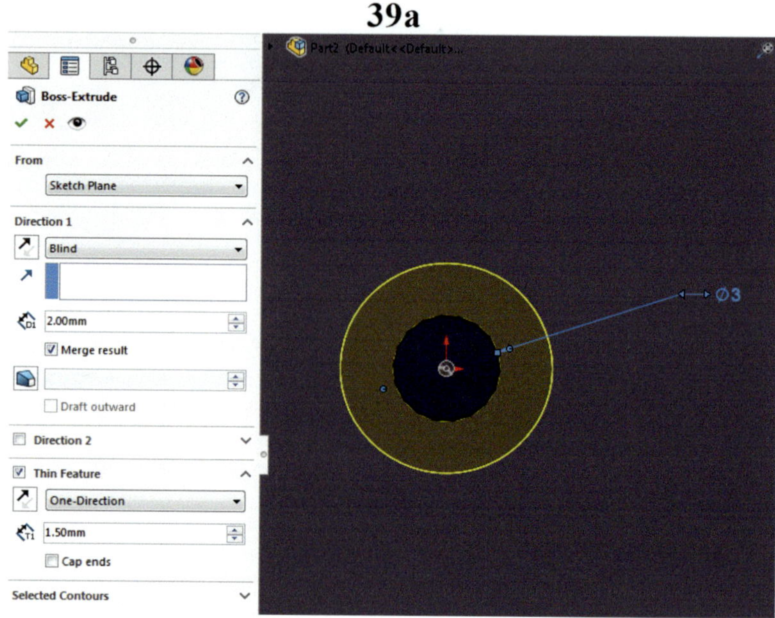

Fig. 3.212 Draw an inner cylinder to serve as support bushing

39b.   Fillet the new cylinder at its base.

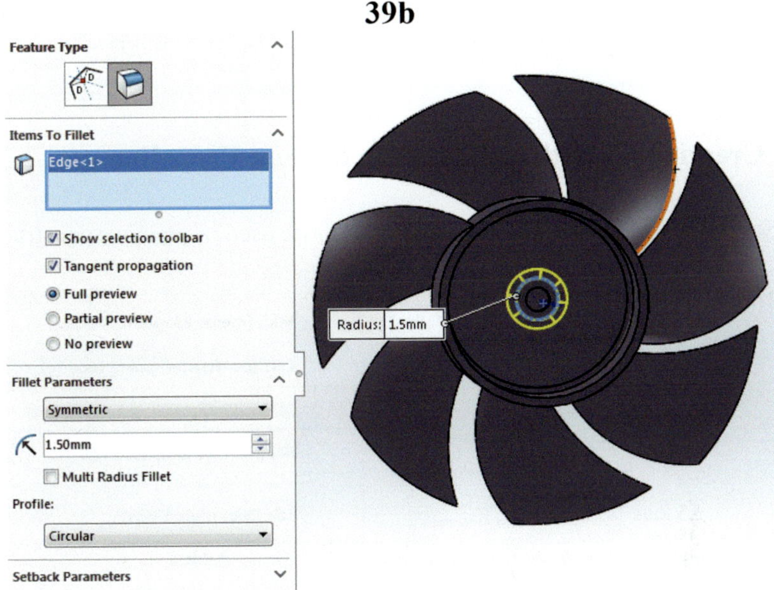

Fig. 3.213   Fillet the inner cylinder edge in contact with the rest of the propeller

39c.   Select the two indicated annular areas and Fillet them.

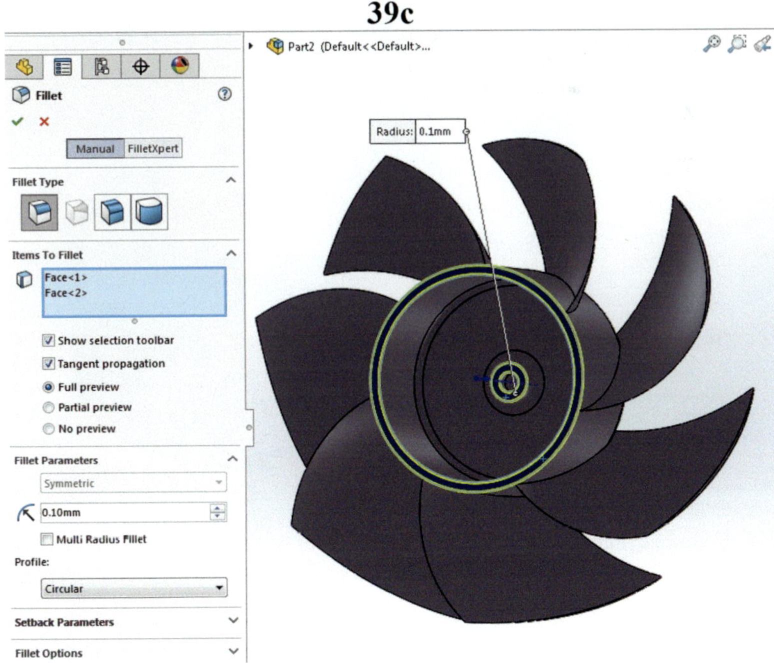

Fig. 3.214   Select the two outer edges to smooth them out

Smaller features and larger elements later on will require a suitable large number of finite elements for computations.

3.4 Creating Cylindrical Bushing for Fan Propeller

50.   Start a new part. The steps are similar to those used in previous construction.

Fig. 3.215 Select sketch plane

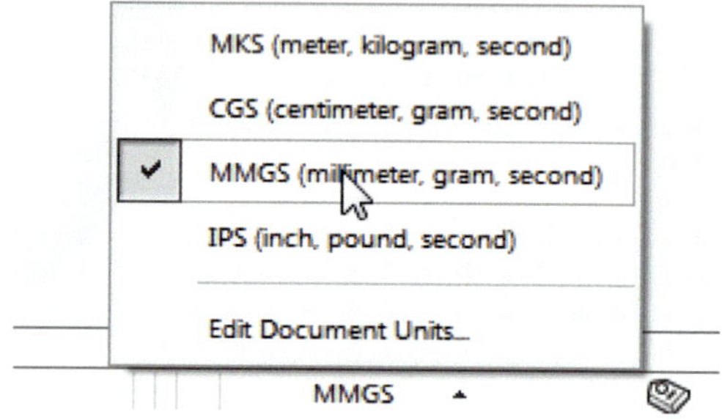

51.   Make the top plane the Sketch plane draw a circle with the indicated dimensions and then Extrude the circle with the option of Thin Feature in order to a cylinder.

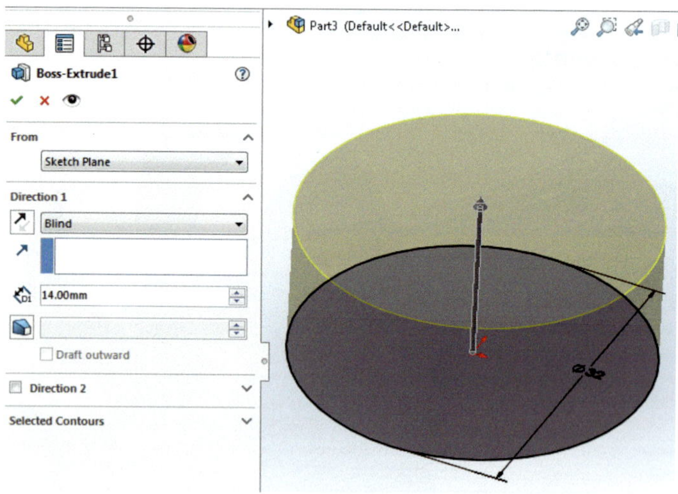

Fig. 3.216 Create a cylinder centered on the drawing area

52.   Create two cylindrical stubs at the ends of the main cylindrical.

52a.   Create a cylinder with a diameter of 2.2 mm and a height of 1.00 mm on one side.

52b.   Create a cylinder with a diameter of 2.1 mm and a height of 2.00 mm on one side.

52a.   Create a cylinder with a diameter of 2.2 mm and a height of 1.00 mm on one side.

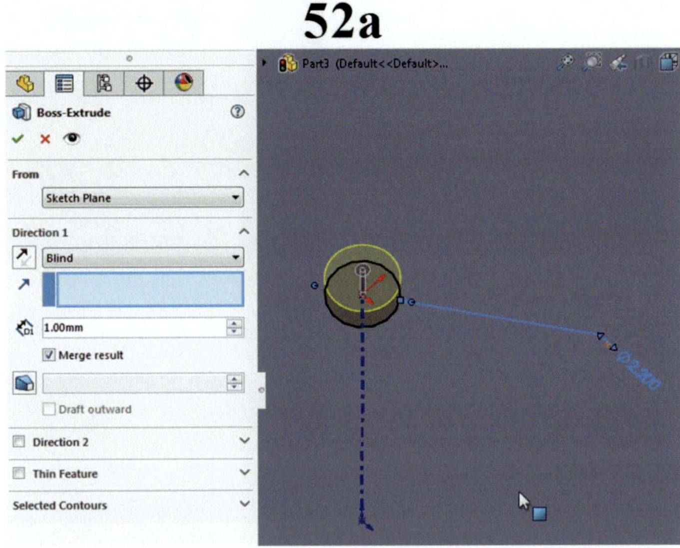

Fig. 3.217   Extrude the first cylinder

52b.   Create a cylinder with a diameter of 2.1 mm and a height of 2.00 mm on one side.

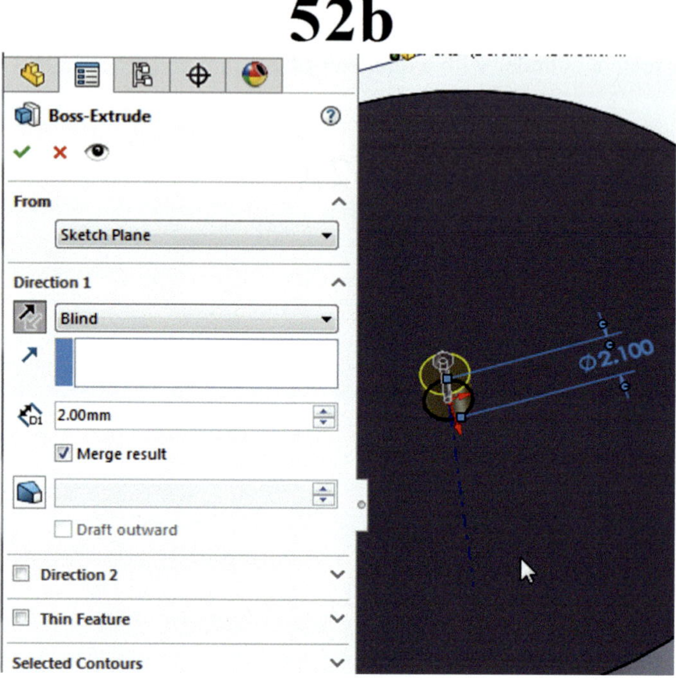

Fig. 3.218 Extrude the second cylinder

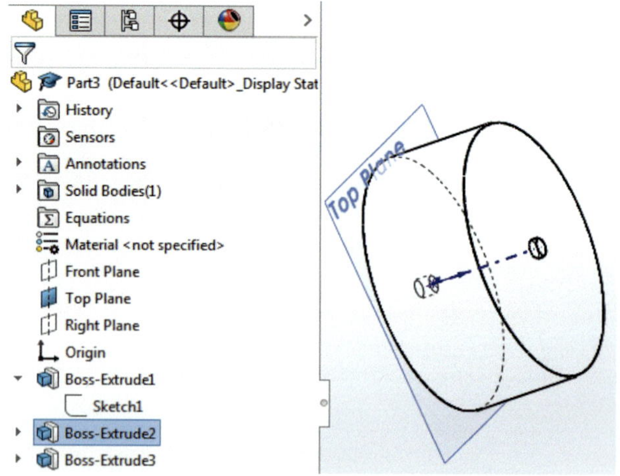

Fig. 3.219 Wire view of the current two cylinder array

3.5 Assembly

53.   There some pieces already created. Now an Assembly is made to put the pieces in the right place.

Fig. 3.220 Select Assembly from the Solidworks menu

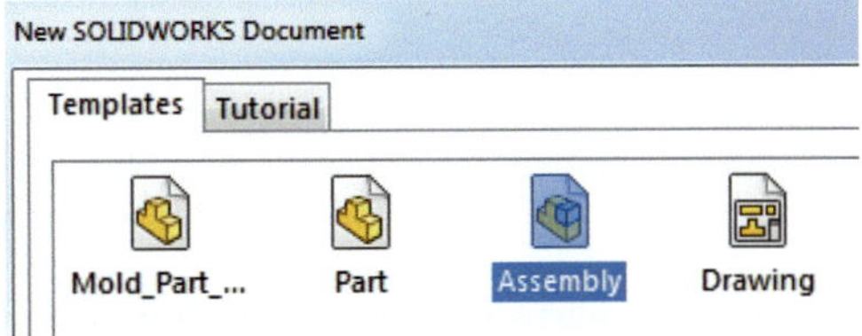

54.   The parts are now assembled at the origin what is needed for the simulations.

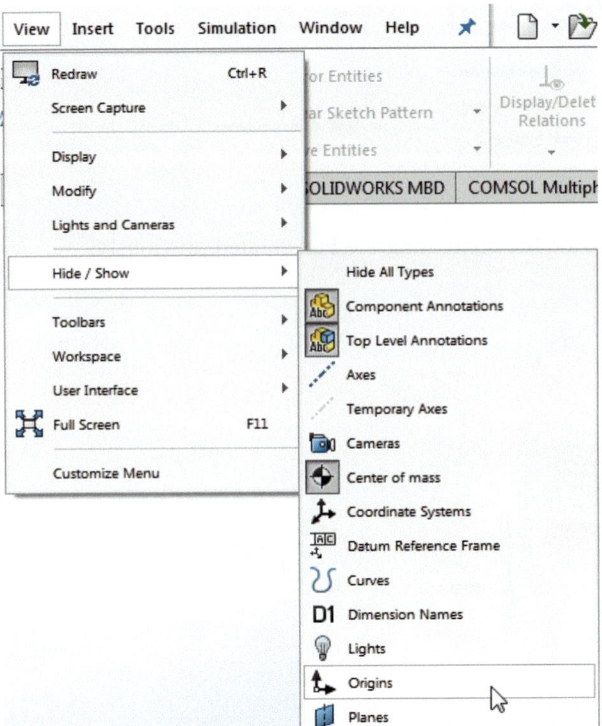

Fig. 3.221 Activate visualization of Origins

55.   The unit sets are defined at the start.

Fig. 3.222 Select the
MMGS set of units

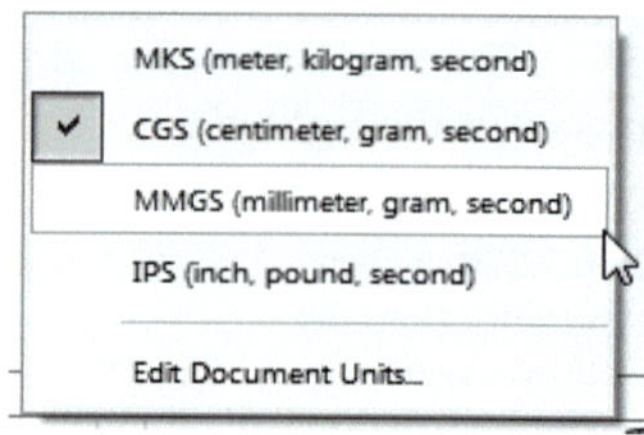

56.   The parts available are found using the Insert tab followed by the Existing
Part/Assembly Selection.

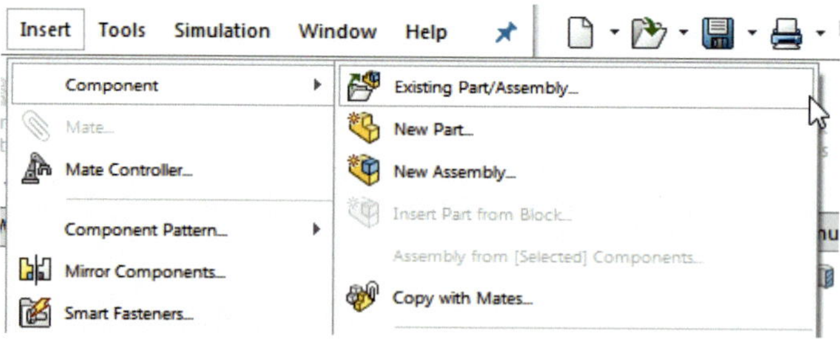

Fig. 3.223 Select the MMGS set of units

57.   Select the file containing the Fan created previously.

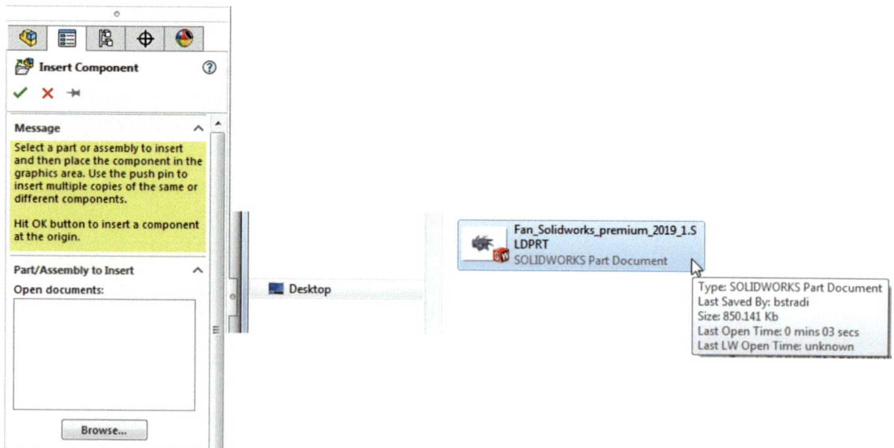

Fig. 3.224 Select the fan propeller as the first element to insert

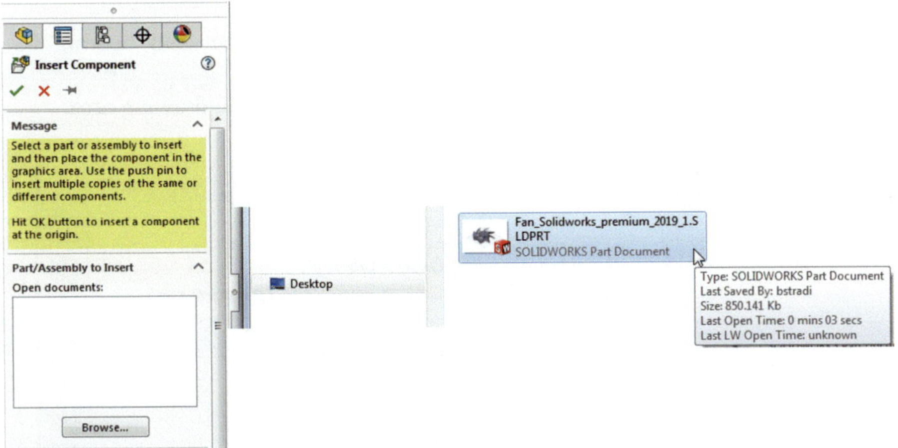

Fig. 3.225 Insert first part of the assembly on the drawing area

58.   It is important for the part to be centered with origin of the canvas.

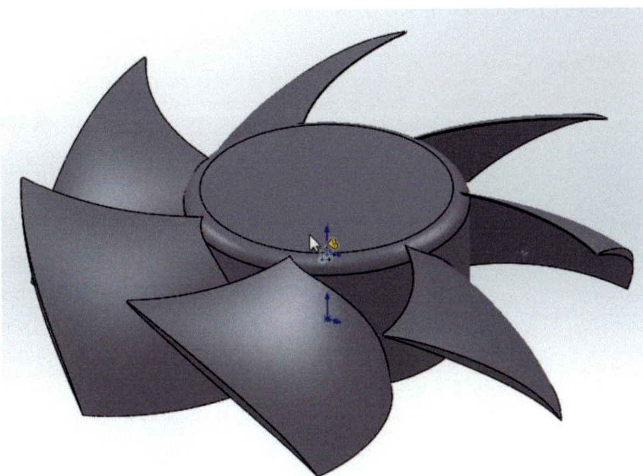

Fig. 3.226 Verify location of centers of canvas and propeller

59.   The Fan needs to have the capacity to rotate. It is possible to build that property using mate properties. It is more illustrative to use a ball bearing inserted into the Fan.

59a.   Go to de Design Library on the right-hand side of the drawing area.

59b.   Find the Toolbox in the Design Library. If not enabled, proceed to Add the library. It is possible that for this step a Premium version of Solidworks be needed.

59c.   Find the Selection for ISO measurements.

59d.   Find the Selection for Ball Bearings.

59e.   Find the Selection for Ball Bearings: instrument ball bearing_68_iso.sldprt

59a.   Go to de Design Library on the right-hand side of the drawing area.

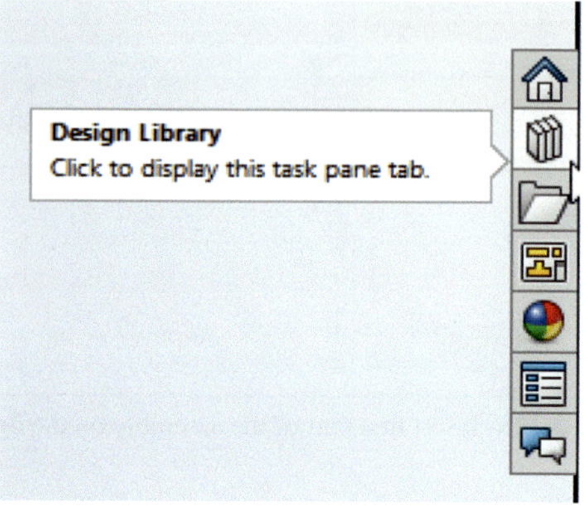

Fig. 3.227 Locate the Design Library for a ball bearing

59b.   Find the Toolbox in the Design Library. If not enabled, proceed to Add the library.

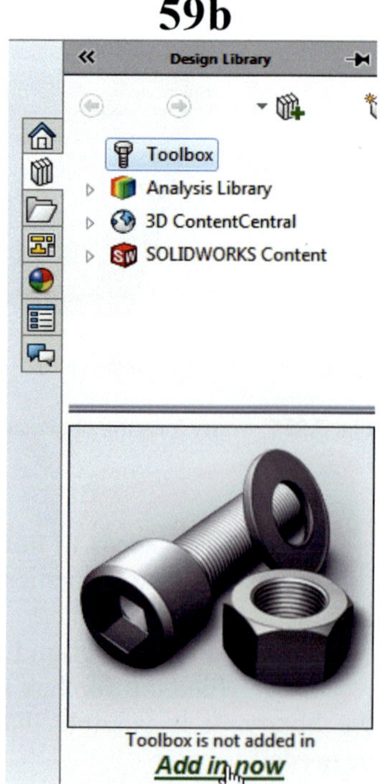

Fig. 3.228 Enable Toolbox in the Design Library

59c.   Find the Selection for ISO measurements.

Fig. 3.229 Select ISO
measurement for parts

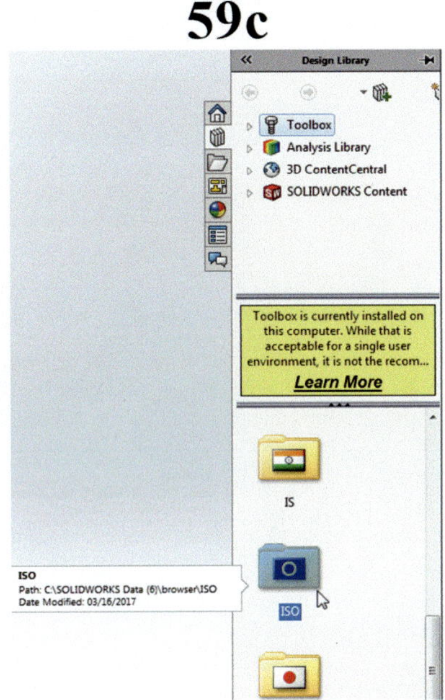

59d.   Find the Selection for Ball Bearings.

59d

Fig. 3.230 Select folder for Bearings

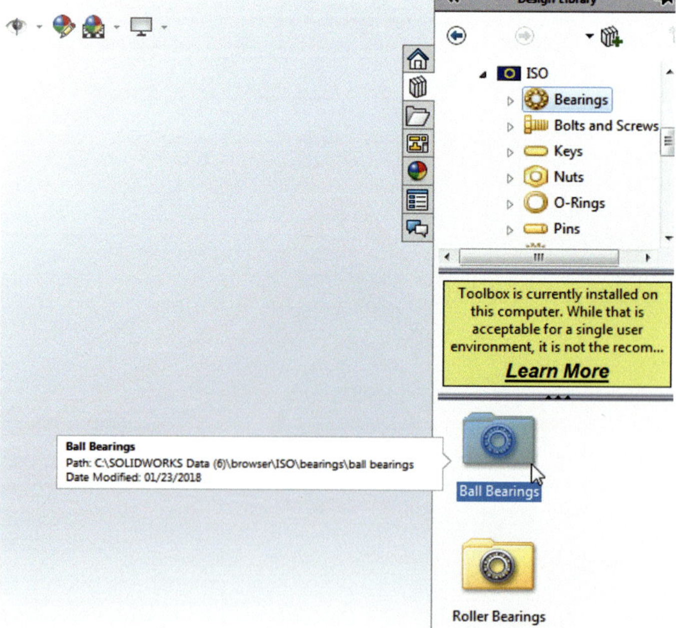

Fig. 3.231 Select folder for Ball Bearings

59e.   Find the Selection for Ball Bearings: instrument ball bearing_68_iso.sldprt

Fig. 3.232 Select ball
bearing_68_iso.sldprt

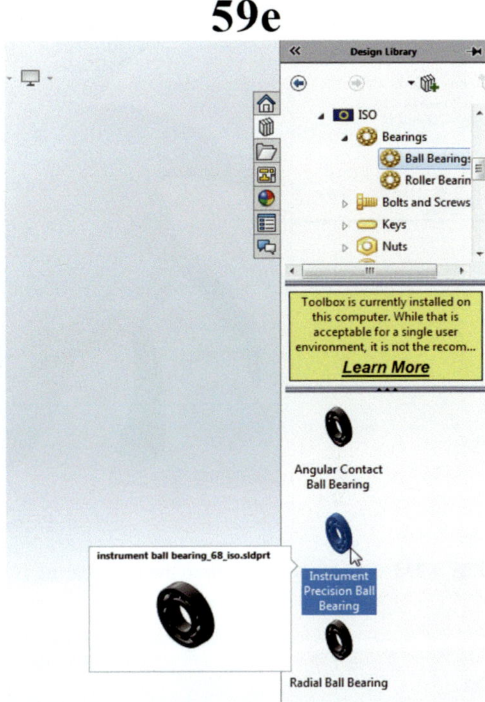

60.   The ball bearing is selected from the menu and dragged into the drawing area. Get it close to the inner circumference in order to auto resize the ball bearing.

60a.   Use the wheel on the mouse to magnify the view and get the ball bearing close to the hole where it is going to be pushed in. The ball bearing will resize automatically, if not, manual resizing will be needed.

60b.   From the Ball Bearing menu, select the serial number 170204 that has the right outside diameter for the orifice on the Fan.

60c.   Select from the menu the Detailed Display in order to see the inner arrangement of the ball bearing.

60a.   Use the wheel on the mouse to magnify the view and get the ball bearing close to the hole where it is going to be pushed in. The ball bearing will resize automatically, if not, manual resizing will be needed.

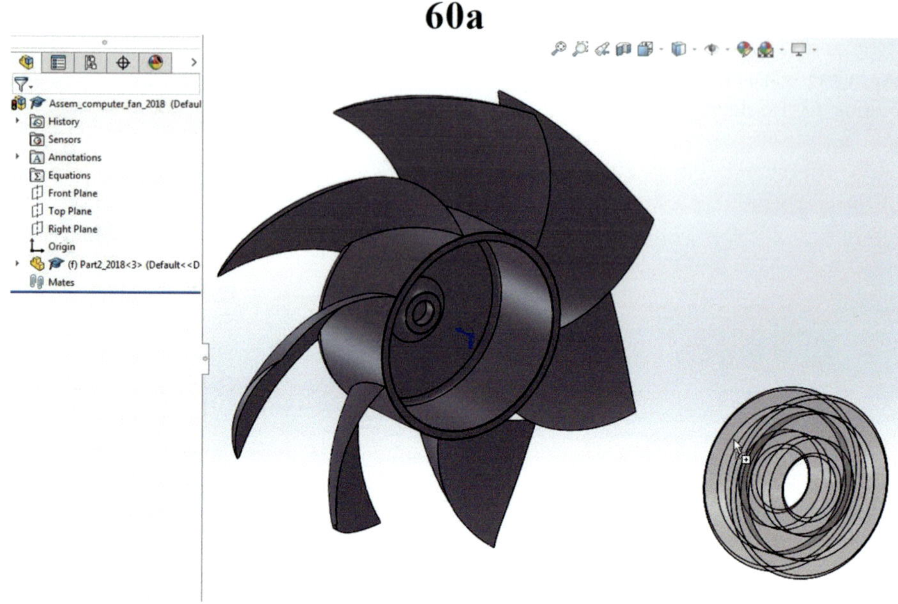

Fig. 3.233 Select the area where the ball bearing is to be located

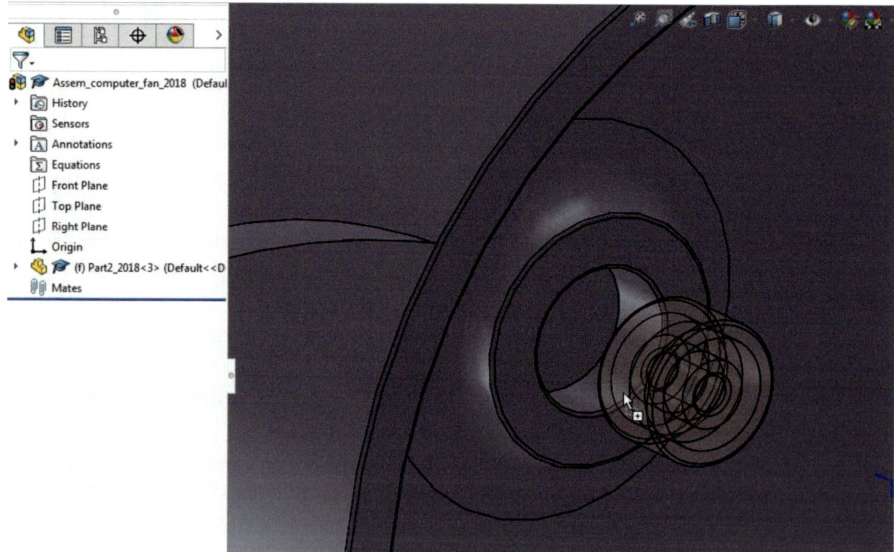

Fig. 3.234 Move the ball bearing close to cavity to automatically resize

60b.   From the Ball Bearing menu, select the serial number 170204 that has the right outside diameter for the orifice on the Fan.

Fig. 3.235 Select the ball
bearing with appropriate
measurements

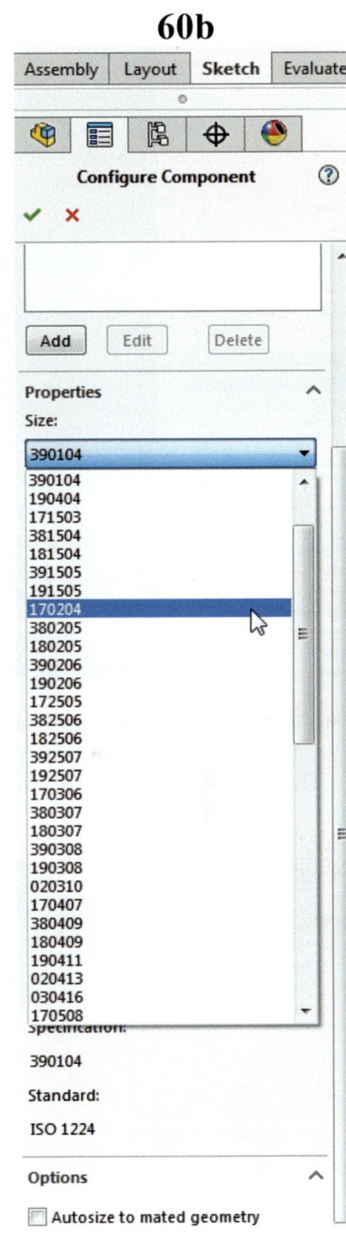

60c.   Select from the menu the Detailed Display in order to see the inner arrangement of the ball bearing.

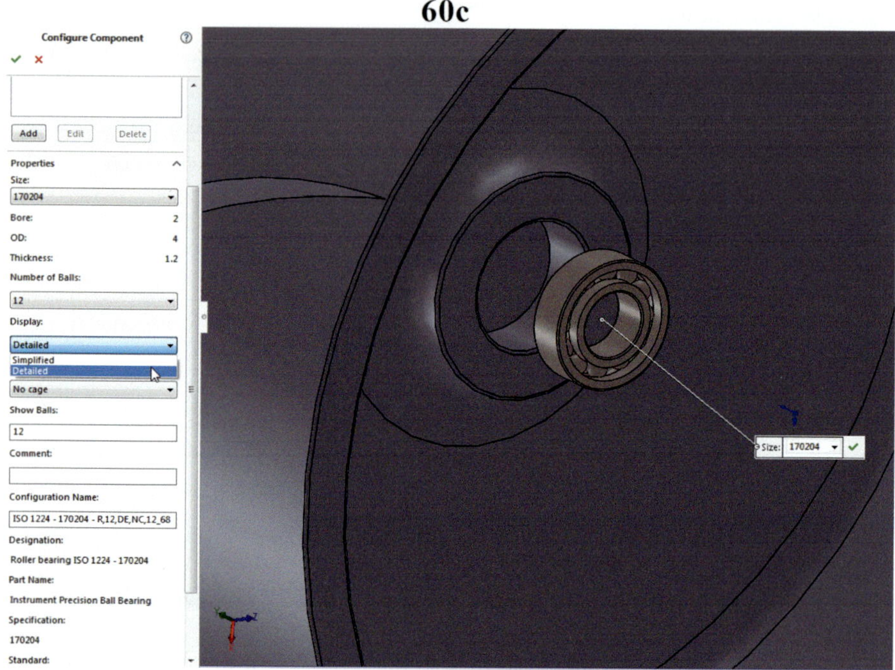

Fig. 3.236 Select a Detailed description of the Ball Bearing

The drawing should look like this figure at this point.

Fig. 3.237 Locate the Ball Bearing close to the aperture

61.   Proceed to match the orifice with the ball bearing. This process requires the operation Mate.

61a.   Left-click on the outer surface of the ball bearing, afterward right-click and choose the Mate option from the menu.

61b.   The two surfaces to match are the inner surface of the fan seen through the orifice and the outer flat circular regions on the side of the ball bearing. In this case, the inner circular area is chosen.

61c.   The inner circular area is shadowed by bodies in the line of sight form outside the figure. Select Other in order to list and choose the area of interest.

61d.   The area of interest is shown in orange color and when chosen appears on the detailed menu on the left of the drawing area.

61e.   Select Coincident from the Standard Mates menu. This selection makes the two surfaces to touch. This in turn makes the Ball bearing to be flush to the inner wall.

61f.   The ball bearing in this position properly allow for movement of the fan while axially connected to the inner motor assembly and to the frame.

61a.   Left-click on the outer surface of the ball bearing, afterward right-click and choose the Mate option from the menu.

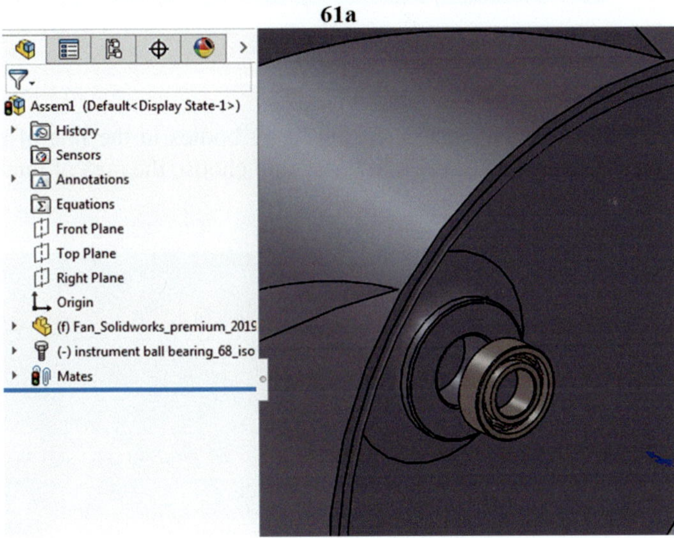

Fig. 3.238 Take a larger view of the two elements

61b. The two surfaces to match are the inner surface of the fan seen through the orifice and one of the flat circular regions on the side of the ball bearing. In this case, the inner circular area is chosen.

61b

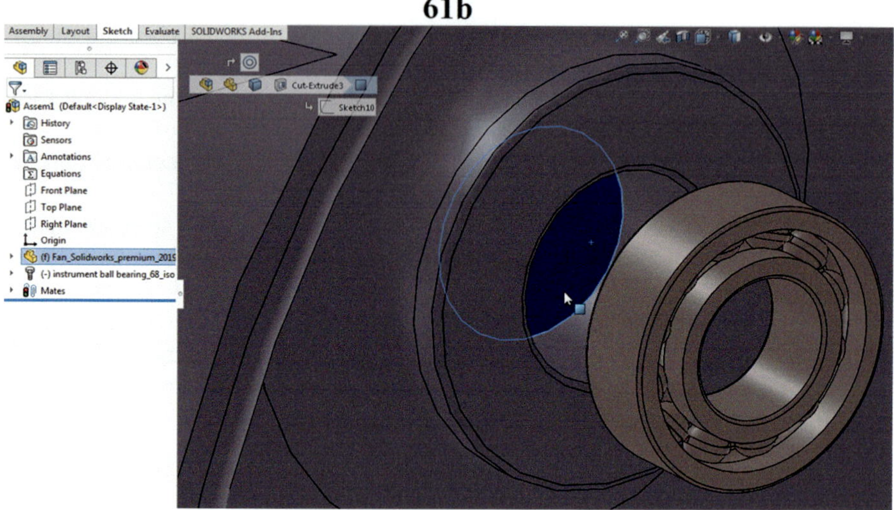

Fig. 3.239 Select inner surface seen through the hole

61c. The inner circular area is shadowed by bodies in the line of sight form outside the figure. Select other in order to list and choose the area of interest.

61c

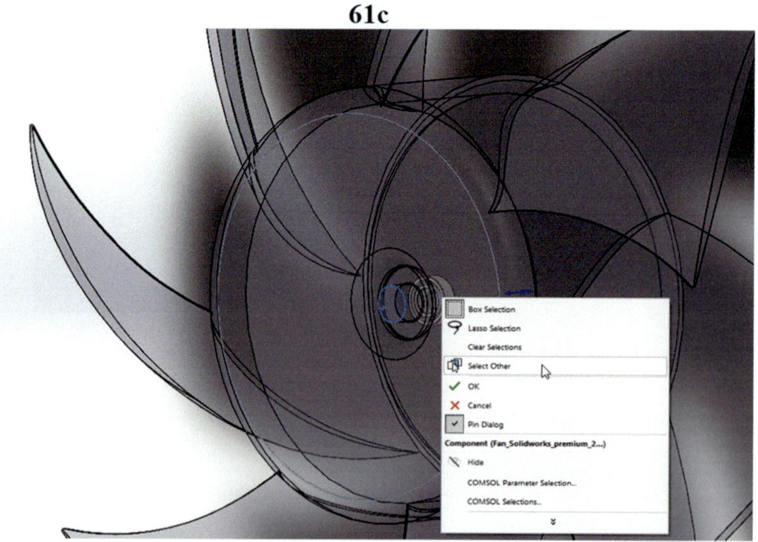

Fig. 3.240 Use Select Other to select the side circular rim of the ball bearing

61d.   The area of interest is shown in orange color and when chosen appears on the detailed menu on the left of the drawing area.

61d

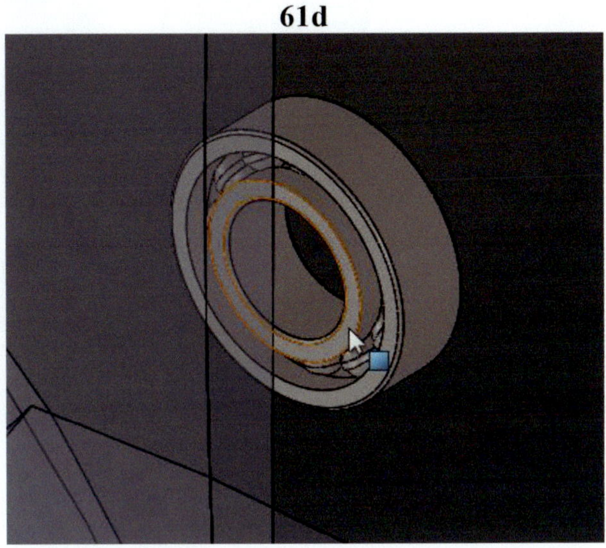

Fig. 3.241 The inner circular ring is the area of contact

61e.   Select Coincident from the Standard Mates menu. This selection makes the two surfaces to touch. This in turn makes the Ball bearing to be located flush to the inner wall.

61e

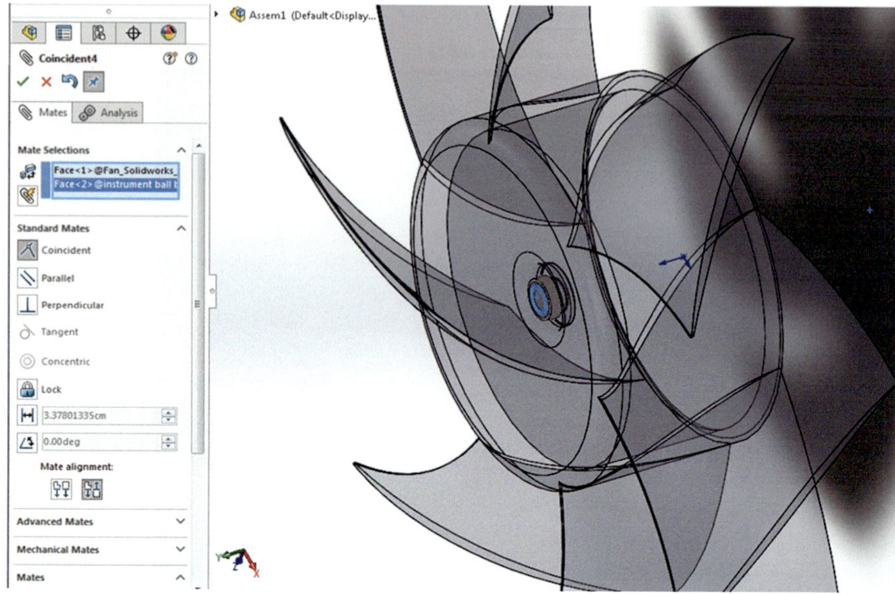

Fig. 3.242 Select Coincident from the Mate Selections menu

61f.   The ball bearing in this position properly allow for movement of the fan while axially connected to the inner motor assembly and to the frame.

61f

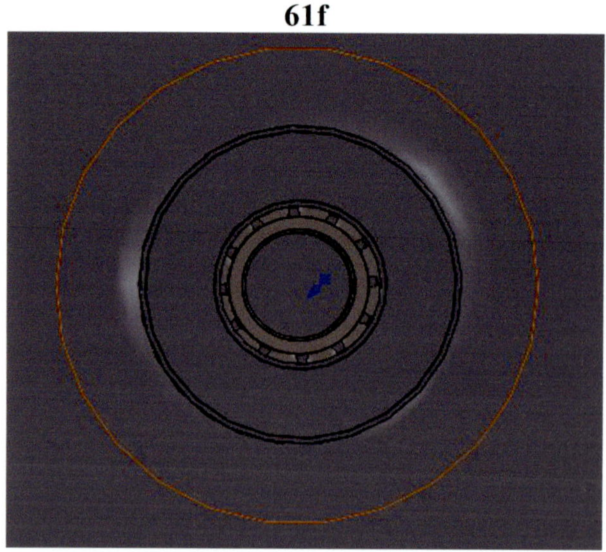

Fig. 3.243 Ball bearing positioned in well

The Fan should look like this prior to continuing assembly.

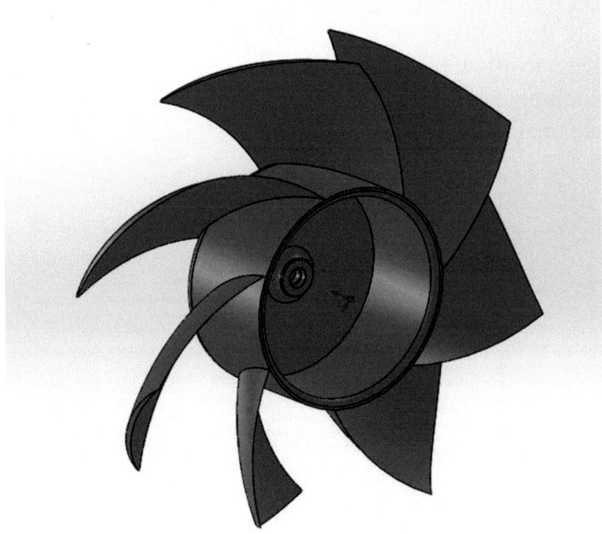

Fig. 3.244   Fan propeller with added ball bearing

The second element in the assembly is the piece that represents the motor. Traditionally, a motor powers the rotation of the fan. In this case, that level of detail is not necessary whereby a current is passed through the motor coils to generate an electromagnetic field, and make the fan turn. As a substitute, a cylinder is added that makes up for the motor volume and serves to connect the fan to the frame.

62.   Insert the cylindrical piece into the fan.

62a.   Locate the cylindrical part on the menu.

62b.   Drag the part to the drawing area. A double set of axis appears on the drawing when the center of the part is coincident with the center of the canvas or drawing area.

62c.   On the menu on the left-hand side (the f letter on the part indicates that it is fixed in its current position), select the part (left-click on it) and open the menu for that part (right-click on it), select Float so the part can move freely.

62d.   The part is properly placed when the stub with a diameter of 2.1 mm stay out, the other stub of 2.2 mm in diameter locks into the ball bearing.

62e.   Before proceeding further, the command Measure can be used to verify the diameter size of the external stub to make sure the part is positioned properly.

62f.   While holding the Crtl key down, select both: the circumference of the stub that will be in contact with the fan's ball bearing assembly and the inner wall of the fan.

62g.   Mate the surfaces as Coincident.

62h.   Sometimes, the matching is imperfect and a second Mate is needed to make the elements Concentric.

62i.   Notice that the Fan (there is a small f letter next to it on the left-hand menu to indicate that is fixed in its current position; this will have to be changed later in the process to allow for rotational motion), the ball bearing, and the cylinder part are now assembled.

62a.   Locate the cylindrical part in the menu.

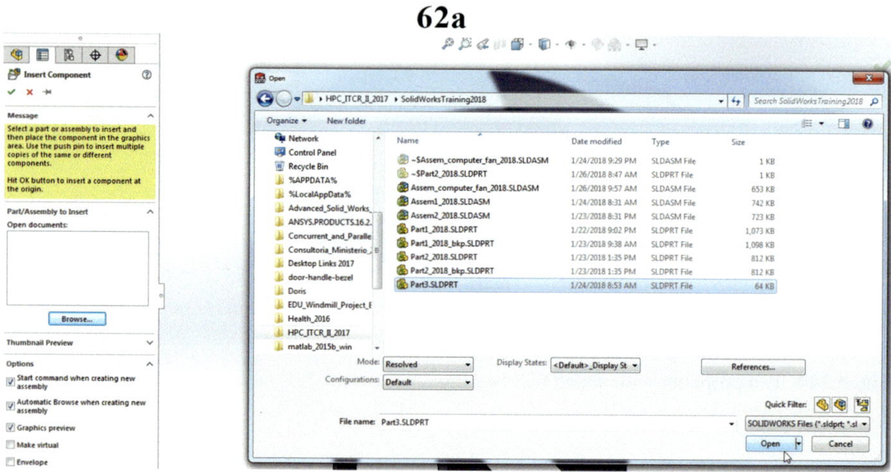

Fig. 3.245 Select the cylindrical part proxy for the electric motor

62b.   Drag the part to the drawing area. A double set of axis appears on the drawing when center of the part is coincident with the center of the canvas or drawing area.

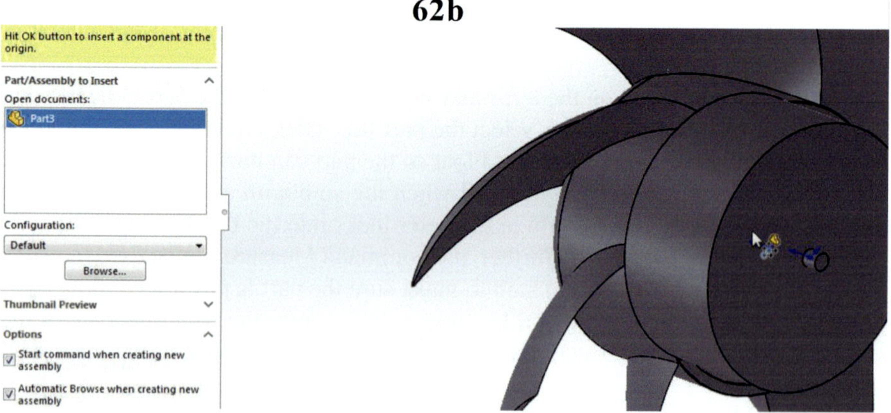

Fig. 3.246 Place the center of the cylindrical part with the center of the drawing area

62c.   On the menu on the left-hand side (the f letter on the part indicates that it is fixed in its current position), select the part (left-click on it) and open the menu for that part (right-click on it), select Float so the part can move freely.

62c

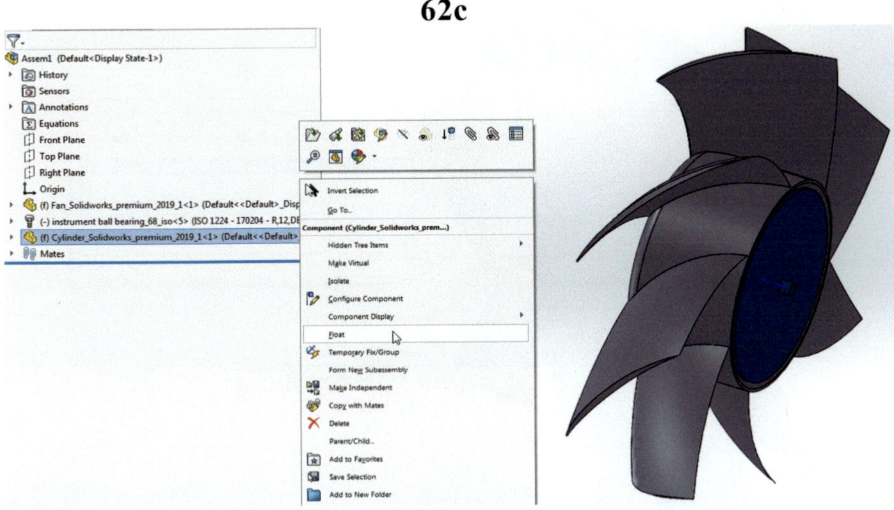

Fig. 3.247 Place the cylinder into the fan with Float

62d.   The part is properly placed when the stub with a diameter of 2.1 mm stay out, the other stub of 2.2 mm in diameter locks into the ball bearing.

62d

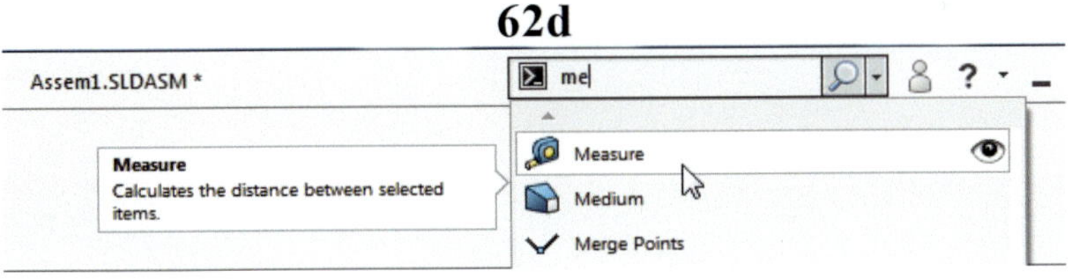

Fig. 3.248 Locate the Measure option on the drawing menu

62e.   Before proceeding further, the command Measure can be used to verify the diameter size of the external stub to make sure the part is located properly.

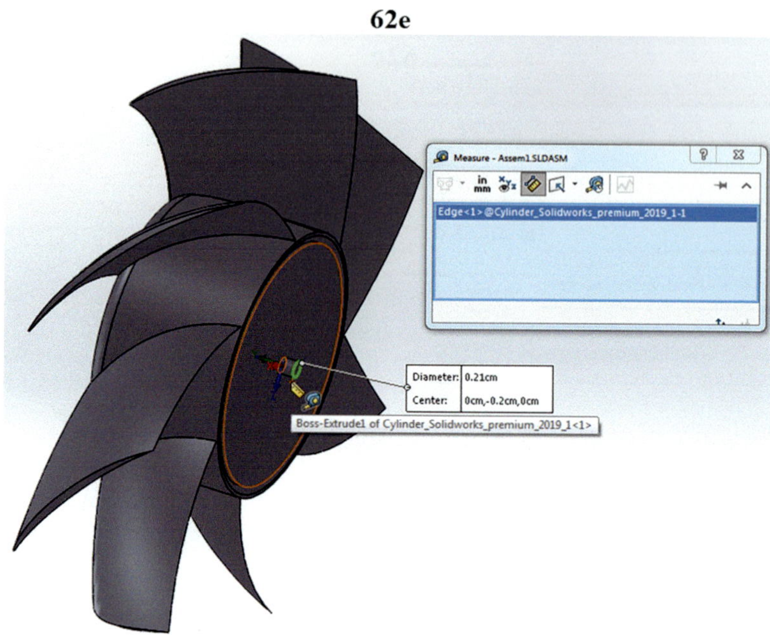

Fig. 3.249 Verify the orientation of the cylindrical part

62f. While holding the Crtl key down, select both: the circumference of the stub that will be in contact with the fan's ball bearing assembly and the inner wall of the fan.

10f

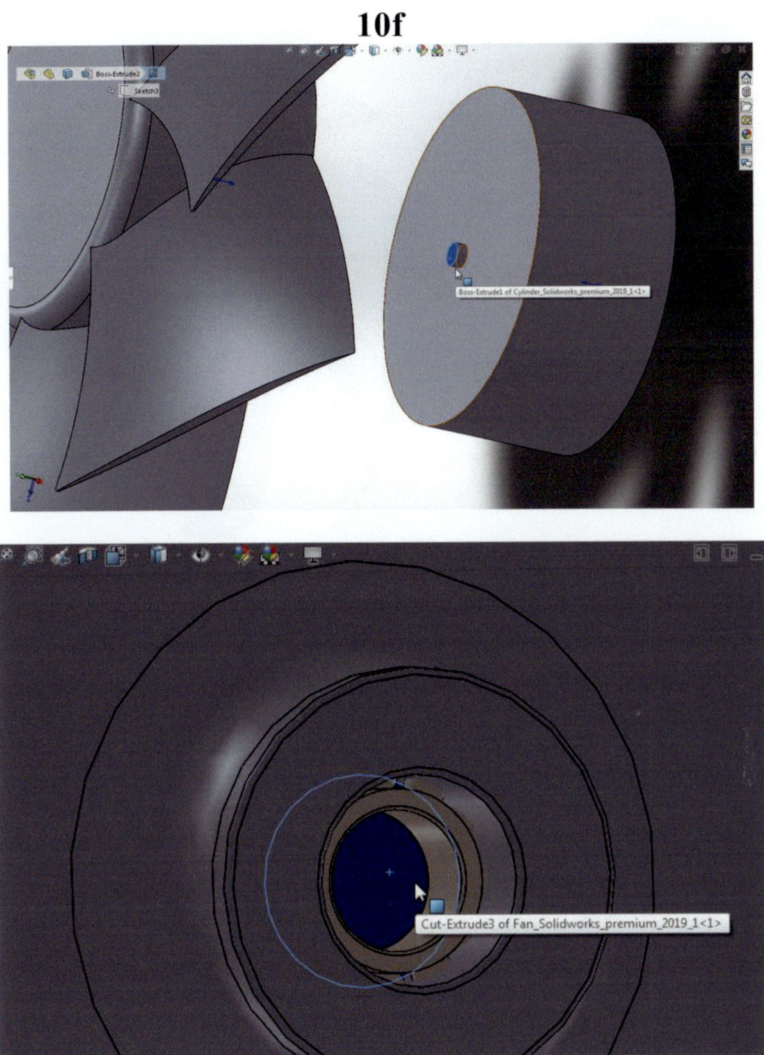

Fig. 3.250 Select the tip of the protruding stub and the inner wall of the fan's propeller

62g.   Mate the surfaces as Coincident.

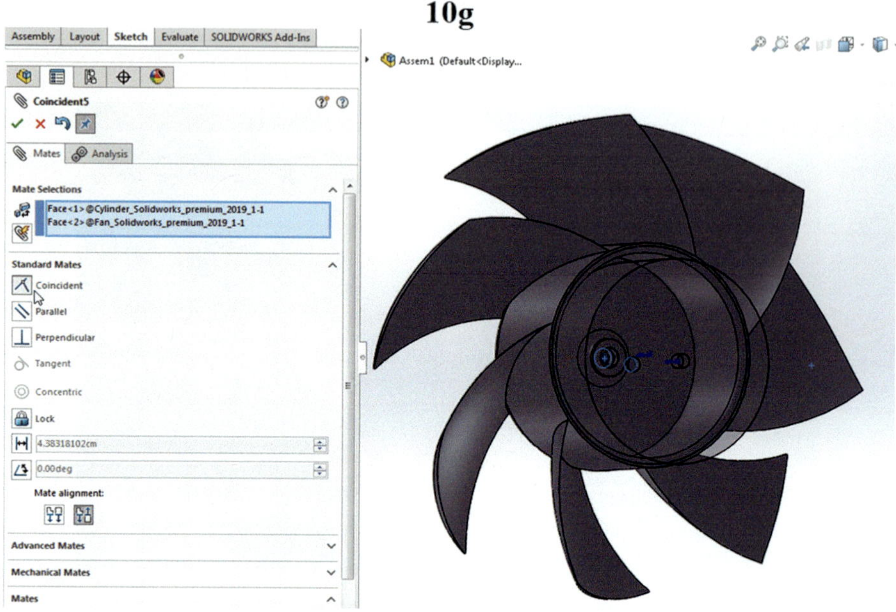

Fig. 3.251 Make the two faces coincident

62h.   Sometimes, the mating is imperfect and a second Mate is needed to make
the elements Concentric.

Fig. 3.252 Correct mating applying the additional Concentric condition

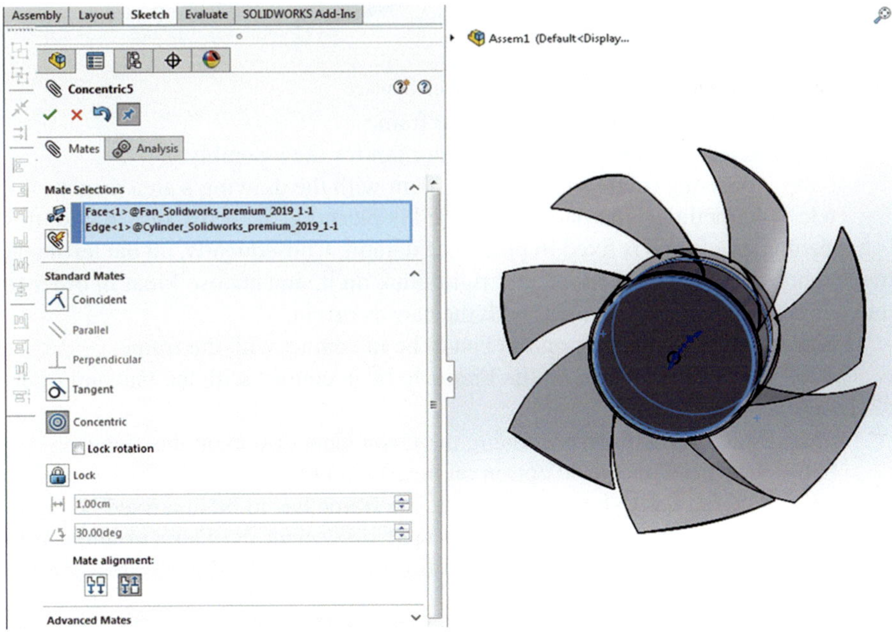

Fig. 3.253 Larger view to align cylinder and ball bearing

62i.   Notice that the Fan (there is a small f letter next to it on the left-hand menu to indicate that is fixed in its current position; this will have to be changed later in the process to allow for rotational motion), the ball bearing, and the cylinder part are now assembled.

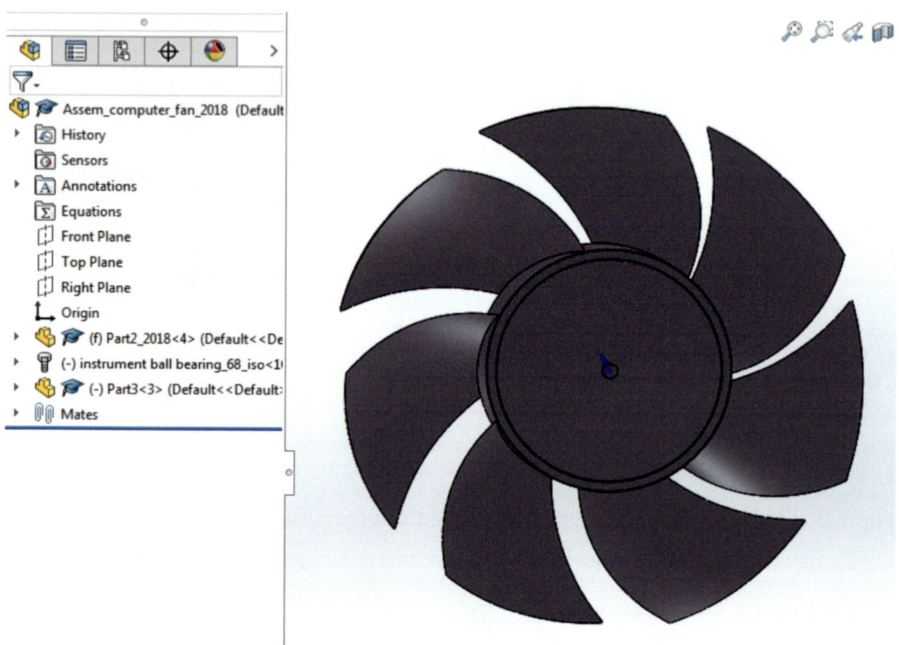

Fig. 3.254 Momentarily the fan's propeller is fixed in place

63.   Attach the current assembly to the frame.

63a.   Select the frame of the computer fan for the assembly.

63b.   Place the frame's origin coincident with the drawing's area origin.

63c.   Sometimes, the part to be added is placed incorrectly. Just in case this is happening, the frame is fixed in place as a default. Consequently, on the left-hand menu, choose the last part added and right-click on it, and choose Float in order to move it into a position coincident with the canvas origin.

63d.   Select the surface on the Fan to be in contact with the frame.

63e.   Select the surface on the Frame to be in contact with the Fan and right-click to Mate.

63f.   Mate the two surfaces using the Coincident option on the menu.

63h.   Use the Concentric option on the Mate menu.

63i.   The fan has to be able to rotate, the frame has to be in a fixed position. Consequently, from the menu on the left-hand side select the Fan and allow it to Float. Similarly, select the Frame and choose the menu selection Fix to keep the frame on a specific location on the canvas.

63j.   Make sure the computer fan can rotate. Left-click on the fan and rotate with the mouse.

63a.   Select the frame of the computer fan for the assembly.

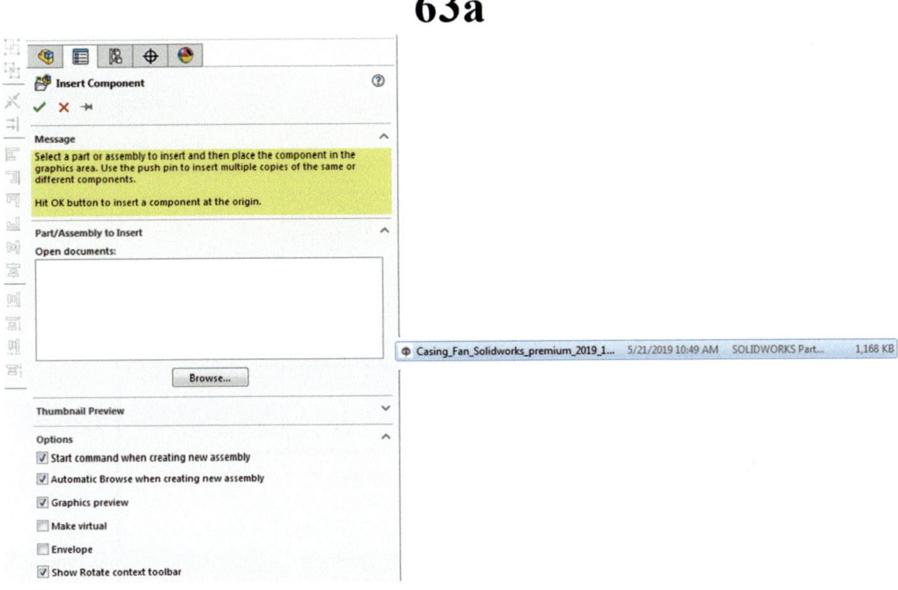

Fig. 3.255 Select the frame of the computer fan

63b.   Place the frame's origin coincident with the drawing's area origin.

Fig. 3.256 Place the
computer fan frame center
to coincide with the canvas
origin

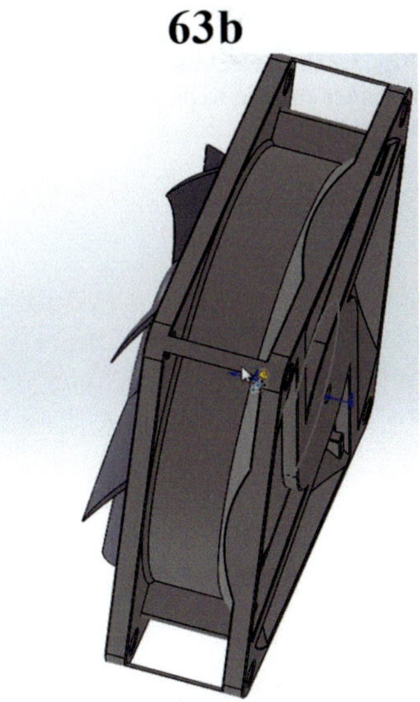

63c.   Sometimes, the part to be added is placed incorrectly. Just in case this is happening, the frame is fixed in place as a default. Consequently, on the left-hand menu, choose the last part added and right-click on it, and choose Float in order to move it into a position coincident with the canvas origin.

63c

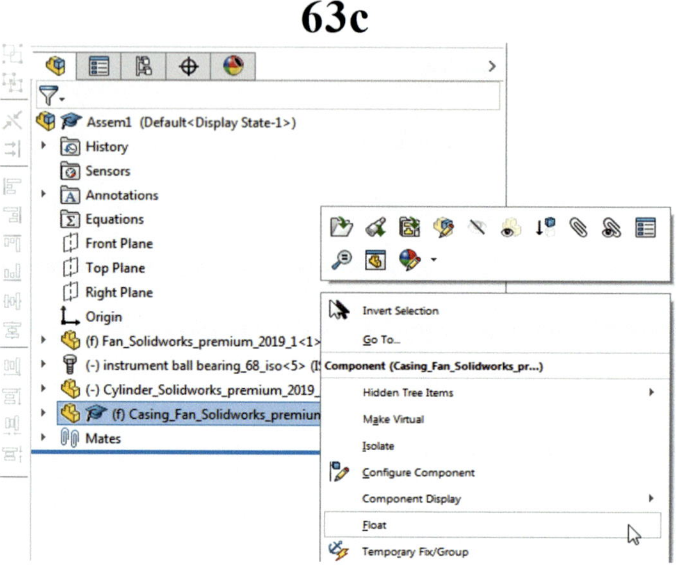

Fig. 3.257 Allow the Frame to Float for proper location

63d.   Select the surface on the Fan to be in contact with the frame.

Fig. 3.258 Select the cylinder surface to prepare for Mate

63d

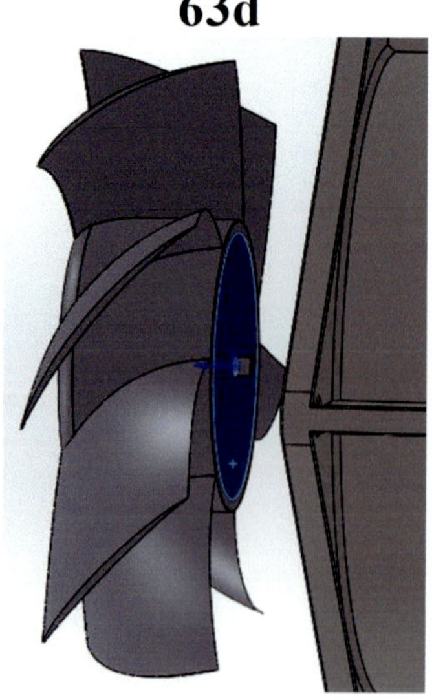

63e. Select the surface on the Frame to be in contact with the Fan and right-click to Mate.

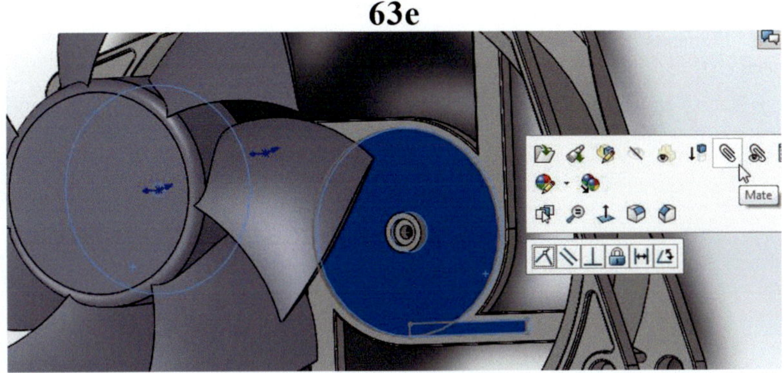

Fig. 3.259 Select the frame surface to prepare for Mate

63f. Mate the two surfaces using the Coincident option on the menu.

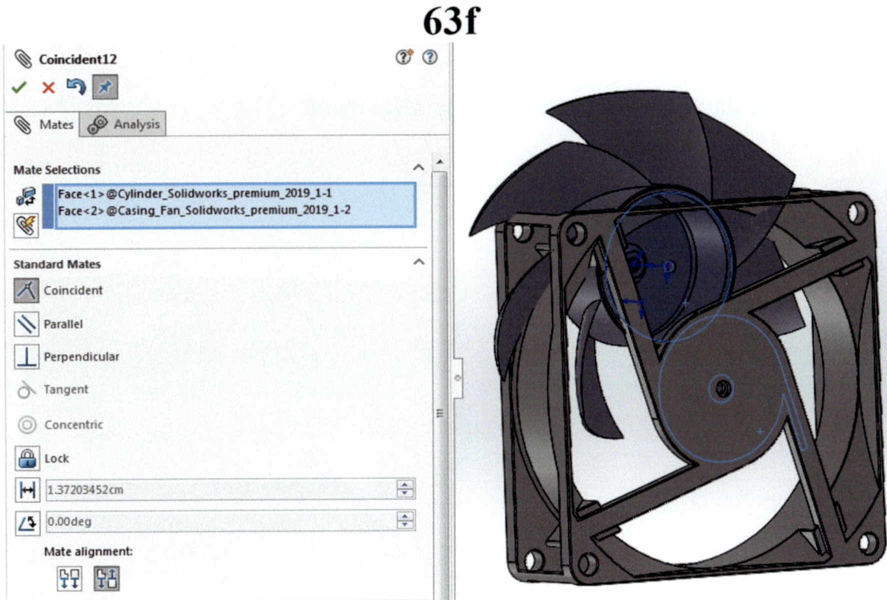

Fig. 3.260 Mate the cylinder's and frame's surfaces using Coincident

63g.   Notice the two centers are not aligned. Hold the Crtl key and left-click on the surface of the circular stub on the fan and the inner cylindrical area on the frame. Right-click and choose to Mate both entities.

Fig. 3.261 Elements misaligned

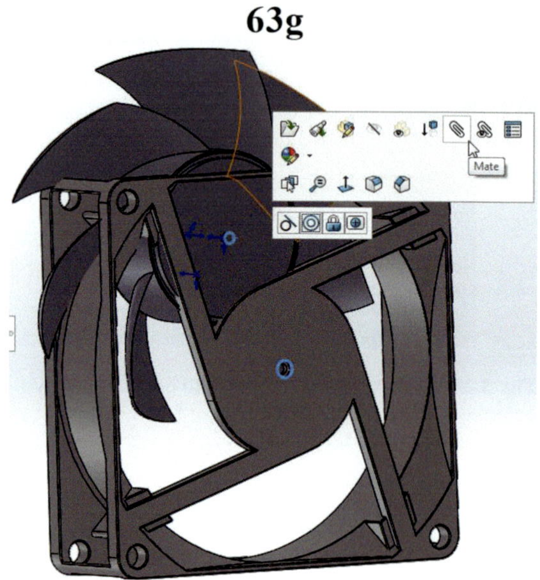

63h.   Use the Concentric option on the Mate menu.

Fig. 3.262 Mate the two cylindrical edges

63i.   The fan has to be able to rotate, the frame has to be in a fixed position. Consequently, from the menu on the left-hand side select the Fan and allow it to Float. Similarly, select the Frame and choose the menu selection Fix to keep the frame on a specific location on the canvas.

63i

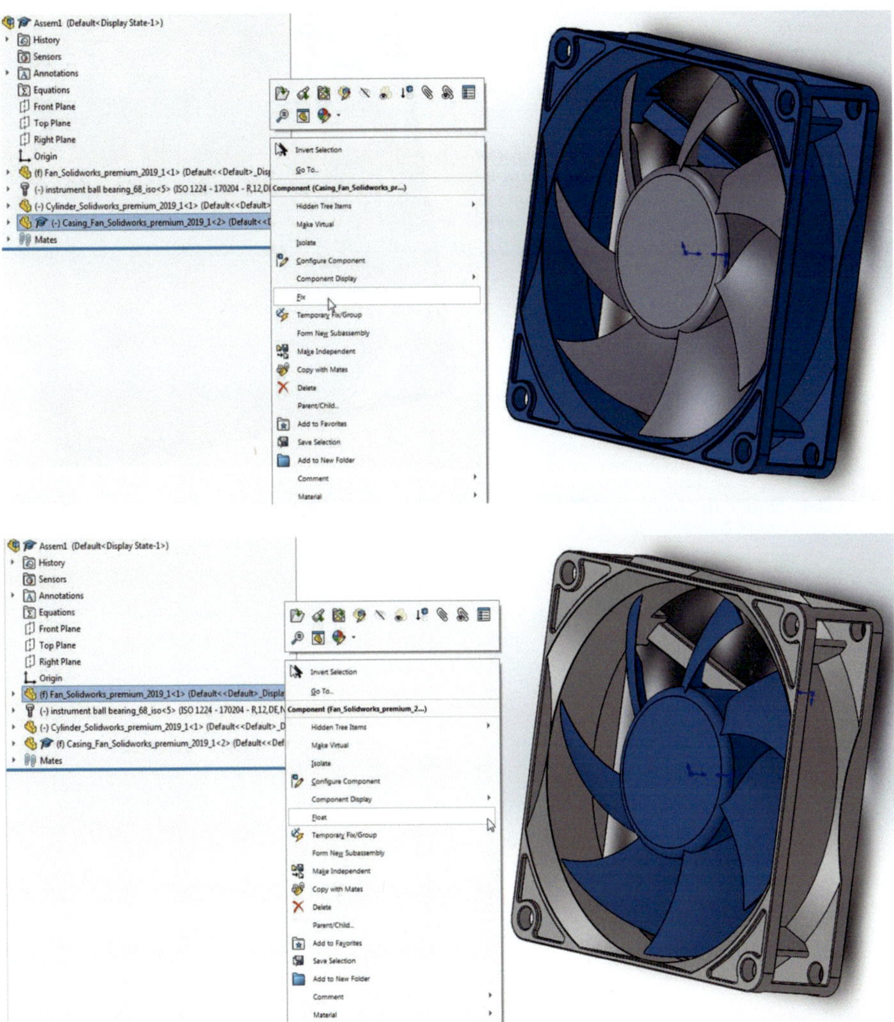

Fig. 3.263 Fix the position of frame and release position of fan's Propeller

63j. Make sure the computer fan can rotate. Left-click on the fan and rotate with the mouse.

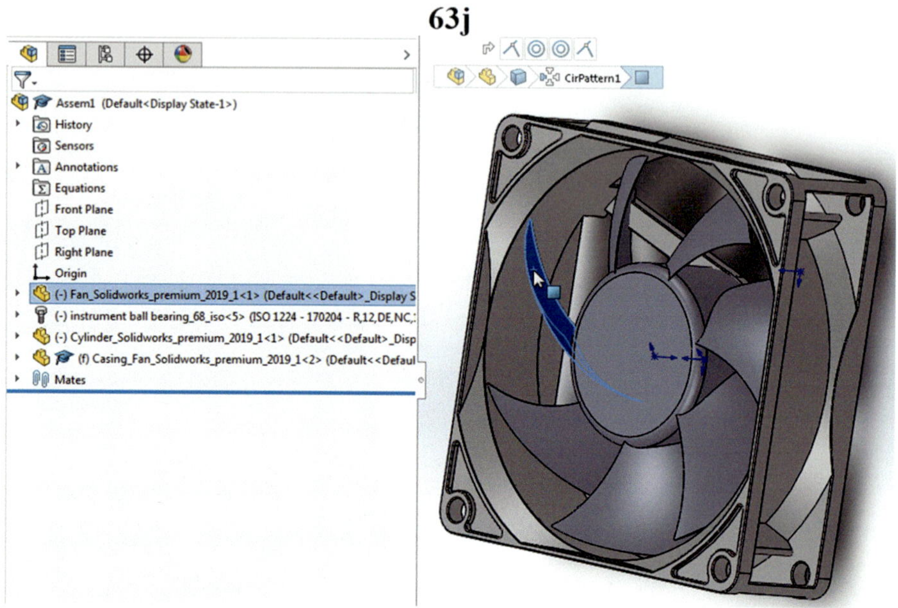

Fig. 3.264 Rotate the fan's propeller to verify freedom of movement

Chapter 4
Motion Studies

4.1 Building a Motion Study

The Fan has to rotate in order for the simulation to have any resemblance to reality. Consequently, the motion study will set up the conditions that allow for the rotation of the fan. Notice that it is without importance the source of power to make the fan rotate. As long as it rotates, the computational fluid dynamics computations will work properly.

1. Click on Motion Study 1 at the bottom of the screen.

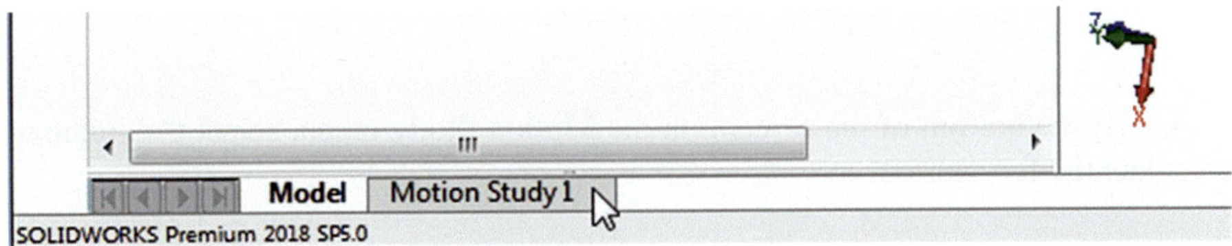

Fig. 4.1 Select Motions Study 1 from the menu

2. Select the Motor option from ribbon enabled by the Motion Study.

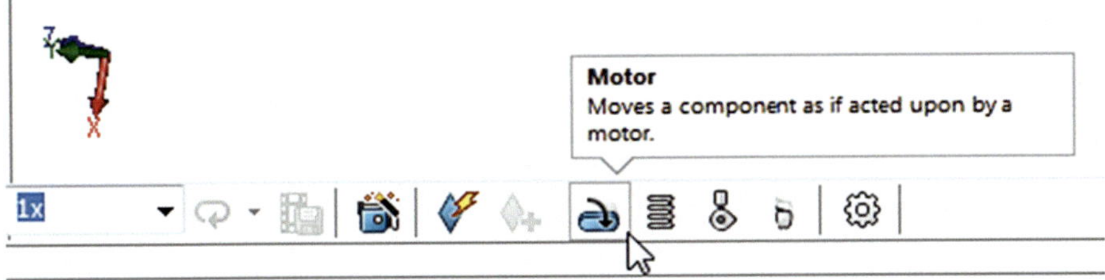

Fig. 4.2 Select the Motor option on the menu ribbon

© Springer Nature Switzerland AG 2020
B. A. Stradi-Granados, *Cloud Computing for Engineering Applications*,
https://doi.org/10.1007/978-3-030-40445-1_4

3.   Select the circular area of the Fan. Assign a clockwise rotation and a rotational speed of 400 rpm. The direction of rotation is varied using the arrows on the Component/Direction option menu. Click the Check mark.

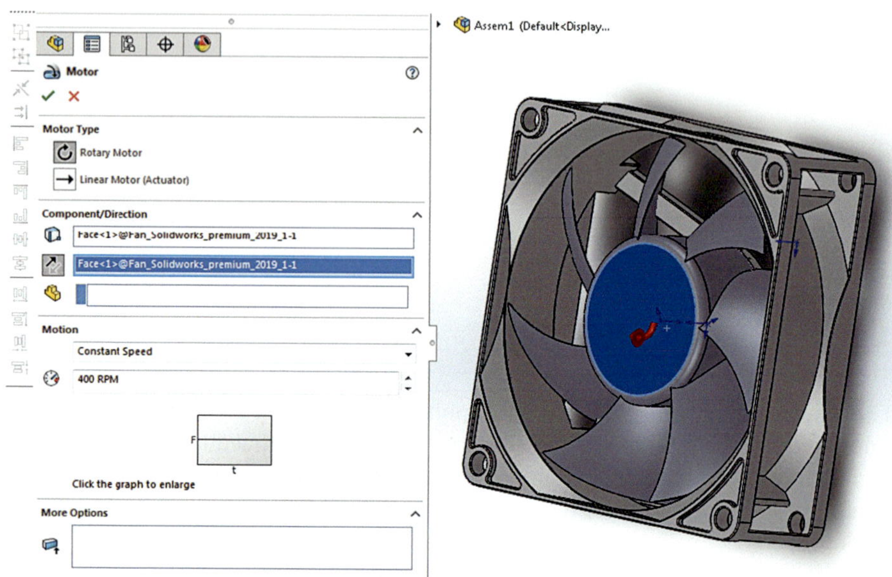

Fig. 4.3 Select area, direction, and speed of rotation

4.   At the bottom of the screen, on the Motion Study menu, select Calculate to calculate the fan motion.

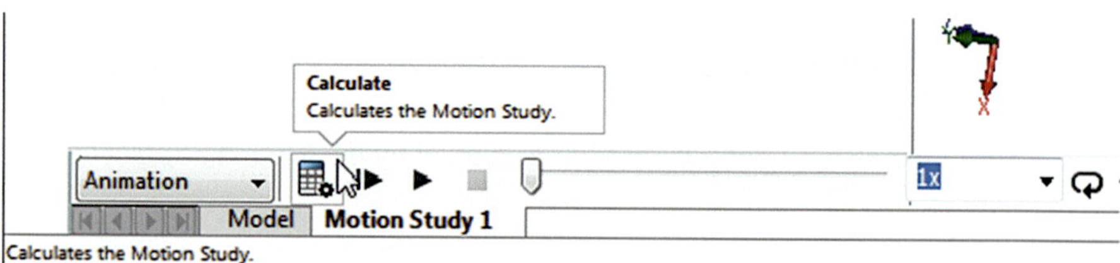

Fig. 4.4 Calculate the fan motion

5.   Once calculations are finished, modify the speed to 5× and afterward press Play.

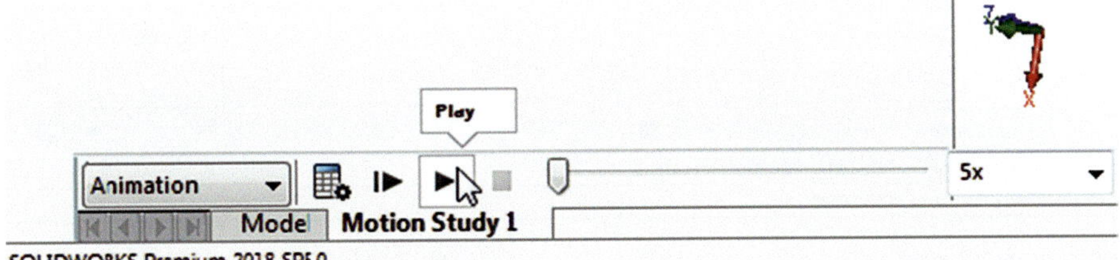

Fig. 4.5   Increase playing speed and test Play function

6.   Create a playing Loop to simulate the continuous motion of the fan.
   6a.   Open the Playback menu using the tab to the right of the horizontal arrow.
   6b.   Select Loop as the Playback mode for the fan Motion Study.
6a. Open the Playback menu using the tab to the right of the horizontal arrow.

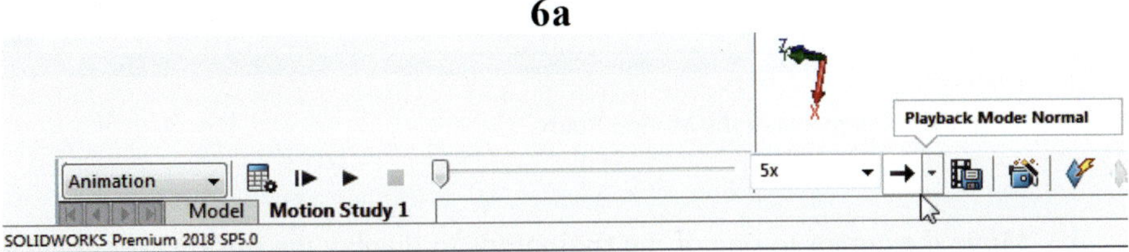

Fig. 4.6   Find the Playback menu on the menu ribbon

6b. Select Loop as the Playback mode for the fan Motion Study.

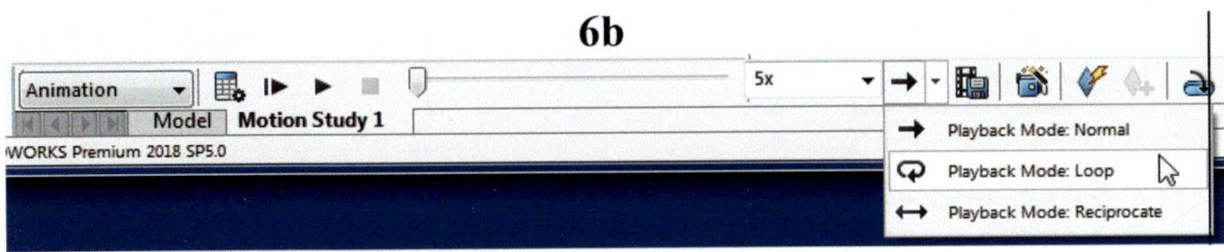

Fig. 4.7   Select Loop as the Playback mode

7.   Run the study under a loop operation

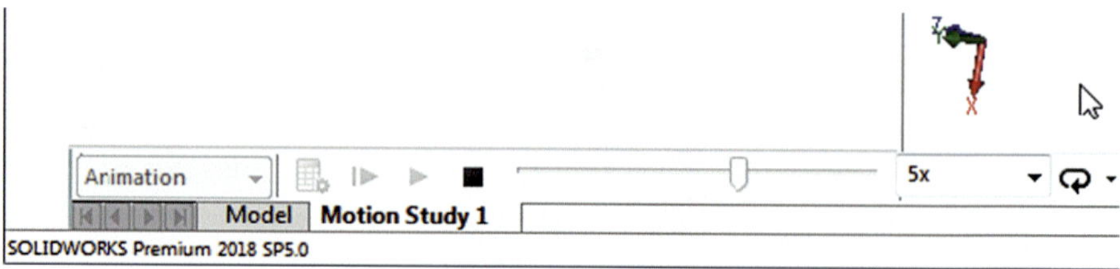

Fig. 4.8 Run the study under a loop operation

8.   At the very right, click on the tab to display the elements of the Motion Study.

Fig. 4.9 Display the elements of the Motion Study

9.   Move the cursor to one of the rhomboids to display the information related to the duration of the event on the same horizontal line.

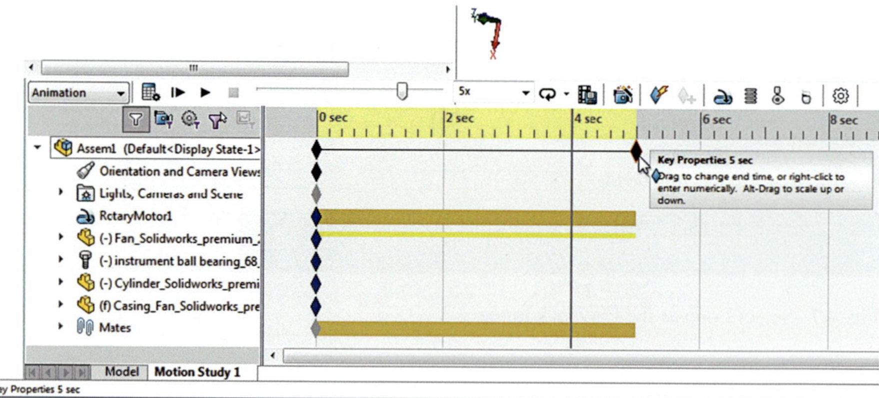

Fig. 4.10 Display the duration of events

10.   Move the upper rhomboid to 15 s. This makes for a longer time of the simulation of 15 s that is recorded and then replayed during the looping period.

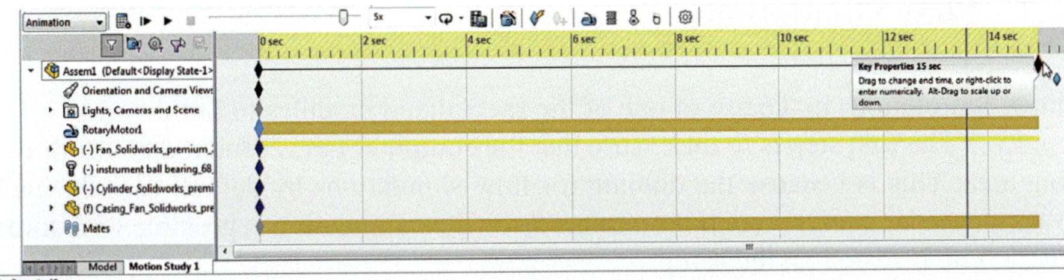

Fig. 4.11   Modify the duration of the motion study using the rhomboid

11.   The updated duration is shown in the inner framed region.

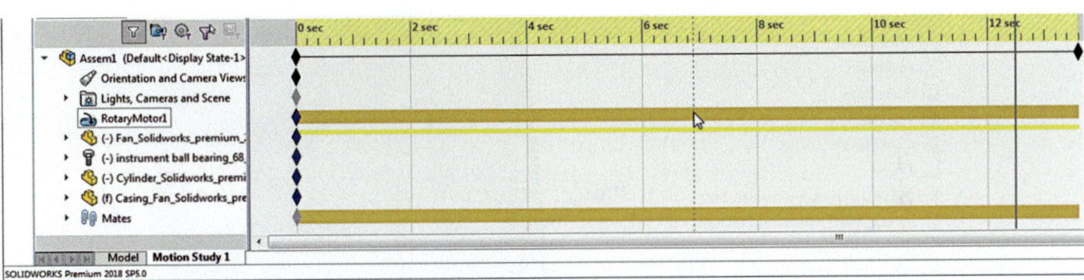

Fig. 4.12   Verify the longer time assign to the Rotary Motor

12.   Play the Motion Study to verify the rotation of the Fan.

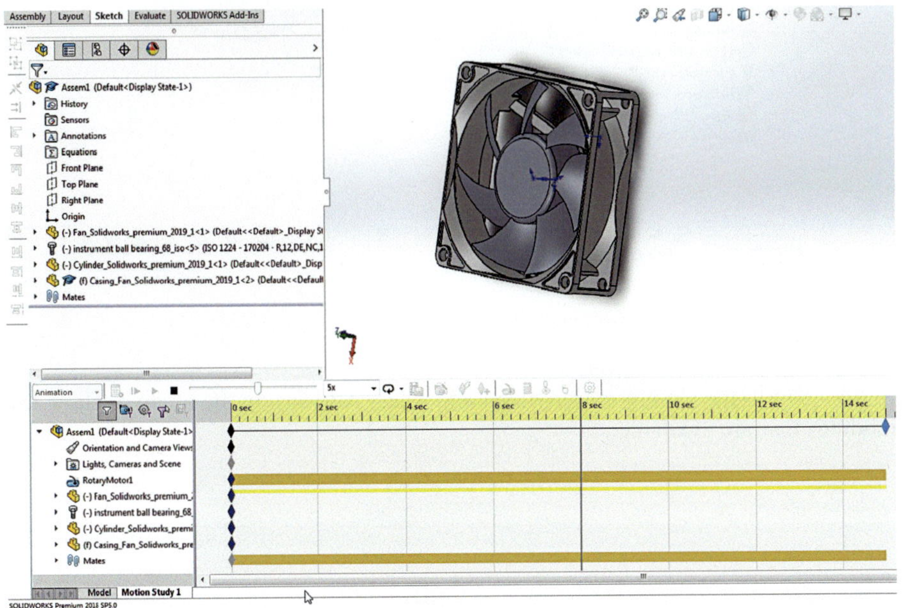

Fig. 4.13   Play the Motion Study to verify the rotation of the Fan

4.2   Flow Simulations

Flow Simulations make use of one of the specialized modules in Solidworks.

13.   The first step is to make sure that the computer fan is centered on the drawing area. This is because the domain for flow simulations by default uses the canvass' center as a reference. If not centered, the flow simulation is possible with some extra time invested on the setup.

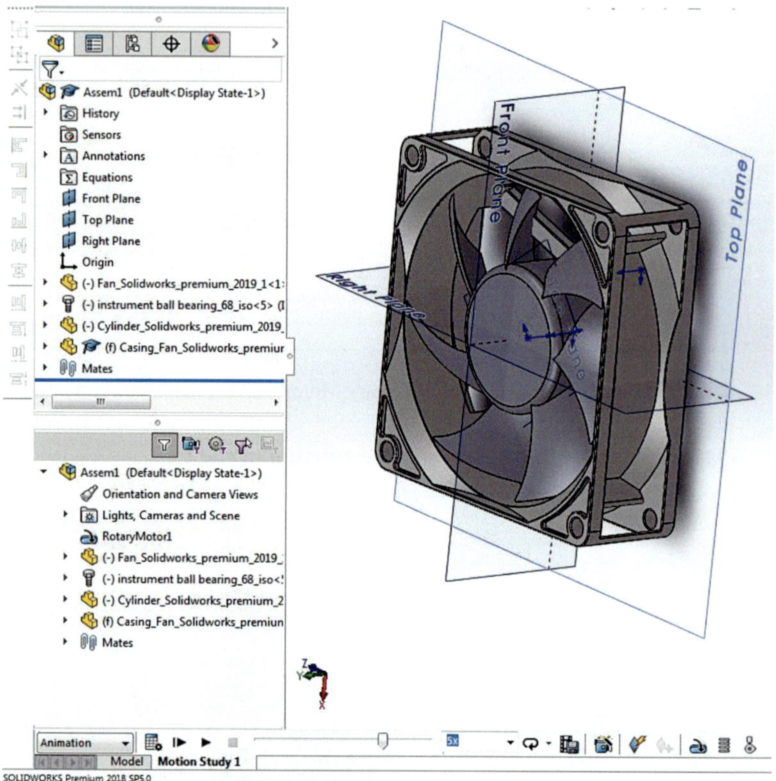

Fig. 4.14 Verify the central position of the computer fan

14.   Go to the Configuration menu and select Add-ins on the menu to show the modules that are available.

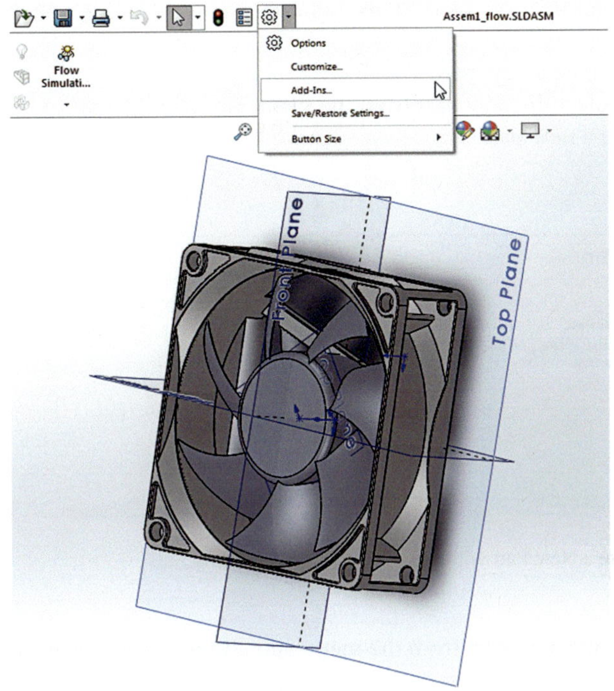

Fig. 4.15 Locate the Add-Ins for additional Modules available in Solidworks

15. Checkmark the Solidworks Flow Simulation 2018 and the Start Up box.

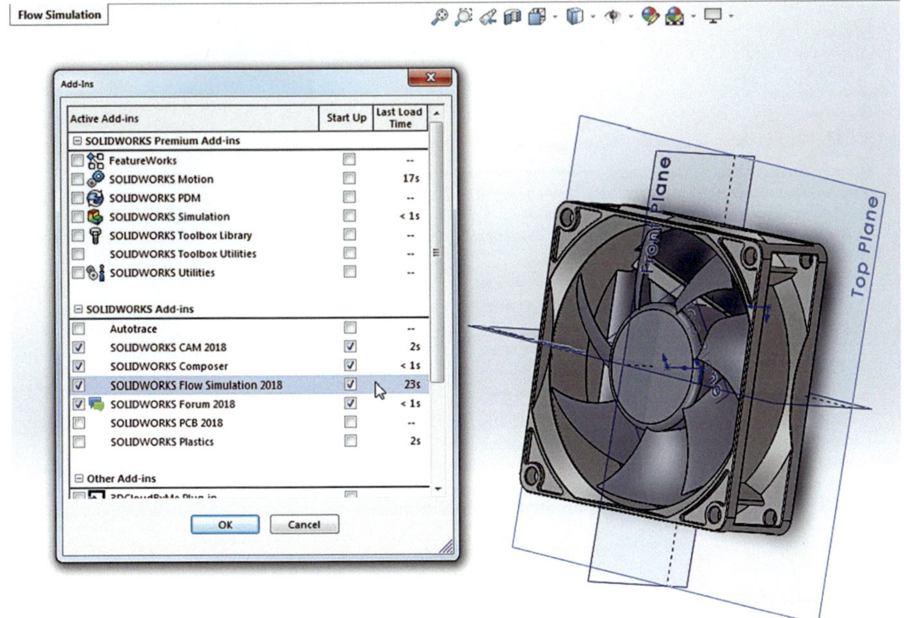

Fig. 4.16 Add Flow Simulation

16.   A volume is created. This volume is the rotating volume during the simulation.

16a.   Create a new part.

16b.   Select the new part from the menu and edit the part to make changes.

16a. Create a new part.

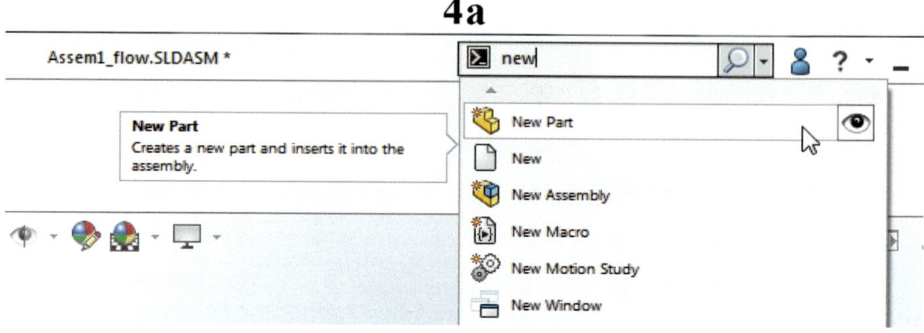

Fig. 4.17   Create a New Part within the Simulation Assembly

16b. Select the new part from the menu and edit the part to make changes.

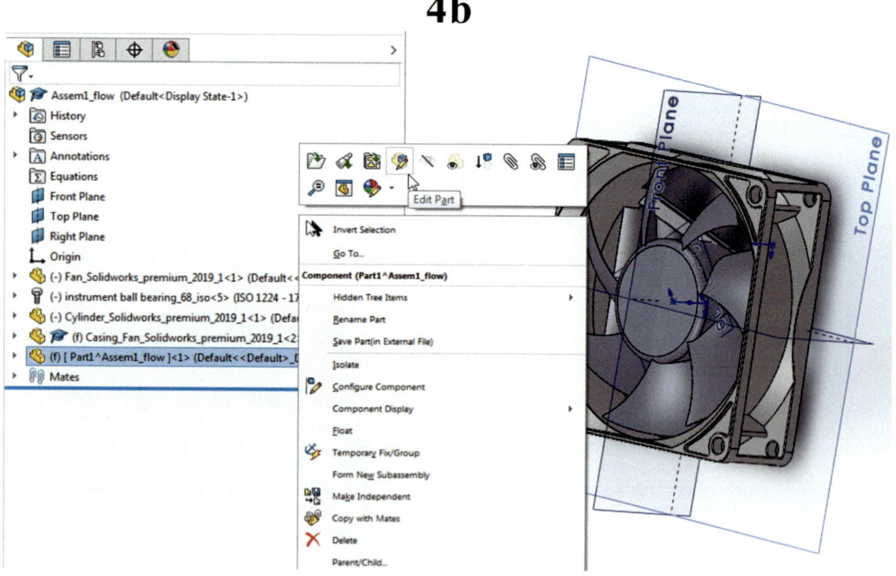

Fig. 4.18   Select the menu icon to create a new element

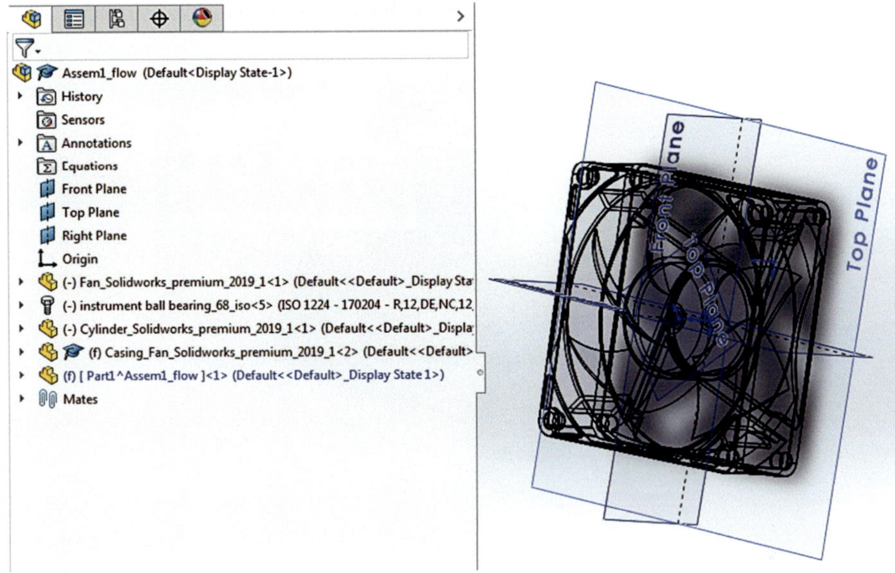

Fig. 4.19 Modify the viewing selection

17.   Select the circular core surface of the fan.

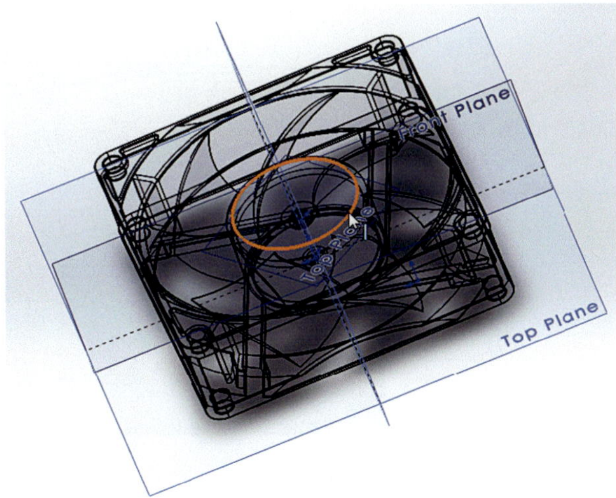

Fig. 4.20 Select the circular center circumference

18. Select the circular core surface of the fan and Sketch on that surface.

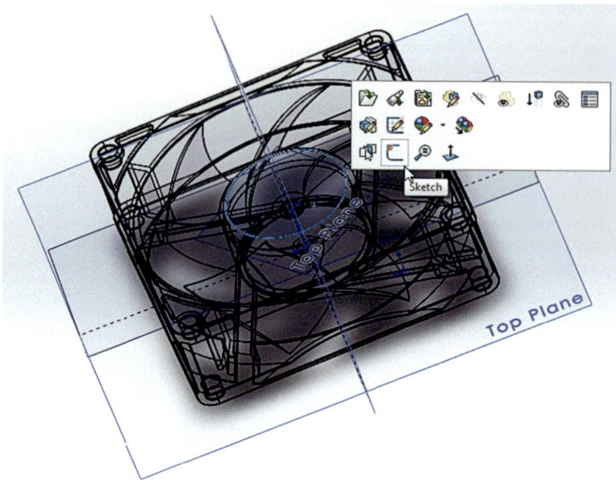

Fig. 4.21 Select the circular center to Sketch a new element

19. Draw a circle on the fan surface. The diameter reaches the outer rim of the fan blades.

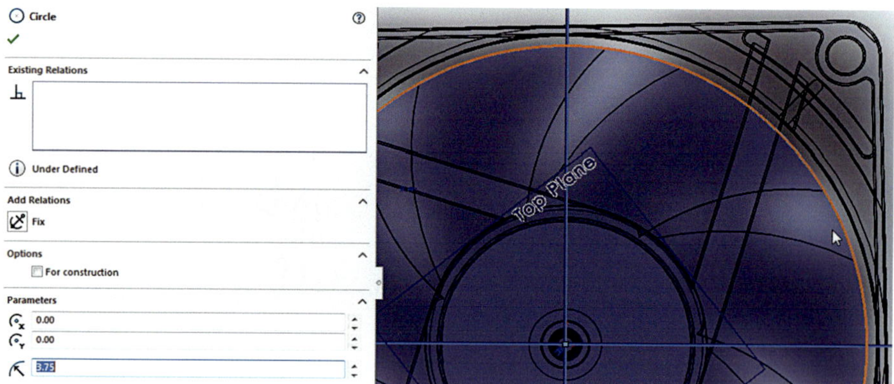

Fig. 4.22 Draw a circle to the edge of the blades

20.   Extrude the circle using the Extruded Boss/Base.

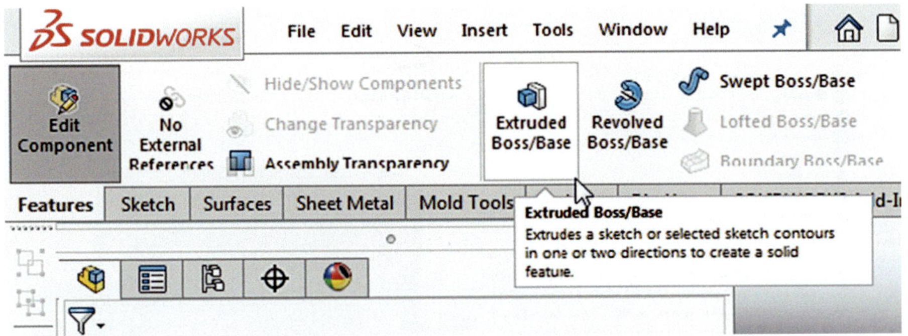

Fig. 4.23 Select to Extrude the sketched circle

21.   The cylindrical region has to reach up to the vertex of the blade as indicated with the orange point.

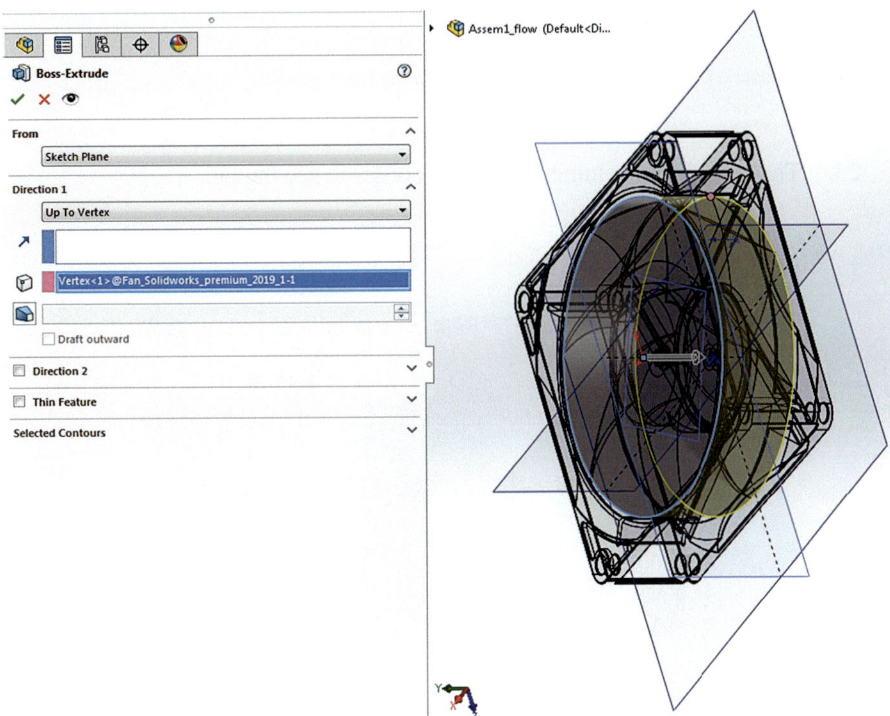

Fig. 4.24 Extrude to a vane corner on the back of the propeller

22.   Cylindrical region is created and overlays upon the fan blades.

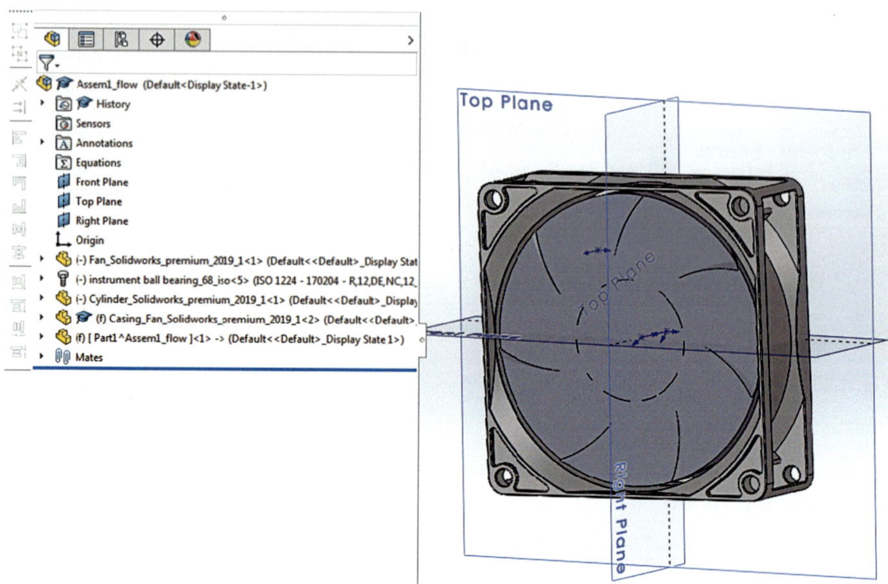

Fig. 4.25   Create a Cylindrical region overlaid upon the fan's vanes

23.   The cylindrical volume is hidden in order to see the fan.

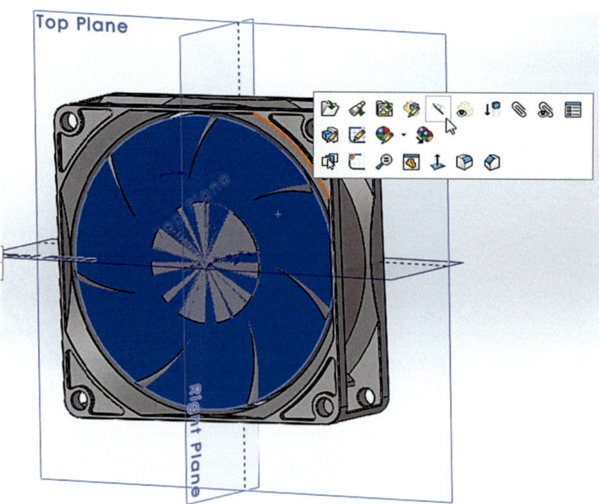

Fig. 4.26   Make the new cylindrical body invisible

24.   Select the tab for Flow Simulation and click on the Wizard.

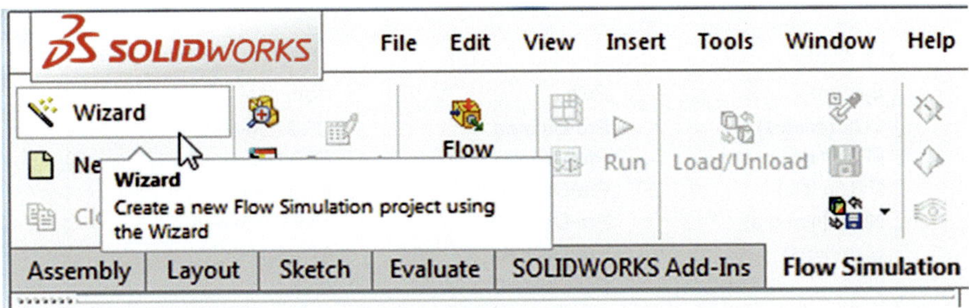

Fig. 4.27 Locate the Wizard option for Flow Simulations

25.   Follow the Wizard, starting with a title for the simulation.

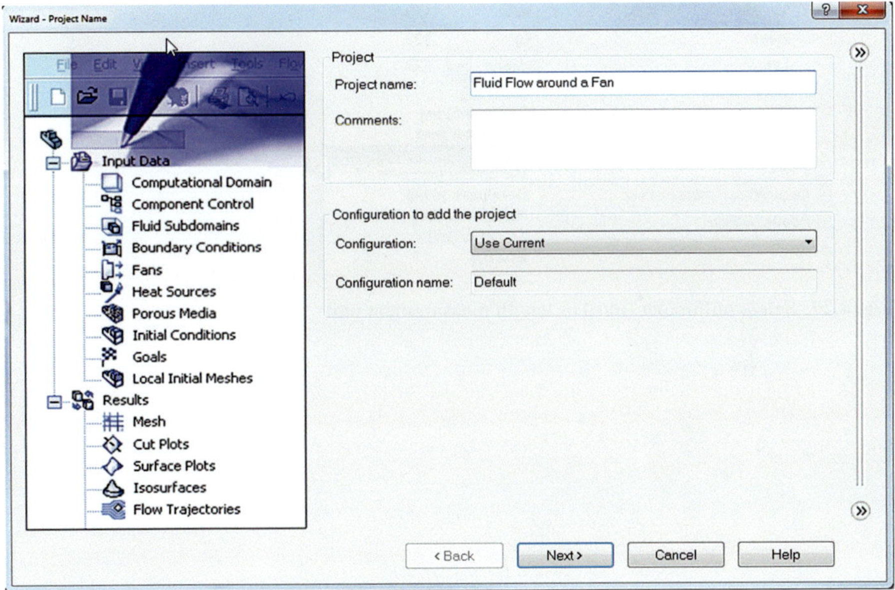

Fig. 4.28 Provide a name for the Project

26.   Select the Unit system with lengths measured in millimeters.

Unit system:

System	Path	Comment
CGS (cm-g-s)	Pre-Defined	CGS (cm-g-s)
FPS (ft-lb-s)	Pre-Defined	FPS (ft-lb-s)
IPS (in-lb-s)	Pre-Defined	IPS (in-lb-s)
NMM (mm-g-s)	Pre-Defined	NMM (mm-g-s)
SI (m-kg-s)	Pre-Defined	SI (m-kg-s)
USA	Pre-Defined	USA

☐ Create new   Name: SI (m-kg-s) (modified)

Parameter	Unit	Decimals in results display	1 SI unit equals
⊟ Main			
Pressure & stress	Pa	.12	1
Velocity	m/s	.123	1
Mass	kg	.123	1
Length	Meter ▾	.123	1
Temperature	Meter [m]		1
Physical time	Decimeter [dm]		1
Percentage	Centimeter [cm]		1
⊞ HVAC	Millimeter [mm]		
⊞ Geometrical Characteristic	Micrometer [micrometer]		
⊞ Loads&Motion	Dekameter [dam]		
⊞ Heat	Hektometer [hm]		
	Kilometer [km]		
	Inch [in]		

Fig. 4.29 Select millimeter (mm) as length measurement unit

27.   The angular velocity units are selected to revolutions per minute (RPM).

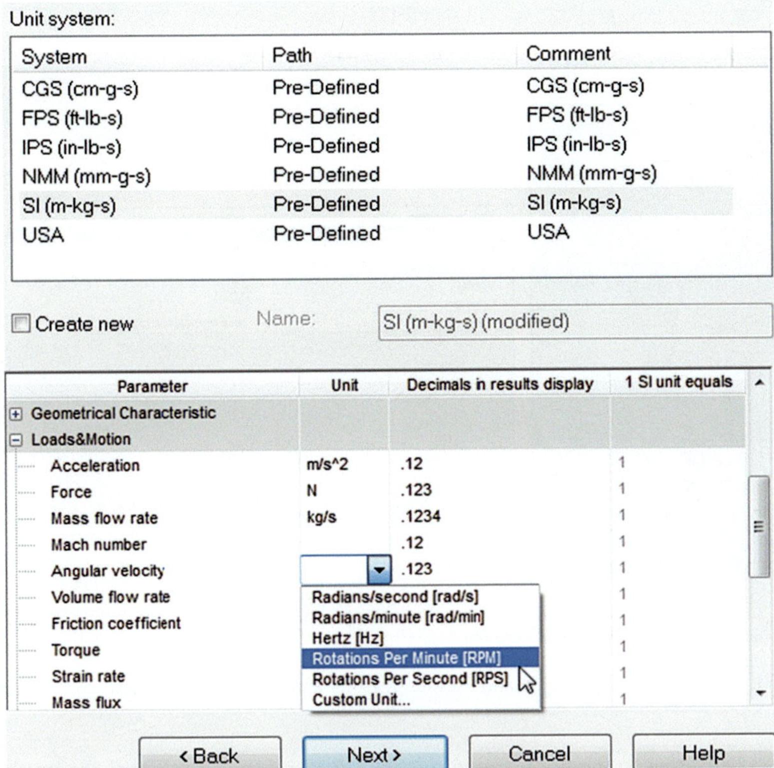

Fig. 4.30 Select the units for the speed of the rotating propeller

28.   The Analysis type is external, no closed cavities are taken into account (because there is no flow), rotation is a Physical Feature and *Y* is the direction of flow.

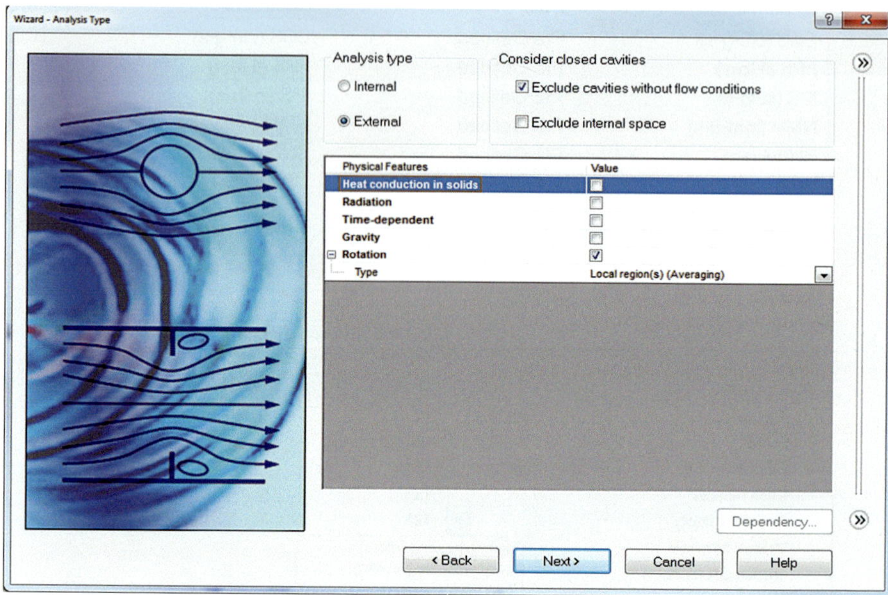

Fig. 4.31 Select the Analysis type and exclude regions without flow conditions

29.   Continue to the next screen to select Air and the Fluid from the predefined compound available.

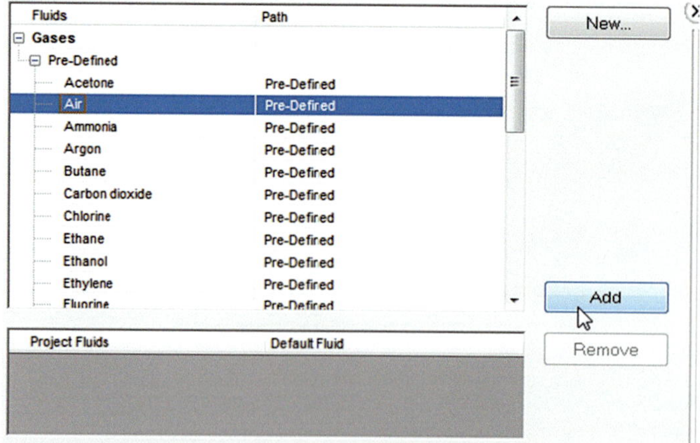

Fig. 4.32 Select Air from the list of the substances available

30.   Air is the Project's fluid.

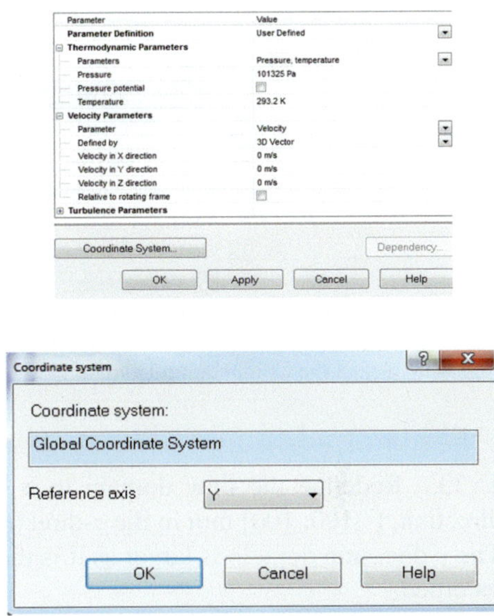

Fig. 4.33 Default fluid and coordinate system

The default thermal conditions and roughness are left at their default values.
Initial ambient conditions are not modified, but the Coordinate System is set to Y.

31.   The default domain is very large (so large that we only see the vertical components of the domain.

Fig. 4.34 Fluid flow domain

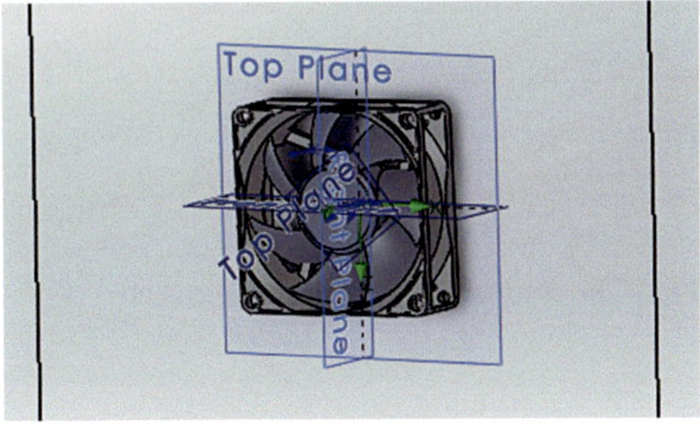

32.   Select the Computational Domain with the cursor and right-click on the selection (or double click on it).

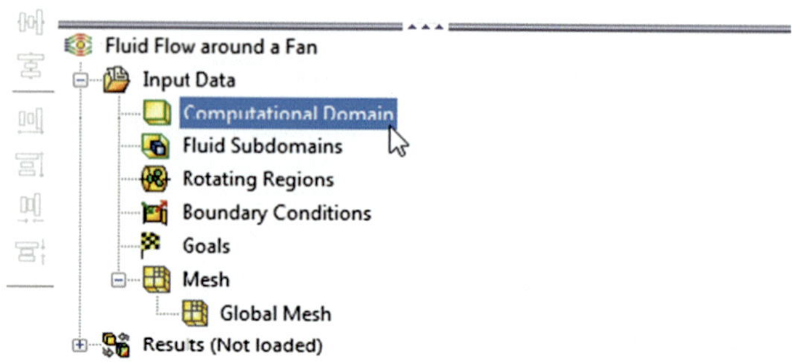

Fig. 4.35 Select the Computation Domain

33.   Redefine the flow domain to a smaller size of [−50,50] mm in the *x*-direction, [−100, 100] mm in the *y*-direction, and [−50, 50] mm in the *z*-direction. The *y*-direction domain is larger as it is the direction of flow and more features are recorded.

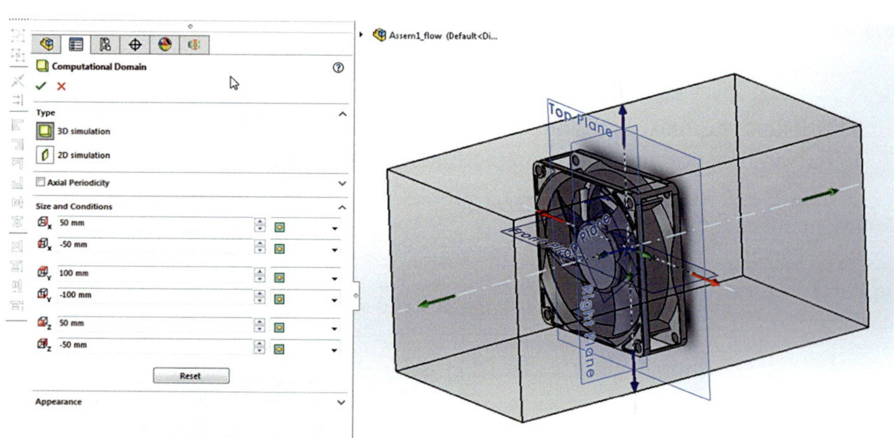

Fig. 4.36 Redefine flow domain dimensions

34.   Hide the computation domain for a clearer view. Hide reference planes and other elements are not needed.

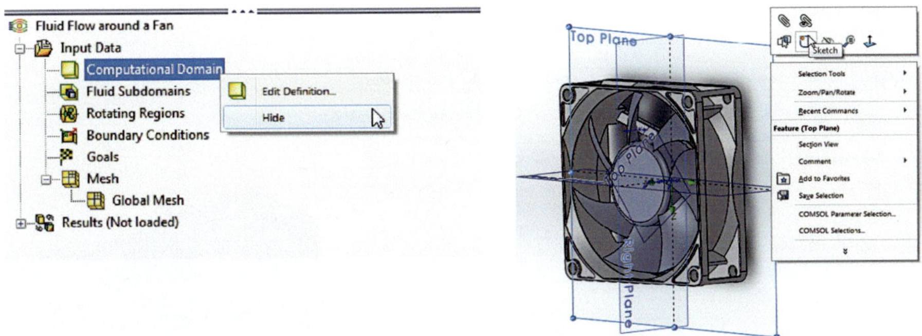

Fig. 4.37 Make invisible computational domain

35.   Select the Rotating Regions label on the Menu. Assign a rotational speed of 600 rpm clockwise to the last region that was created.

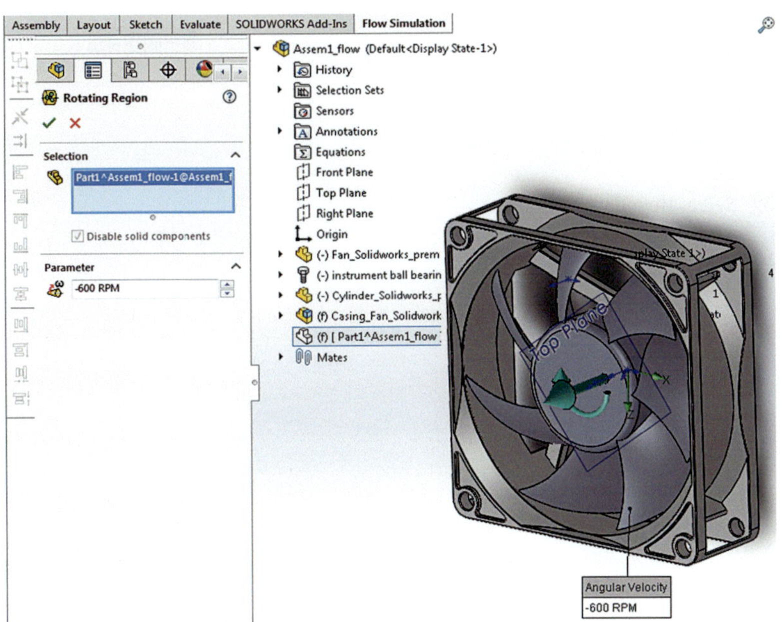

Fig. 4.38 Select cylindrical volume as rotating volume from menu

36.   Continue with boundary conditions. The immediate inner surface to the fan's vanes is Real Walls.

In a motor the part that rotates is the Rotor and the part that stays fixed in position is the Stator.

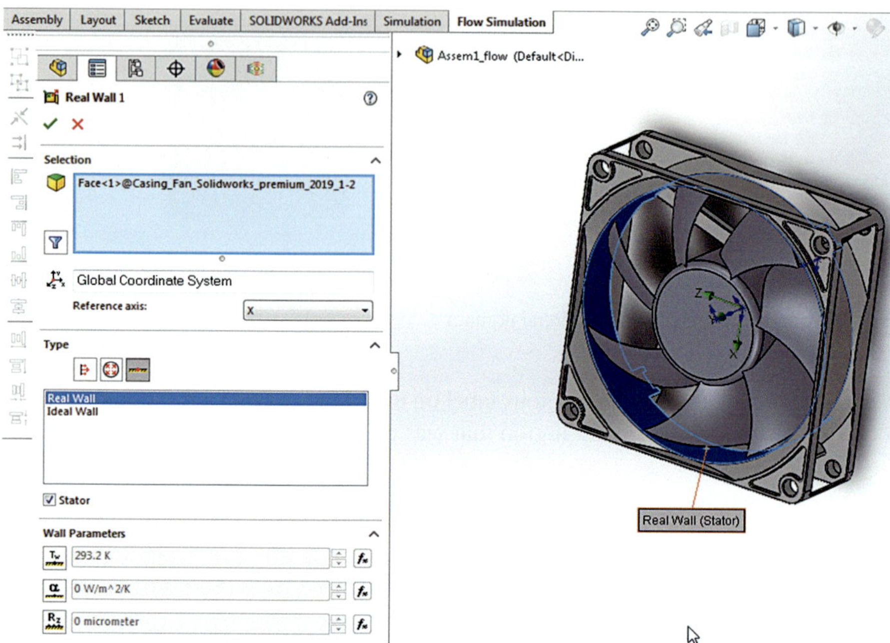

Fig. 4.39 Select Boundary Conditions

37. Lo and Behold proceed to run the Simulation.

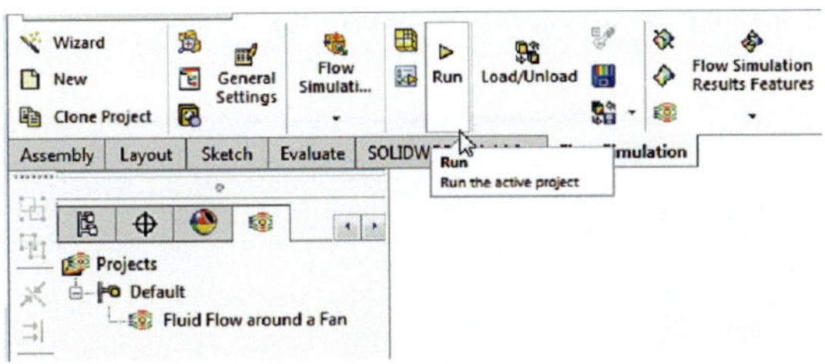

There may be some trouble like the following

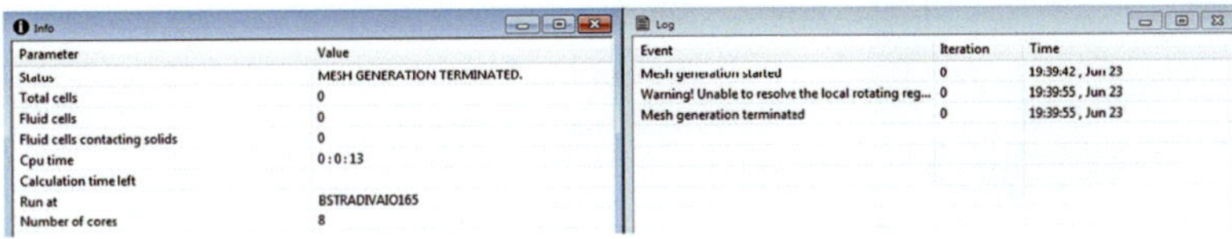

In this case in the Mesh tab open the Global Mesh Settings and make smaller elements

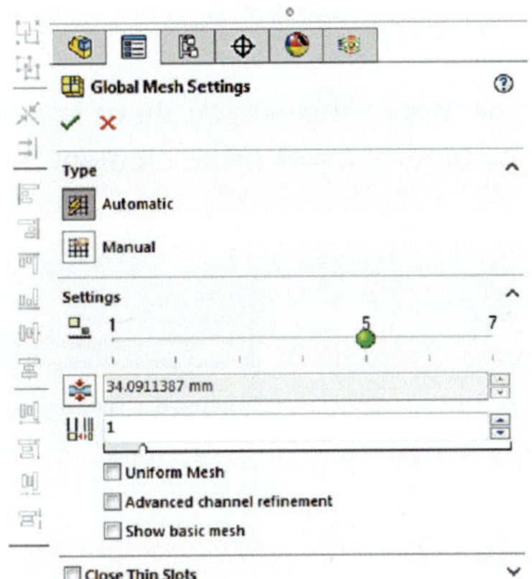

Fig. 4.40 Proceed to Run the active project

38.	The mesh settings are adjusted by the software. An interesting parallel exercise is to open task manager and check how much of the processor power is being used; this is left for as a self-study exercise.

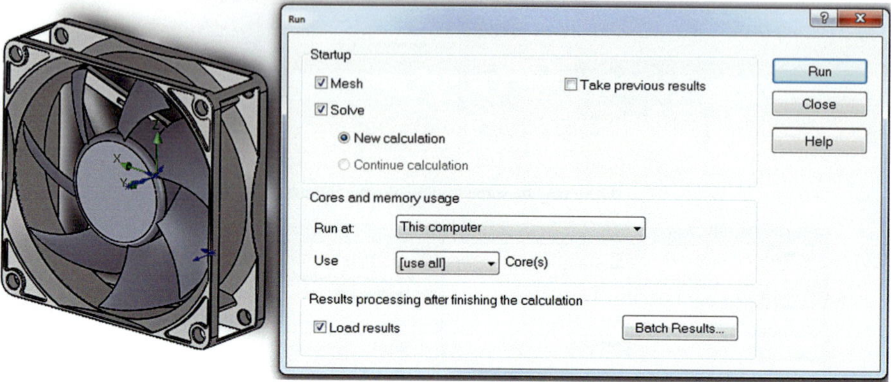

Fig. 4.41 Default settings for simulation

39.	The simulation runs and in this screen there is a summary of the process. Notice the meshing step characteristics of finite element methods.

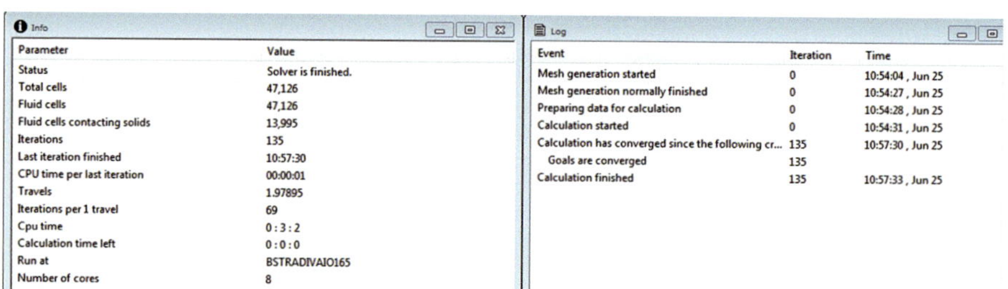

Parameter	Value
Status	Solver is finished.
Total cells	47,126
Fluid cells	47,126
Fluid cells contacting solids	13,995
Iterations	135
Last iteration finished	10:57:30
CPU time per last iteration	00:00:01
Travels	1.97895
Iterations per 1 travel	69
Cpu time	0 : 3 : 2
Calculation time left	0 : 0 : 0
Run at	BSTRADIVAIO165
Number of cores	8

Event	Iteration	Time
Mesh generation started	0	10:54:04 , Jun 25
Mesh generation normally finished	0	10:54:27 , Jun 25
Preparing data for calculation	0	10:54:28 , Jun 25
Calculation started	0	10:54:31 , Jun 25
Calculation has converged since the following cr...	135	10:57:30 , Jun 25
Goals are converged	135	
Calculation finished	135	10:57:33 , Jun 25

Fig. 4.42 Summary of process execution parameters

40.   The results are shown in graphical form by selecting Flow Trajectories.

Fig. 4.43 Inserting Flow
Trajectories

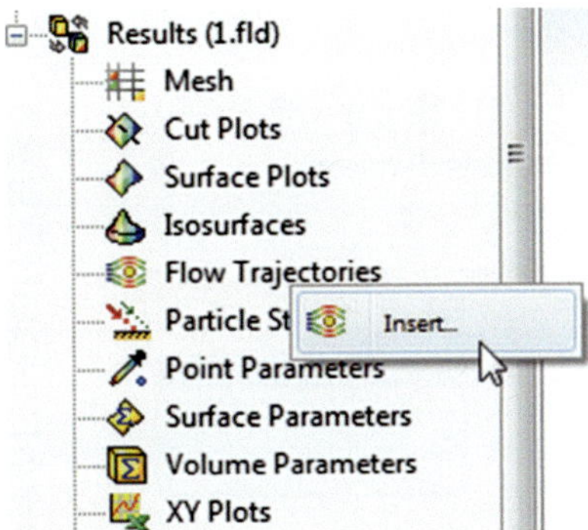

41.   A few elements are needed for the plotting of Flow Trajectories. A plane of
reference in this case is the Top Plane.

Fig. 4.44 Selecting the
plane perpendicular to the
flow

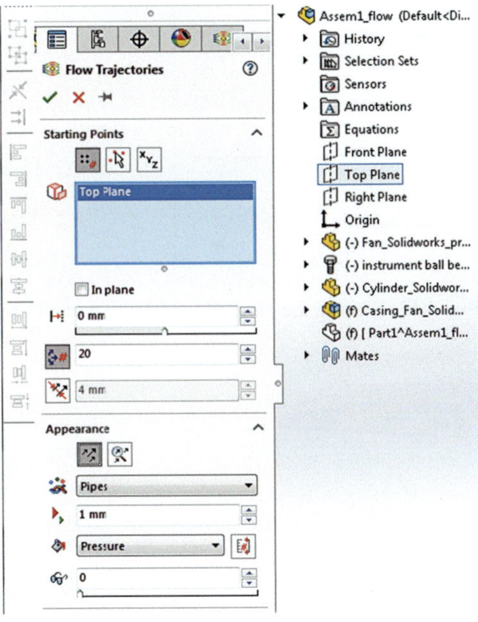

42.   A distance for the particles to enter the computational domain indicated by the light green plane.

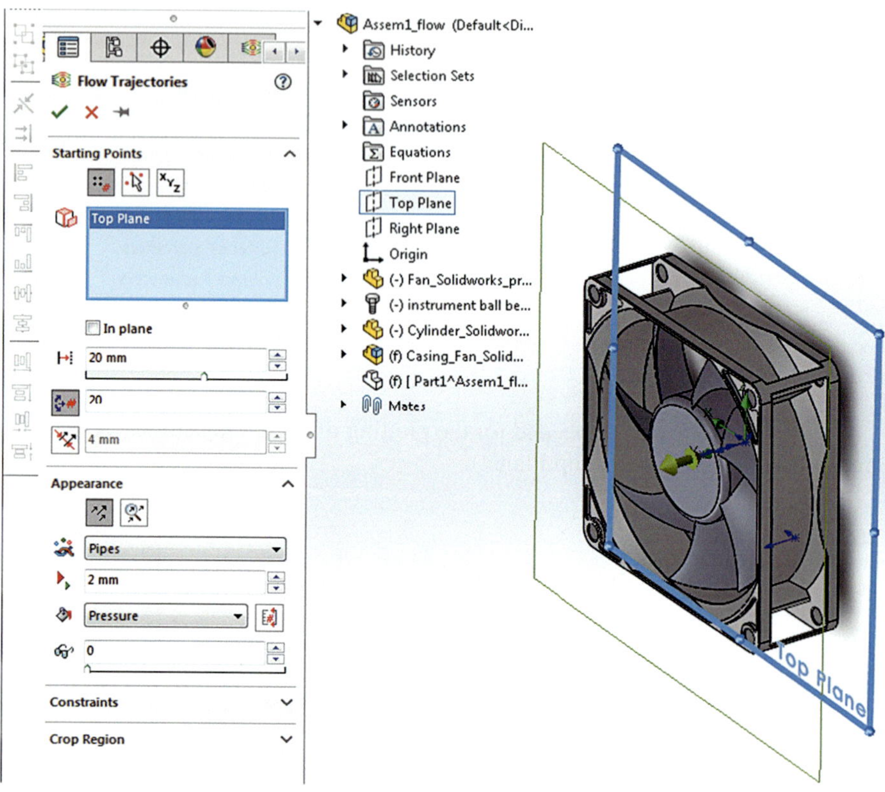

Fig. 4.45 Placing the entry plane for the particles to enter the computational volume

43.   The setting for the Appearance is Lines and Arrows and not many of them to get started. Velocity is the quantity illustrated on the figure and quantified on the scale.

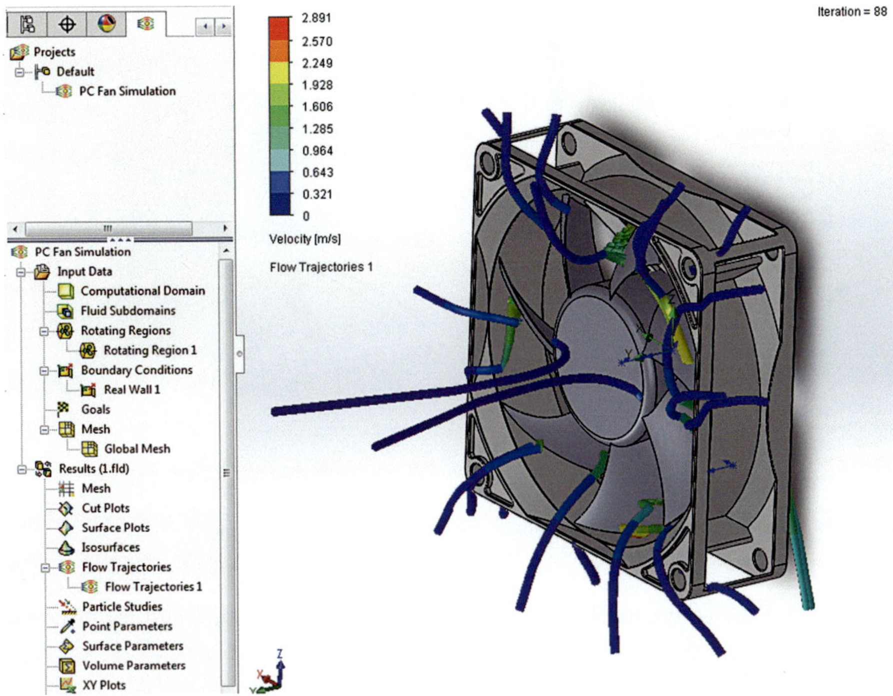

Fig. 4.46 Tubes depict flow lines

44.   To animate the flow trajectories, open Flow Trajectories 1 and press Play. Take advantage of the opportunity and change the trajectory tubes for 100 lines with a line width of 4.

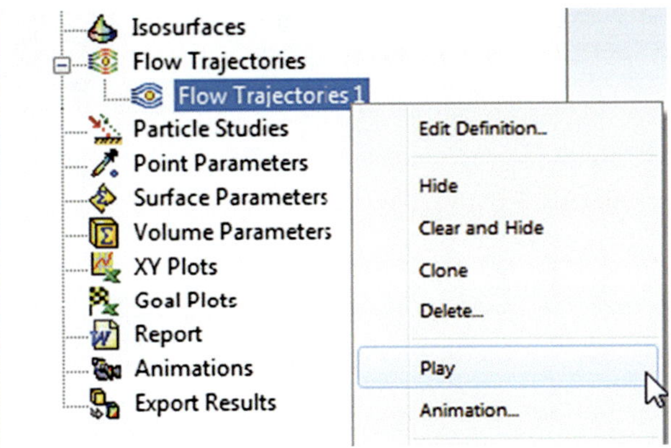

Fig. 4.47 Proceed to Play the animation of Flow Trajectories

45.   The variation of the particle velocity with time is animated on the screen.

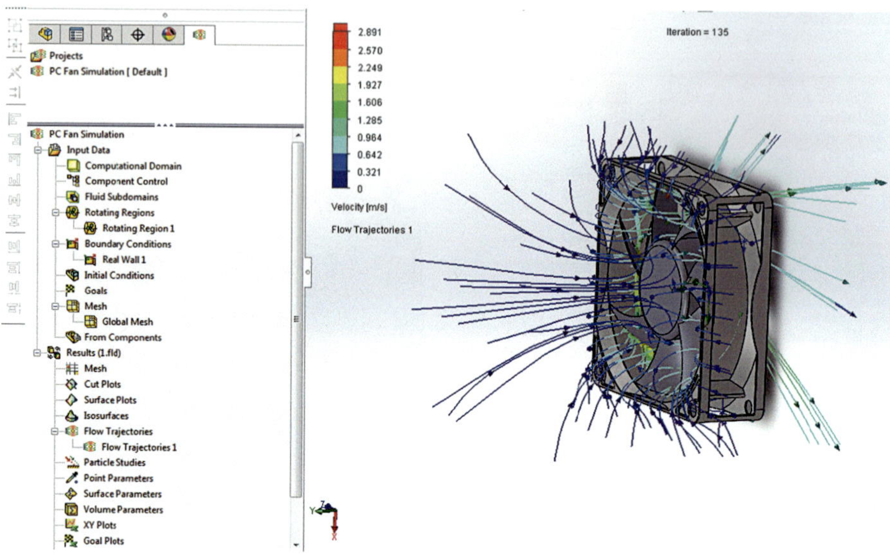

Fig. 4.48 On the results generated, Animate the particle velocity with time

46.   Use the command Clone to copy the current simulation and try different settings.

Fig. 4.49 Clone the current simulation to experiment on the copy of the model

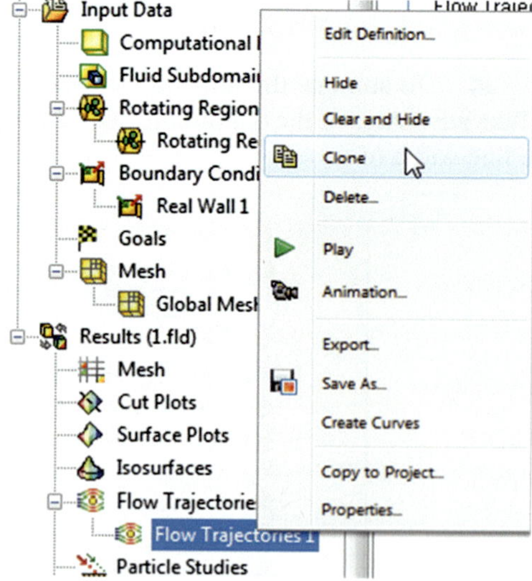

47.   Run the new Simulation.

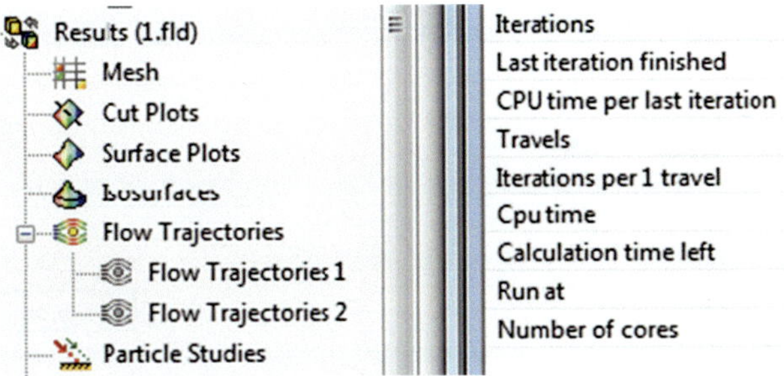

Fig. 4.50 Run the new simulation

48.   Redefine the flow trajectories, this time with 500 particles and lines with arrows.

Fig. 4.51 Experiment by adding more particles to the model

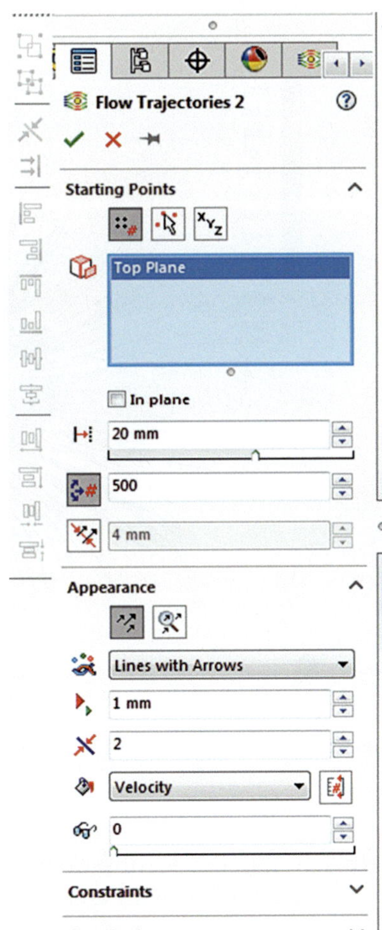

49.   Some computations are needed.

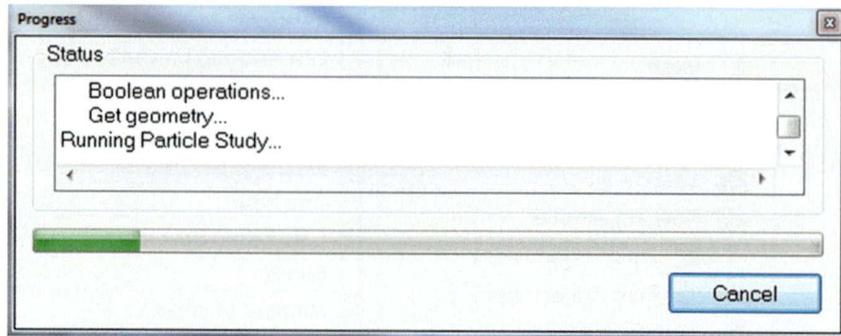

Fig. 4.52 Simulation progress

50.   The graph generated is somewhat more crowded with so many lines still very illustrative.

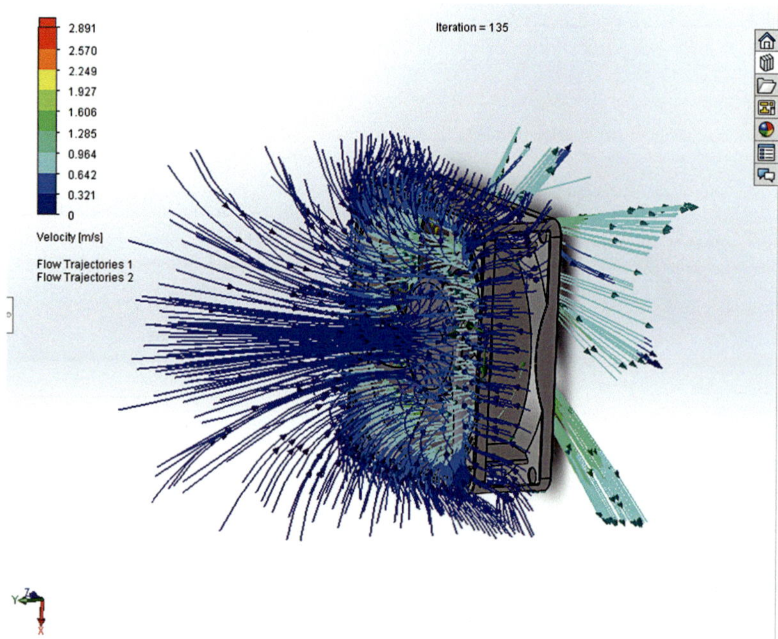

Fig. 4.53 Flow lines according to Solidworks

51.   The maximum on the scale can be varied to a specific maximum value, if necessary. It is also possible to slice the graph to a specific plane using Section View at the top of your drawing area.

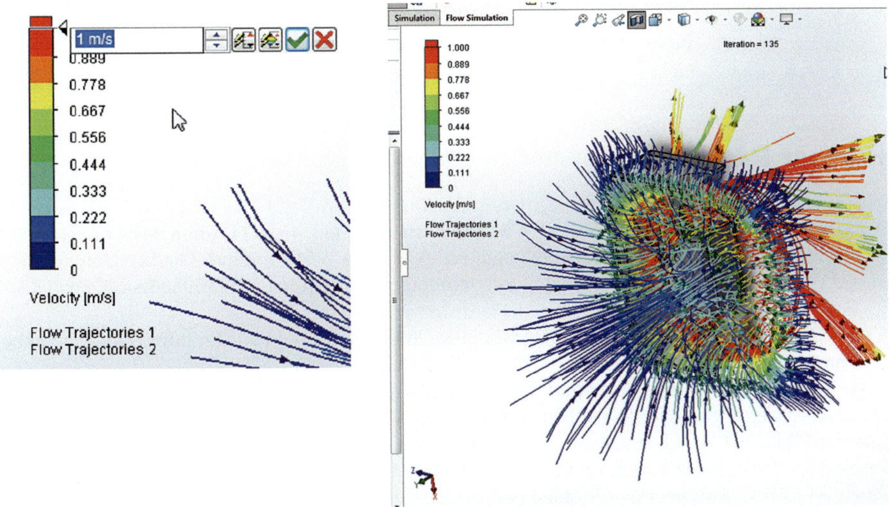

Fig. 4.54 Modify the maximum on the scale

52. Cross section with detail of fluid flow with 100 lines.

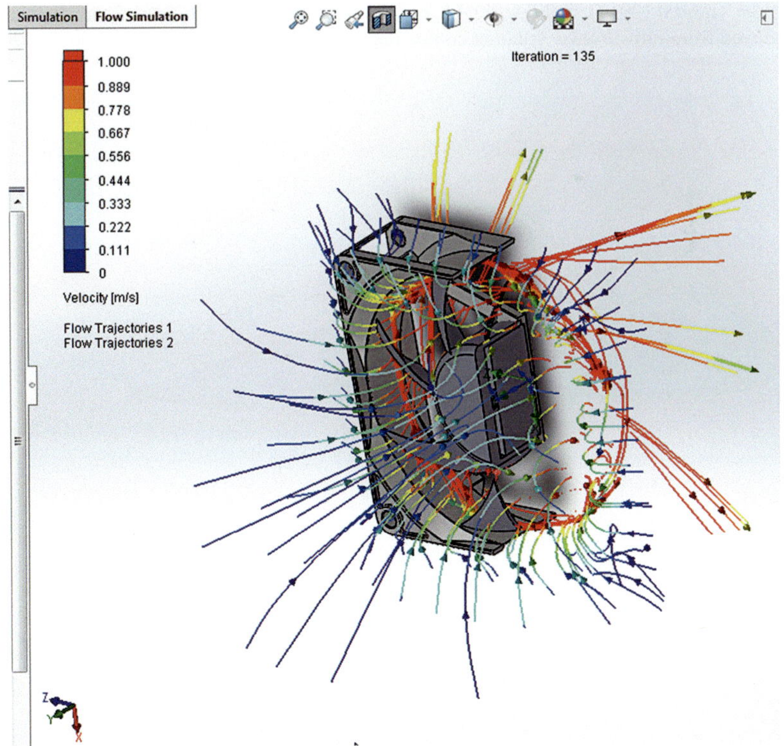

Fig. 4.55 Cross section view with 100 lines

Long exercise
Now time to apply to your very own necessities

Further Readings

Amazon (2019). Noctua NF-A12x25 PWM, premium quiet fan, 4-pin (120mm, brown). Retrieved from https://www.amazon.com/Noctua-NF-A12x25-PWM-Premium-Quality-Quiet/dp/B07 C5VG64V/ref=sr_1_11_sspa?keywords=computer+fan&qid=1563850986&s=gateway&sr= 8-11-spons&psc=1

CAD CAM Tutorial | Sketch Computer Fan in Solidworks. Retrieved from https://www.youtube. com/watch?v=bspA1quUf9s

CAD Training. Solidworks. Essential training videos and manuals 1–16.

Dassault Systems. Solidworks student edition DVD. Academic year 2016–2017.

Lombard, M. (2013). Advanced solidworks DVD, Training, Infinite skills.

Perez, M. (2013). Solidworkes mold tools DVD. Infinite skills.

Perez, M (2013). Solidworks—Molded part design. Infinite skills.

Shutterstock (2017). Computer fan images. Retrieved from https://www.shutterstock.com/search/ computer+fan

Solidworks Flow Simulation (2017). Centrifugal pump. Retrieved from https://www.youtube.com/ watch?v=Vr9A2oVy570

Solidworks Fun (2017). 2 Blade, 3 blade and 4 blade propeller CFD in solidworks (tutorial). Retrieved from https://www.youtube.com/watch?v=a4lAhFsls5k

Solidworks Tutorial (2017). Solidworks FL tutorial #282: PC fan with flow simulation analysis. Retrieved from https://www.youtube.com/watch?v=Vr9A2oVy570

Chapter 5
Matlab Primer

5.1 Basic Annotation

Convention *log* is natural log.

- $\log_{10}$ is the logarithm with base 10
- Trigonometric functions use radians not degrees.
- Files in matlab must end in *.mfilename.m*
- Commands: *help* x gives you the help for x

5.2 Examples

Equation 8. Value of *r*

$$r = 2;$$

Equation 9. Volume

$$\text{volume} = \frac{4}{3}\pi r^{3}$$

Equation 10. Volume computation

$$\text{vol} = \left(\frac{4}{3}\right) * \underset{\pi}{pi} \quad \overset{Multiplication}{*} \quad r \quad \underset{exponential}{\wedge} \quad 3 \quad \overset{\substack{Edit/remove \\ display}}{;}$$

© Springer Nature Switzerland AG 2020

B. A. Stradi-Granados, *Cloud Computing for Engineering Applications*,

https://doi.org/10.1007/978-3-030-40445-1_5

Equation 11. Split line into two

$$\mathrm{vol} = \left(\frac{4}{3}\right) * \mathrm{pi} \ldots \text{ellipsis indicates that a long expression}$$

$$* r^3$$

5.3   Arithmetic Operators

+	Addition
−	Subtraction
*	Multiplication
/	Division
\	Reverse division (Gaussian elimination)

Equation 12. Matlab if statement
Example

```
if cond
   something
end
     r=2;
     if  r>0
          vol=(4/3)*pi*r^3;
     end

Logical  equal  ==
     if       (r==2)
          vol=(4/3)*pi*r^3;
     end
```

5.4   Logical Operators

~=	Not equal
>	Greater than
<	Less than
>=	Equal or greater than

<=	Equal or less than
&	And

Equation 13. Logical or

```
if (g>3 | g<0)
     a=6
     end
```

5.5   Naming Variables, Programs, and Subroutines

- Do not use names of reserved functions.
- Names may not start with numbers, include special characters (underscore permitted) or include blanks.
- The same for programs, subroutines, and functions.

Table 5.1 Special numbers and variable names

Variable name	Meaning	Value
eps	Machine epsilon	$2.220 * 10^{-16}$
pi	π	3.14159
i and j	Unit imaginary	$\sqrt{-1}$
inf	Infinity	∞
NaN	Not a number	
date	Date	
flops	Floating-point operation count	

5.6 Loops

Equation 14. Loops: for/end

- for/end

$$\text{for} \qquad \overbrace{r = 1:5}^{\substack{\textit{notice increments} \\ \textit{of one unit}}}$$
$$vol = \left(\frac{4}{3}\right) * pi * r \wedge 3$$
$$disp\left([r, vol]\right)$$
$$\text{end}$$

Equation 15. Loops: while/end

- while/end

$$\text{while } r < 5$$
$$r = r + 1$$
$$vol = \left(4/3\right) * pi * r \wedge 3$$
$$disp\left(\left[r, vol\right]\right)$$
$$\text{end}$$

Equation 16. Stepping down index with two colons

- Stepping down the index:

$$\text{for } r = s \ \overset{\textit{two}}{:} \ - 1 \ \overset{\textit{colons}}{:} \ 1$$
$$vol = (4/3) * pi * r^3$$
$$disp\left([r, vol]\right)$$
$$\text{end}$$

$$\text{for } r = 1:s$$
$$\quad \text{for } s = 1 : r$$
$$\quad\quad vol = (4/3) * pi * (r \wedge 3 - s \wedge 3)$$
$$\quad\quad disp\left([r, vol]\right)$$
$$\text{end}$$

5.7 Matlab Number Format

- 15 decimal places   *format long*
- 6 decimal places   *format short*

- Matlab has no *goto* command but has a *break* command that would terminate program execution.
- To clear variables in memory *clear x, y, z*
- To clear the window *clc*

Equation 17. Input from Keyboard

while r < 10

$$r = input\ ('type\ radius\ \overbrace{(or-1\ to\ stop):\ ')}^{input\ a\ value\ from\ keybord}$$
$$if\ r < 0$$

this would kill the
while and stop the if

$$\overbrace{break}$$
$$end$$
$$vol = \frac{4}{3} * pi * r^{\wedge}3;$$
$$fprintf\ ('volume = \%7.3\ f\backslash n, vol')$$

end

$$\underbrace{-\ -\ -}_{3\ digits}\quad\underbrace{\cdot}_{period}\quad\underbrace{-\ -\ -}_{3\ digits}$$

3 digits, period, 3 digits

And then skip to the next line because

5.8   Reading and Writing

Three ways:

- Keyboard
- Datafile (reading or writing to a data file)
- Using save or load

Reading input from keyboard:

- Z = input ('type radius:')

Output format:

- fprintr('the volume of the sphere %12.5f\n',vol)

Writing into a specific file:

- fprintf('file_x', 'volumen=%12.5 f\n', vol) will write the output in the file named file_x

5.9   Array Variables

Row form:

- $x = [0, 0.1, 0.2, 0.3, 0.4, 0.5]$
- Particular element $x(3) = 0.2$

Array with fixed increment

- $x = 2{:}{-}0.4{:}{-}2$

Column array:

- $z = [0; 0.1; 0.2; 0.3; 0.4; 0.5]$ (row)
- $z = [0, 0.1, 0.2, 0.3, 0.4, 0.5]'$ (column)

Operations: when two vectors x and y have the same dimensions.

Table 5.2 Operations in array variables

Operation	
+	Standard element by element subtraction
−	
.*	Element (i) of x by element (i) of y
./	Element (i) of x divided by element (i) of y

Equation 18. Two-dimensional array
   Two-dimensional array

$$X = \begin{bmatrix} 0,0.1,0.2; 0.3,0.4,0.5 \end{bmatrix} \quad = \quad X = \begin{bmatrix} 0 & 0.1 & 0.2 \\ 0.3 & 0.4 & 0.5 \end{bmatrix}$$

Equation 19. Display rows and columns

$$\underbrace{x(1,i)}_{\substack{Display \\ an\ entire \\ row}} \quad , \quad \underbrace{x(i,1)}_{\substack{Display \\ an\ entire \\ column}}$$

5.10   Getting into Matlab

- Info
- Demo
- Help
- Helpbrowser
- Lookfor

5.11   Variable Names

- The name must begin with a letter of the alphabet. After that, the name can contain letter, digits, and the underscore character (e.g., value_I), but it cannot have a space.
- MATLAB is case-sensitive. That means that there is a difference between upper and lowercase letters.
- Names of built-in functions can, but should not, be used as variable names, these are reserved words.
- **who** shows variables that have been defined in this Command Window (this just shows the names of variables).
- **whos** shows variables that have been defined in this Command Window (this shows more information on the variables, similar to what is in the Workspace Window).
- **clear** clears out all variables, so they no longer exist.
- **clear** *variablename* clears out a particular variable.
- Double precision: sign, integer, 15 decimal places.
- *format compact,* tighter on the screen, you can fit more
- Ellipsis …
- \ division (Gaussian elimination), / division
- (), ^, -, *, /, \, +, − highest to lowest priority
- The expressions are evaluated form left to right.
- Type *script* shows the contents of the file name script.m

5.11.1   Matlab Uses

Matlab can be used to solve different problems, some of them are:
Parallel Computing
Math, Statistics, and Optimization
Control Systems
Signal Processing and Communications
Image Processing and Computer Vision
Test and Measurements
Computational Finance
Computational Biology
Code Generation and Verification
Application Deployment
Database Connectivity and Reporting

5.12   Exercise 1: Taylor Series

Consider the function $f(x) = e^x \sin (x)$ and a Taylor series for this function around $x_0 = \Pi/4$.

1. How many terms are needed in the Taylor series to approximate the value of $f(1) = e^1 \sin (1)$ with relative error no larger than $1e^{-6}$
2. What is the final computational error ε_a.

```
%
clear
clc
%
        ea=1
        es=1e-6
        iter=0
        itermax=100
        x0=pi/4
        x1=1
        f0=exp(x0)*sin(x0)
        total=0
        fN=exp(x1)*sin(x1)
        %
        syms x
        fx=exp(x)*sin(x)

        while ea>es
            iter=iter+1
            coeff=subs(diff(fx,x,iter),x,x0)
            term=coeff/factorial(iter)*(x1-x0)^iter
            total=total+term
            ea=abs((total+f0-fN)/fN)
        end
clc
%
format compact
fN
double(total)
iter
double(ea)
        %
```

Fig. 5.1 Results from
Taylor series

```
fN =
  2.287355287178843
ans =
  0.736473770563126
iter =
  5
ans =
  7.346048594154421e-07
```

5.13   Exercise 2: Newton–Raphson Method

Consider the following set of equations

$$y = 3x^4 + 2x^2 + 1$$

$$y = 3x + 2$$

determine the intersections in the domain where $x \in [-2, 2]$.

Fig. 5.2 Results from the
Newton–Raphson method

x0=2 (first initial point)

```
x0 =
   0.911704251339163
iter =
   7
ea =
   4.677273935635659e-08
```

x0=-2 (second initial point)

```
x0 =
   -0.276513913845809
iter =
   8
ea =
   3.263209087728432e-09
```

```
%
clear
clc
        x=-2:0.01:2;
        fx1=3*x.^4+2*x.^2+1
        fx2=3*x+2
%
%
        fx_sum=3*x.^4+2*x.^2+1-(3*x+2)
%
        plot(x,0,x,fx1,'b-',x,fx2,'r-',x,fx_sum,'g-')
        xlabel('x')
        ylabel('y')
        title('fx1 fx2 fx_sum vs x')
%
%                   fx_sum: in a single equation to find the roots
%
        es=1e-6
        ea=1
        iter=0
        itermax=100
        x0=2       %each solution needs a different starting point
%         x0=-2
%
while ea>es && iter<itermax
%
        FX0=3*x0^4+2*x0.^2+1-(3*x0+2)
        DFX0=12*x0^3+4*x0-3
        x1=x0-FX0/DFX0
        iter=iter+1
        ea=abs((x1-x0)/x1)
        x0=x1
end
%
%
clc
x0
iter
ea
%
```

5.14   Exercise 3: Gauss Elimination

Consider the following set of equations
   Solve the following set of equations using Gaussian elimination

$$\begin{bmatrix} 1 & 2 \\ 3 & 1 \end{bmatrix}\begin{bmatrix} x_1 \\ x_2 \end{bmatrix} = \begin{bmatrix} 2.5 \\ 3 \end{bmatrix}$$

$A = [1\ 2;3\ 1];$
$b = [2\ 3]';$
$x = A \backslash b$

Fig. 5.3 Result from the
Gauss elimination
procedure

```
x =

   0.800000000000000
   0.600000000000000
```

5.15   Exercise 4: LU Decomposition

Determine the inverse matrix using LU decomposition for Matrix A.

$$A = \begin{bmatrix} 1 & 2 \\ 3 & 1 \end{bmatrix}$$

format short
$L = [1\ 0;3\ 1];$
$U = [1\ 2;0\ -5];$
$d1 = L \backslash [1\ 0]';$
$d2 = L \backslash [0\ 1]';$
%
$c1 = U \backslash d1;$
$c2 = U \backslash d2;$
$Ainv = [c1\ c2]$

Fig. 5.4 Result from the
LU decomposition
procedure

```
Ainv =

   -0.2000    0.4000
    0.6000   -0.2000
```

5.16   Exercise 5: Trajectories and Tracking

Two circles intercept $x^2 + y^2 = 16$, $(y - 4)^2 + (x - 4)^2 = 5$. Show the trajectories from the initial point to the intersection point.

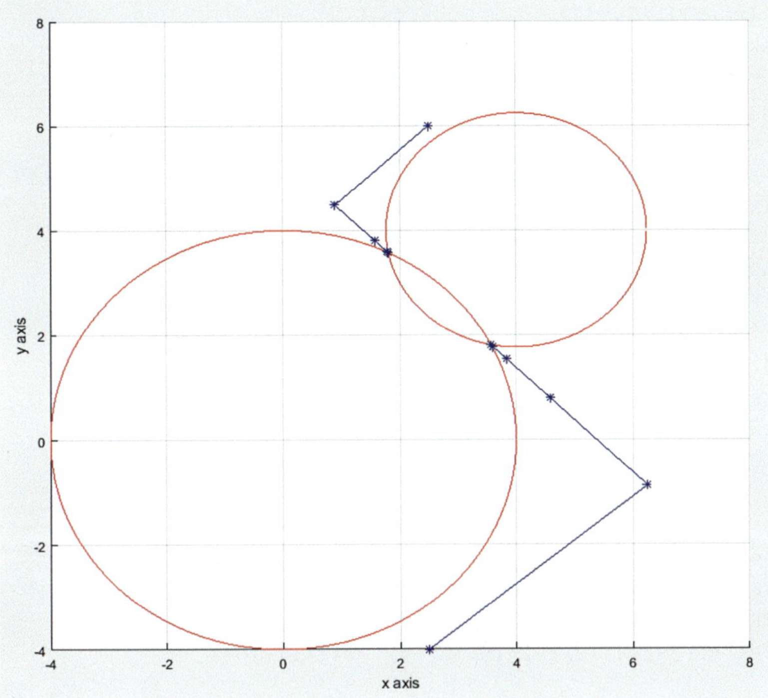

Fig. 5.5 Result from trajectories and tracking

```matlab
clear
clc
hold on
x1=-4:0.01:4
y11=sqrt(16-x1.^2)
y12=-y11
%plot(x1,y11,x1,y12)
%break
x2=4-sqrt(5):0.001:4+sqrt(5)
y22=sqrt(5-(x2-4).^2)+4
y21=-sqrt(5-(x2-4).^2)+4

plot(x1,y11,'r-',x1,y12,'r-',x2,y22,'r-',x2,y21,'r-')
grid on
%break
% initial point
X0=2.5
Y0=-4
counter=0
modulus=1
XI=[X0 Y0]'

while modulus>0.01 & counter <20
counter=counter+1
XTRACK(counter)=XI(1)
YTRACK(counter)=XI(2)
J=[-2*XI(1)  -2*XI(2);-2*(XI(1)-4)  -2*(XI(2)-4)]

F10=16-XI(1)^2-XI(2)^2
F20=5-(XI(1)-4)^2-(XI(2)-4)^2
F0=[F10 F20]'

DX=inv(J)*(-F0)
%DX=[J\-F0]'
XNEW(1)=XI(1)+DX(1)
XNEW(2)=XI(2)+DX(2)
temp=(XNEW'-XI)./XNEW'
modulus=sqrt(temp'*temp)
%modulus=sqrt((XNEW'-XI)'*(XNEW'-XI))
XI=[XNEW]'
end
XTRACK(counter+1)=XI(1)
YTRACK(counter+1)=XI(2)

plot(XTRACK,YTRACK,'b*-')
```

```
%break
clear
X0=2.5
Y0=6
counter=0
modulus=1
XI=[X0 Y0]'

while modulus>0.01 & counter <20
counter=counter+1
XTRACK(counter)=XI(1)
YTRACK(counter)=XI(2)
J=[-2*XI(1)  -2*XI(2);-2*(XI(1)-4)  -2*(XI(2)-4)]

F10=16-XI(1)^2-XI(2)^2
F20=5-(XI(1)-4)^2-(XI(2)-4)^2
F0=[F10 F20]'
%DX=inv(J)*(-F0)
DX=[J\-F0]'
XNEW(1)=XI(1)+DX(1)
XNEW(2)=XI(2)+DX(2)
modulus=sqrt((XNEW'-XI)'*(XNEW'-XI))
XI=[XNEW]'
end
XTRACK(counter+1)=XI(1)
YTRACK(counter+1)=XI(2)

plot(XTRACK,YTRACK,'b*-')
xlabel('x axis')
ylabel('y axis')
hold off
```

5.17   Exercise 6: Trajectories and Tracking

Consider the values of x from 0 to 500 for which the function **rand** is used to generate y-values using the following code:

```
clear
clc
for i=1:500
    x(i)=i
    y(i)=rand(1)*100
end
plot(x,y,'b*')
```

15.16.1	Determine the average (ymean) of the $\{y(i)\}$.
15.16.2	Plot error(i) vs x(i), where error(i)=$y(i)$-ymean.
15.16.3	Determine the average of $\{$error(i)$\}$.
15.16.4	Determine the least-squares linear fit for $y(i)$ as a function of $x(i)$.
15.16.5	Determine r^2.

You may use *polyfit* and *polyval*.

```
clear
clc
    N=1;                        % order of regression
    npoints=500;        %number of points
    %
    for i=1:npoints
        x(i)=i;
        y(i)=rand(1)*100;
    end
    %
    [P,S]=polyfit(x,y,N);
    %     P=-0.01628396623921*x + 54.47236166026033
    %
    yline=polyval(P,x);
    figure(1)
    plot(x,0,0,y,x,y,'b*',x,yline,'b-')
    xlabel('x value')
    ylabel('y value')
    title('Figure 1 Semester I')
    %
    ymean=mean(y);
    %ymean =50.39322811733798
    error=y-ymean;
    %
    figure(2)
    plot(x,0,0,y,x,error,'r*')
    axis([0 500 -60 60])
    errormean=mean(error);
    %errormean=5.341860287444433e-014
    %
    xlabel('x value')
    ylabel('error value')
    title('Figure 2 Semester I')
    %
    tots=0;
    rss=0;
    for i=1:npoints
    tots=(y(i)-ymean)^2+tots;
    rss=(y(i)-P(1)*x(i)-P(2))^2+rss;
    end
    R2=(tots-rss)/tots;
    %
    %R2= 0.00676765756237
    %
```

.

```
ymean =50.39322811733798
errormean=5.341860287444433e-014
y=-0.01628396623921*x + 54.47236166026033
R²= 0.00676765756237
```

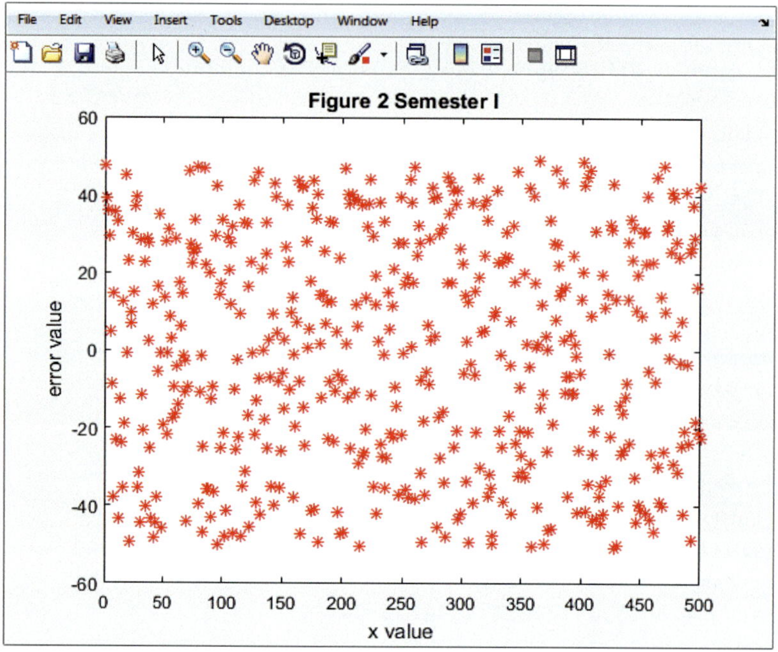

Fig. 5.6 Plot error(i) versus $x(i)$

5.18   Exercise 7: Parallel Processing

Consider the values of x from 0 to 500 for which the function **rand** is used to generate y-values using the following code:

```
clear
clc
tic
parfor i=1:500                    % for i=1:500
    x(i)=i;
    y(i)=rand(1)*100;
end
toc
% plot(x,y,'b*')
delete(gcp)
```

5.18.1   Determine the Processing Time for the Traditional **for** Loop.

Elapsed time is 0.016891 s.

5.18.2   Determine the Processing Time for the Parallel Processing **parfor** Loop.

Starting parallel pool (parpool) using the 'local' profile ... connected to 4 workers.
Elapsed time is 25.758693 seconds.
Parallel pool using the 'local' profile is shutting down.

5.18.3   How Many Workers Perform the Parallel Process?

...connected to 4 workers.

The parallel processing feature in the form of *parfor* requires that all the variables and counters related to the calculation be inside the loop and that may not be possible in all cases.

The number of workers corresponds to the number of physical processors.

The time needed to raise the workers in parallel mode is significant and takes place at the beginning of the calculations. If the computation is repeated using parfor but the parallel pool is already operational the computational time is smaller.

Elapsed time is 0.165430 s.

However, this would indicate that the parallel processing is useful when the workload is massive.

There is some increase in CPU utilization when the program is run with the traditional sequential *for* not reaching saturation at any time.

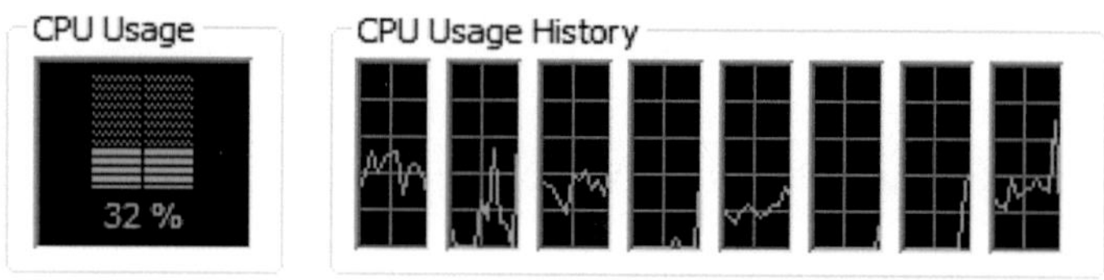

Fig. 5.7 CPU usage with traditional *for* with 50,000,000 iterations

The use of the parallel processing parfor makes a more intense use of processing power. That would be a factor to consider in the design of massive computations to better use computational power.

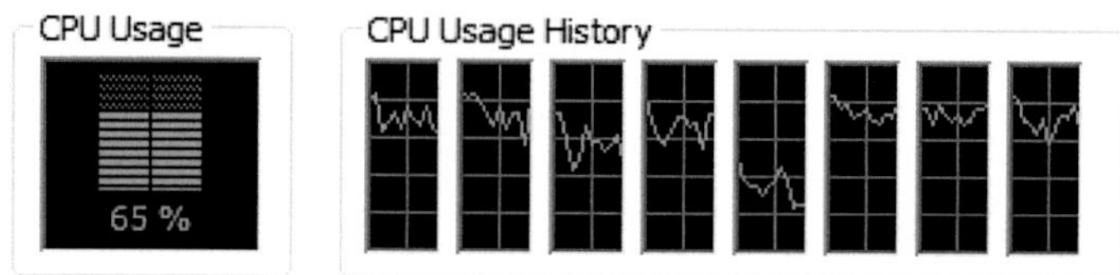

Fig. 5.8 CPU usage with parfor with 50,000,000 iterations

There are other processes running in the background, but their effect is not significant on the observations annotated above.

There are important examples that cannot be parallelized. One that is used in interval mathematics comes from computing all the possible combinations between two arrays and save them into an array for further processing. Take a look and try to understand the flow of information from the main program to the function and the recursive calls performed there in.

```
clear
clc
%
global combinationscounter stack totalrows totalcolumns m array_new
%
m=1;
combinationscounter=0;
totalcolumns=2;
totalrows=2;
%
array_new=repmat('0',totalrows,2^totalrows);

i=1;
j=1;
k=1;
pila=['z'];
stack=['A' 'B'
       '1' '2'
       'p' 'w'
       '3' '4'
       '*' '@'];
%
%display(' in 1')
generatecombination(i,j,k,pila)
%display('out 1')
combinationscounter
array_new
%
```

Fig. 5.9 Main program recursive counting

```
function total=generatecombination(i,j,k,pila)
    %
    global combinationscounter stack totalrows totalcolumns m array_new
                totalcolumns;
                totalrows;
                if i==totalrows+1
                        return
                else
                i
                j
                k
                pila(k)=stack(i,j);
                k=k+1;
    %           display('in 2')
                generatecombination(i+1,1,k,pila)
    %           display('out 2')
                k=k-1;
            if i==totalrows
    %           display('in 3')
                pila';
                m;
                array_new
                array_new(:,m)=pila'
                m=m+1;
    combinationscounter=combinationscounter+1; % -- combination counter
    %           display('out 3')
            end
    %
            if j<totalcolumns
    %           display('in 4')
                generatecombination(i,j+1,k,pila)
    %           display('out 4')
            end
    %
        end
end
```

Fig. 5.10 Function generate combination

```
combinationscounter =
    4
array_new =
AABB
1212
```

Fig. 5.11 Combinations

Table 5.3 Flow of information in counting algorithm

Main Program	Function Call from Main Program(i,j,k,pila) →	First Recursive Call to Function(i,j,k,pila) →	Second Recursive Call to Function(i,j,k,pila) →
Stack ['A' 'B' '1' '2'] i=1, j=1, k=1,pila='z' generatecombination(i,j,k,pila) generatecombination(1,1,1,'z')	i=1, j=1, k=1,pila=['z'] pila(k)=stack(i,j) pila(1)=['A'] k=k+1 k=2 generatecombination(i+1,1,k,pila) generatecombination(2,1,2,'A')	i=2, j=1, k=2,pila=['A'] i==totalrows+1 ? 2≠3 pila(k)=stack(i,j) pila(2)=stack(2,1) pila(2)=1 pila=['A' '1'] k=k+1 k=3 generatecombination(i+1,1,k,pila) generatecombination(3,1,3,['A' '1'])	i=3, j=1, k=3,pila=['A' '1'] i==totalrows+1 ? 3=3 return
		← Return from Second Recursive Call	

```
k=k-1
k=3-1=2
i=2==totalrows=2 ?
m=1
array_new(:,m)=
array_new(:,1)=['A' '1']=['A' '1']
m=m+1=1+1=2
j=1<totalcolumns=2 ?
generatecombinations(i,j+1.k.pila)
generatecombinations(2,2,2,['A' '1']
```

First element of combination matrix is stored as a column element

←

Third Recursive Call to Function(i,j,k,pila) ←

Third Recursive Call to Function(i,j,k,pila) ←
```
i=2, j=2, k=2,pila=['A' '1']
i==totalrows+1 ?
2≠3
pila(k)=stack(i,j)
pila(2)=stack(2,2)
pila(2)=2
pila=['A' '2']
k=k+1
k=3
generatecombination(i+1,1,k,pila)
generatecombination(3,1,3,['A' '2'])
```

Fourth Recursive Call to Function(i,j,k,pila)
→
```
i==totalrows+1 ?
3=2+1
return
```

← **Return from Fourth Recursive Call**
```
m=2
j=2
k=k-1
k=3-1=2
i=2==totalrows=2 ?
array_new(:,m)=
array_new(:,2)=['A' '2']=['A' '2']
m=m+1=2+1=3
```

Second element of combination matrix is stored as a column element ←

Tab. 5.3 (continued)

j=2<totalcolumns=2 ?
end
return

← **End from Third Recursive Call**

end
return

← **End First Recursive Call to Function(i,j,k,pila)**

k=k-1
k=2-1
k=1
i==totalcolumns ?
1≠2
j<totalcolumns ?
1<2
generatecombination(i,j+1,k,pila)
generatecombination(1,2,1,'A')

Fifth Recursive Call to Function(i,j,k,pila) →

i=1, j=2, k=1,pila=['A']
i==totalrows+1 ?
1≠3
pila(k)=stack(i,j)
pila(1)=stack(1,2)
pila=['B']
k=1+1
k=2
generatecombination(i+1,1,k,pila)
generatecombination(2,1,2,['B'])

Sixth Recursive Call to Function(i,j,k,pila) →

i=2, j=1, k=2, pila=['B']
m=3
k=3-1=2
pila(k)=stack(i,j)
pila(2)=stack(2,1)
pila(2)=['1']
pila=['B' '1']
k=k-1=2-1=1
i==totalrows ?
i==2=totalrows
array_new(:,m)=

Third element of combination matrix is stored as a column element

Tab. 5.3 (continued)

array_new(:,3)=['B' '1']=['B' '1']   ←

m=m+1=3+1=4
j<totalcolumns=?
1<2
generatecombinations(i,j+1.k.pila)
 generatecombinations(2,2,2,['B' '1']

Seventh Recursive Call to Function(i,j,k,pila) ←

i=2, j=2, k=2,pila=['B' '1']
i==totalrows+1 ?
2≠3
pila(k)=stack(i,j)
pila(2)=stack(2,2)
pila(2)=['2']
pila=['B' '2']
k=2+1
k=3
generatecombination(i+1,1,k,pila)
generatecombination(3,1,2,['B' '2'])

Eighth Recursive Call
to Function(i,j,k,pila) ←
i=3, j=1, k=2, pila=['B' '2']
i==totalrows+1 ?
i=3=totalrows+1
return

← Return from Eighth Recursive Call
m=4
i=totalrows ?
i=2=totalrows
array_new(:,4)=['B' '1']=['B' '2']
j<totalcolumns ?
2≤2
end
return

Fourth element of combination matrix is stored as a column element   ←

← End from Seventh Recursive Call
end
return

← End from Sixth Recursive Call
k=k-1
k=2-1=1

Tab. 5.3 (continued)

i==totalrows ?
1≠2
j<totalcolumns
2≤2
end
return

← End from Fifth Recursive Call
end
return

← End from First
Recursive Call from Main
Program

$$array\_new = \begin{bmatrix} A & A & B & B \\ 1 & 2 & 1 & 2 \end{bmatrix}$$

The flow of information is somewhat intricate and it is included here to illustrate the complexity of the problem.

This style of concurrent algorithm is not typically taught in the engineering curriculum and thus fitting to those unfamiliar with this programming strategy.

Further Readings

Attaway, S. (2016). *Matlab: a practical introduction to programming and problem solving* (p. 600). Burlington, MA: Butterworth-Heinemann.

Chapra, S. C., & Canale, R. (2009). *Numerical methods for engineers* (6th ed., p. 960). New York: McGraw-Hill Education.

MATLAB Central (2018). Retrieved from https://www.mathworks.com/matlabcentral/

Nakamura, S. (1995). *Numerical analysis and graphics visualization with MATLAB* (p. 477). Upper Saddle River, NJ: Prentice-Hall.

Chapter 6
Shell Installation

6.1 Linux Systems

Shell installation was used via Powershell in previous years with Azure. Many of the previous commands and downloads are no longer functional. This is an update for Linux users.

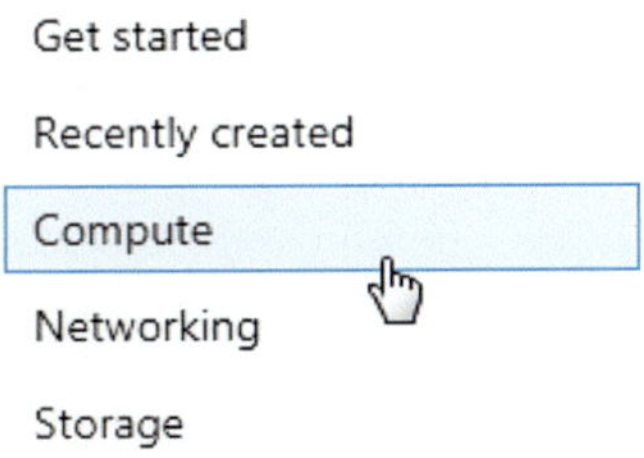

Fig. 6.1 Select Compute from the Microsoft Azure dashboard

Get started

Recently created

Compute

Networking

Storage

Fig. 6.2 Select Ubuntu Server

© Springer Nature Switzerland AG 2020

B. A. Stradi-Granados, *Cloud Computing for Engineering Applications*,

https://doi.org/10.1007/978-3-030-40445-1_6

There are several versions of Linux-based systems and all of them have different quirks. Consequently, Ubuntu, Red Hat Enterprise Linux, SUSE, and CentOS are not exactly the same. Even more, commands vary between versions of the same operating system. The developer should plan ample time to account for these unforeseen variations.

Dashboard > New > Create a virtual machine

Create a virtual machine

PROJECT DETAILS

Select the subscription to manage deployed resources and costs. Use resource groups like folders to organize and manage all your resources.

* Subscription ❶   Azure in Open

        * Resource group ❶   (New) WhiteRibbon
                            Create new

INSTANCE DETAILS

* Virtual machine name ❶   CloudLinux

* Region ❶   West US 2

Availability options ❶   No infrastructure redundancy required

* Image ❶   Ubuntu Server 18.04 LTS
                            Browse all images and disks

* Size ❶   **Standard D2s v3**
                            2 vcpus, 8 GB memory
                            Change size

ADMINISTRATOR ACCOUNT

Authentication type ❶   ● Password   ○ SSH public key

* Username ❶   userlinux

* Password ❶   ···········

* Confirm password ❶   ···········

Login with Azure Active Directory (Preview) ❶   ○ On   ● Off

INBOUND PORT RULES

Fig. 6.3 Initial configuration of workstation

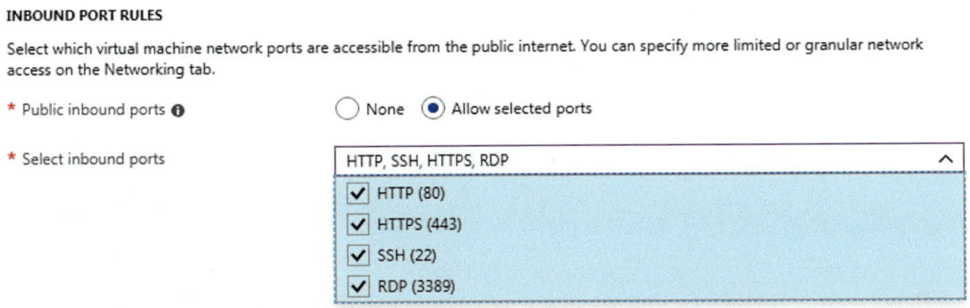

INBOUND PORT RULES

Select which virtual machine network ports are accessible from the public internet. You can specify more limited or granular network access on the Networking tab.

* Public inbound ports ●   ○ None ● Allow selected ports

* Select inbound ports

> HTTP, SSH, HTTPS, RDP   ∧
>> ☑ HTTP (80)
>> ☑ HTTPS (443)
>> ☑ SSH (22)
>> ☑ RDP (3389)

Fig. 6.4 Choose a computer size with affordable performance D2_V3 HDD

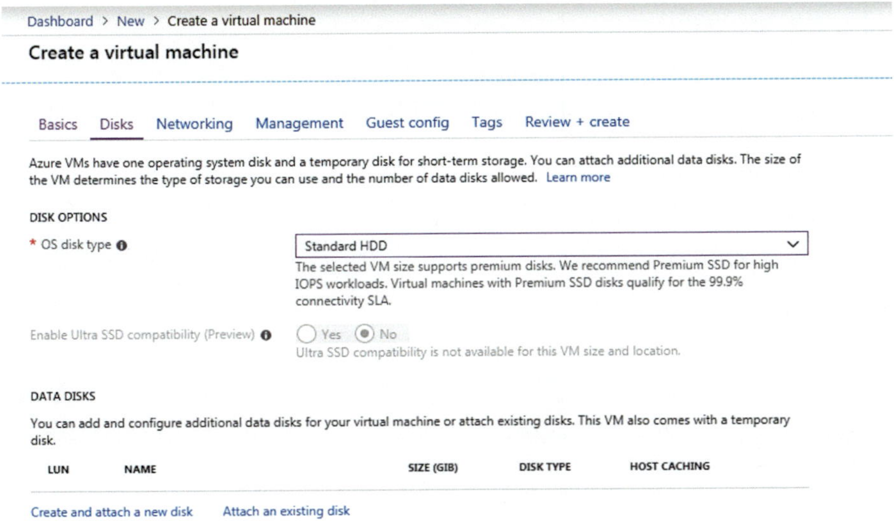

Dashboard > New > Create a virtual machine

Create a virtual machine

Basics **Disks** Networking Management Guest config Tags Review + create

Azure VMs have one operating system disk and a temporary disk for short-term storage. You can attach additional data disks. The size of the VM determines the type of storage you can use and the number of data disks allowed. Learn more

DISK OPTIONS

* OS disk type ●

> Standard HDD   ∨

The selected VM size supports premium disks. We recommend Premium SSD for high IOPS workloads. Virtual machines with Premium SSD disks qualify for the 99.9% connectivity SLA.

Enable Ultra SSD compatibility (Preview) ●   ○ Yes ● No
Ultra SSD compatibility is not available for this VM size and location.

DATA DISKS

You can add and configure additional data disks for your virtual machine or attach existing disks. This VM also comes with a temporary disk.

LUN	NAME	SIZE (GIB)	DISK TYPE	HOST CACHING

Create and attach a new disk   Attach an existing disk

Fig. 6.5 Network settings

The allocation of ports for communication at this point of the configuration process is fairly new as of the writing of this document in the summer of 2018.

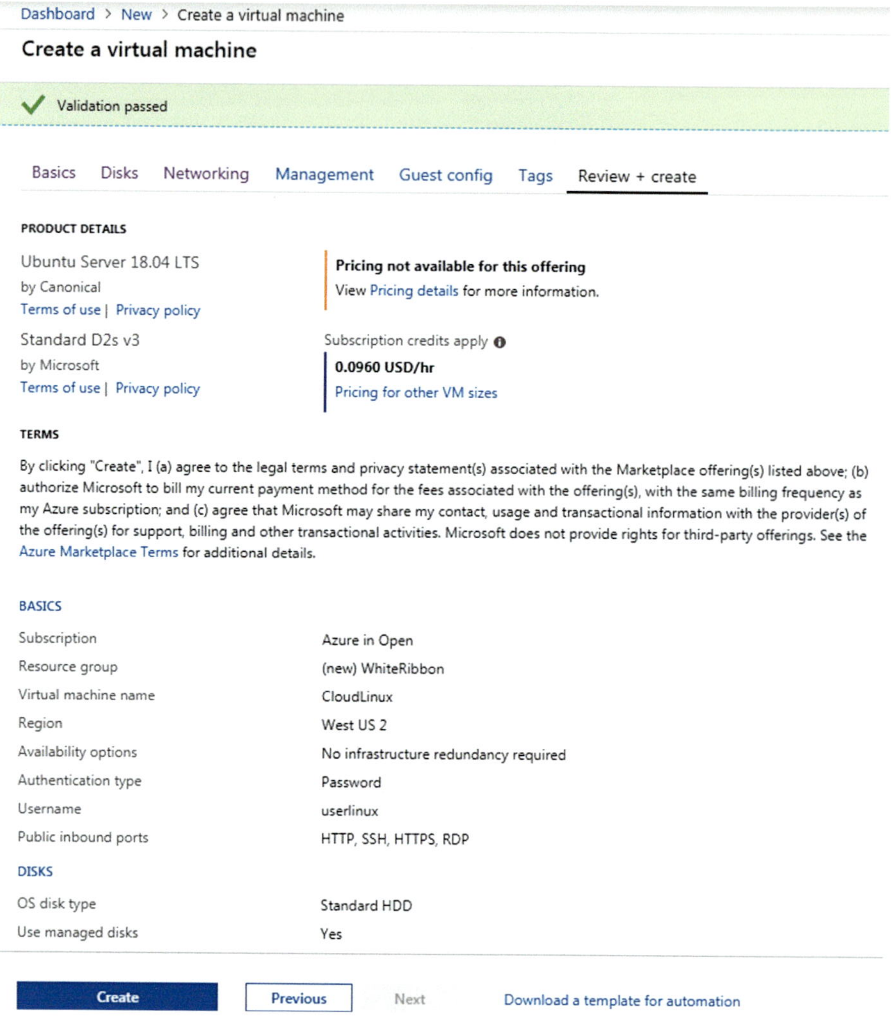

Fig. 6.6 Workstation configuration summary

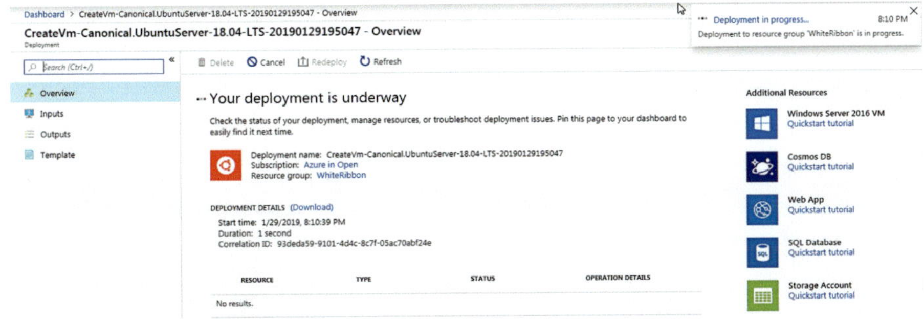

Fig. 6.7 Deploy the new machine

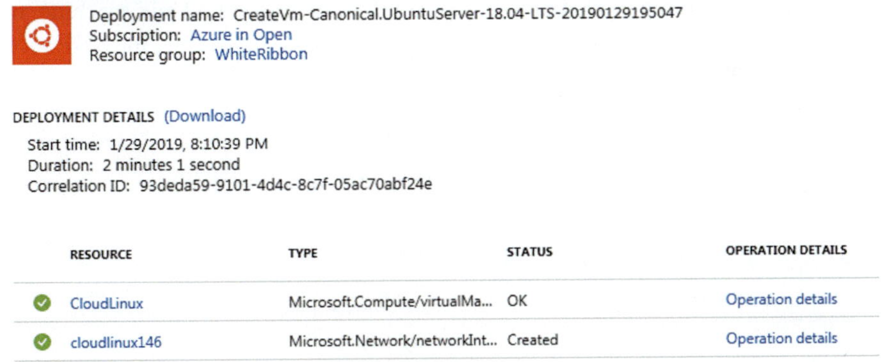

Fig. 6.8 Deployment completed and new Ubuntu workstation is available

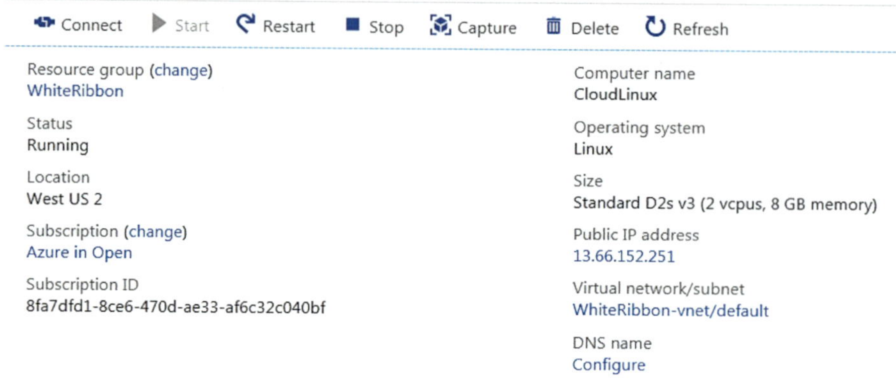

Fig. 6.9 Ubuntu-based workstation

Traditionally, the remote desktop protocol (RDP) is used in Windows to communicate with the workstation in the cloud. In the case of a Unix machine, two parts are needed: the RDP client software in the user's computer and the RDP server software in the cloud workstation.

Try to connect to the cloud workstation using the port 3389.

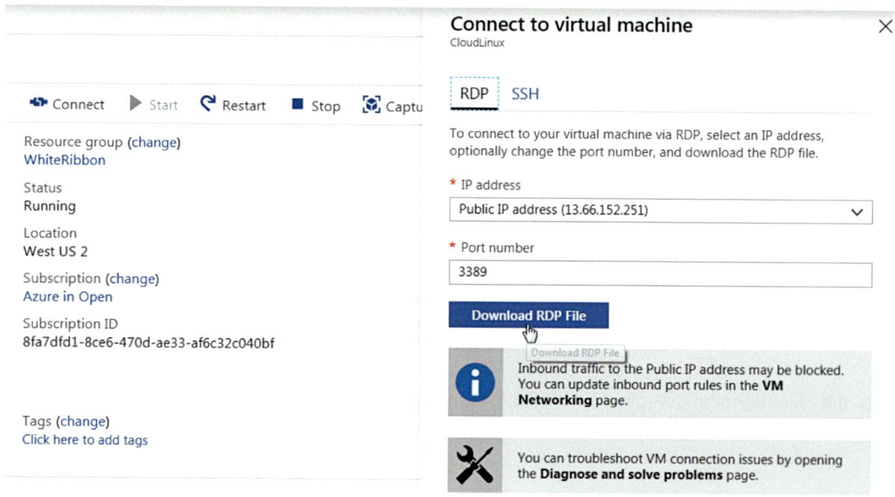

Fig. 6.10 Connection to remote machine

Proceed to download the RDP file and notice that no connection is established.

Fig. 6.11 Connection to remote machine

This is the scripting part where to start the Bash (Linux) or Powershell (Windows). In this case, Bash is used.

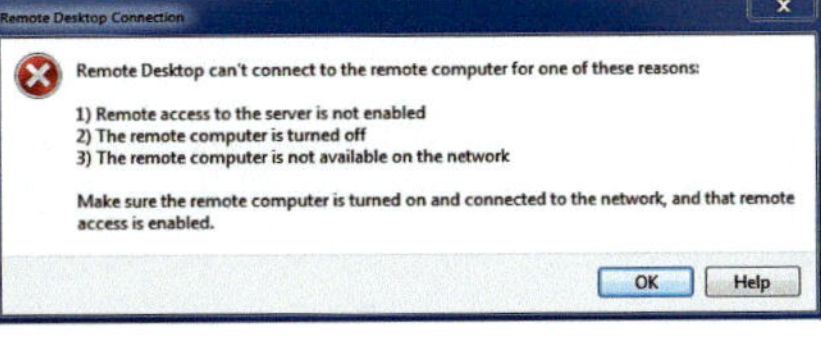

Fig. 6.12 Activating command line in Bash (Linux)

There are charges to use this utility that are necessary.

Fig. 6.13   Create a command storage account

The Bash command screen appears for those familiar with Linux. However, a graphical user interface is traditionally preferred.

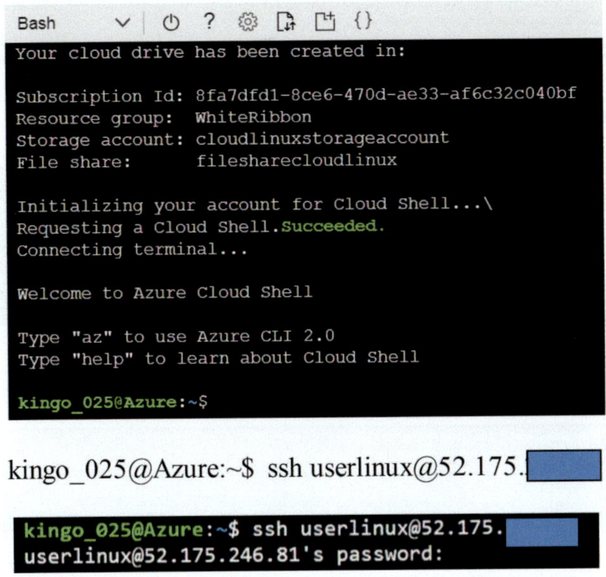

kingo_025@Azure:~$ ssh userlinux@52.175.

Fig. 6.14 Bash command screen

There are a few steps; those steps generate a series of messages.

1. sudo apt-get install -y ubuntu-desktop

```
aspell-autobuildhash: processing: en [en_GB-variant_1].
aspell-autobuildhash: processing: en [en_US-w_accents-only].
aspell-autobuildhash: processing: en [en_US-wo_accents-only].
Processing triggers for initramfs-tools (0.130ubuntu3.6) ...
update-initramfs: Generating /boot/initrd.img-4.15.0-1036-azure
Processing triggers for libgdk-pixbuf2.0-0:amd64 (2.36.11-2) ...
userlinux@CloudLinux:~$
```

2. sudo apt-get install -y gnome-session

```
Setting up fonts-cantarell (0.0.25-4) ...
Setting up gnome-session (3.28.1-0ubuntu3) ...
Processing triggers for fontconfig (2.12.6-0ubuntu2) ...
Setting up adwaita-icon-theme-full (3.28.0-1ubuntu1) ...
userlinux@CloudLinux:~$
```

3.   sudo apt-get install -y tightvncserver

```
Setting up tightvncserver (1.3.10-0ubuntu4) ...
update-alternatives: using /usr/bin/tightvncserver to provide /usr/bin/vncserver (vncserver) in auto mode
update-alternatives: using /usr/bin/Xtightvnc to provide /usr/bin/Xvnc (Xvnc) in auto mode
update-alternatives: using /usr/bin/tightvncpasswd to provide /usr/bin/vncpasswd (vncpasswd) in auto mode
Processing triggers for man-db (2.8.3-2ubuntu0.1) ...
userlinux@CloudLinux:~$
```

4.   sudo apt-get install -y xrdp

```
Processing triggers for man-db (2.8.3-2ubuntu0.1) ...
Setting up xorgxrdp (0.9.5-2) ...
Processing triggers for libc-bin (2.27-3ubuntu1) ...
Processing triggers for ureadahead (0.100.0-20) ...
userlinux@CloudLinux:~$
```

5.   sudo apt-get update

```
Get:40 http://security.ubuntu.com/ubuntu bionic-security/multiverse DEP-11 48x48 Icons [29 B]
Get:41 http://security.ubuntu.com/ubuntu bionic-security/multiverse DEP-11 64x64 Icons [2638 B]
Fetched 18.6 MB in 3s (6128 kB/s)
Reading package lists... Done
userlinux@CloudLinux:~$
```

6.   sudosed -i 's/allowed_users=console/allowed_users=anybody/'/etc/X11/ Xwrapper.config

```
userlinux@CloudLinux:~$ sudo sed -i 's/allowed_users=console/allowed_users=anybody/' /etc/X11/Xwrapper.config
userlinux@CloudLinux:~$
```

7.   sudopasswd root

```
userlinux@CloudLinux:~$ sudo passwd root
Enter new UNIX password:
Retype new UNIX password:
passwd: password updated successfully
userlinux@CloudLinux:~$
```

This is the first screen and finally the remote workstation desktop appears.

Fig. 6.15 Update Ubuntu later

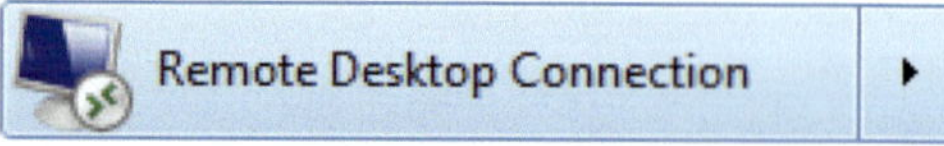

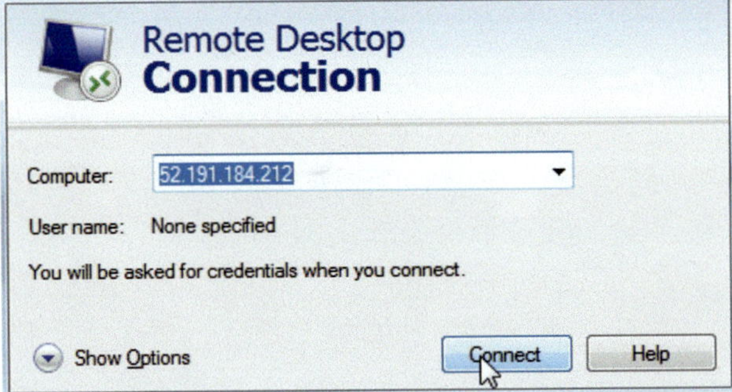

Fig. 6.16 Disregard warning for now

Fig. 6.17 Try connecting
to Linux workstation

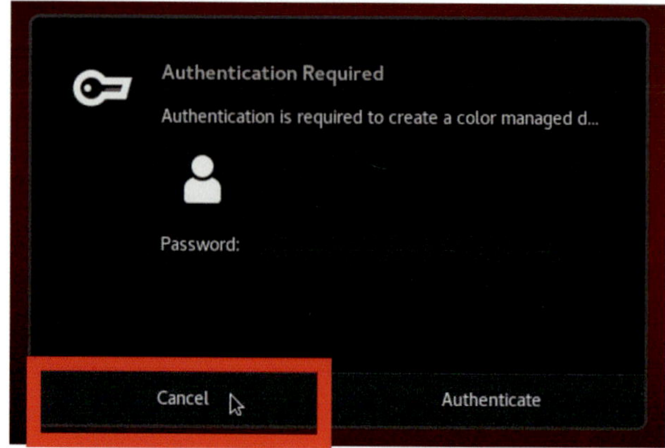

Fig. 6.18 *Root log-in* in Ubuntu

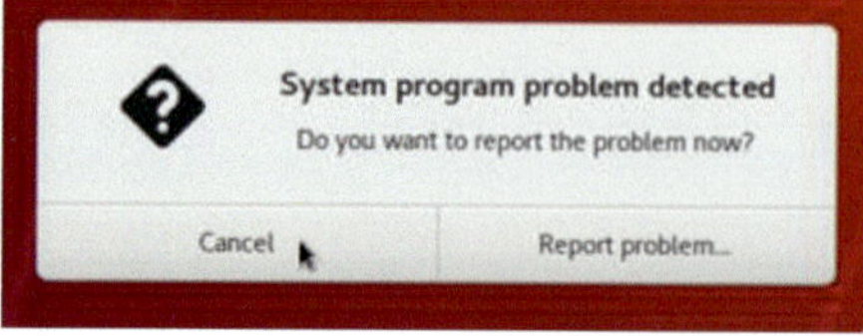

Fig. 6.19 No software updates at this time

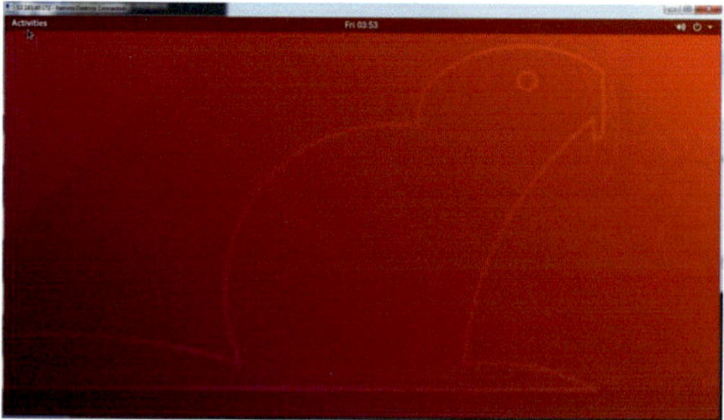
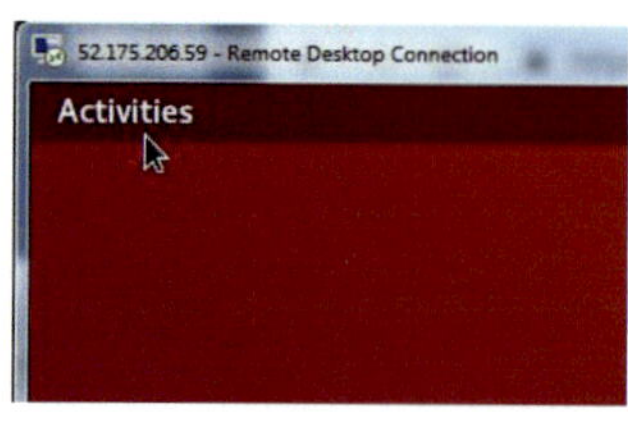

Fig. 6.20 Ubuntu desktop

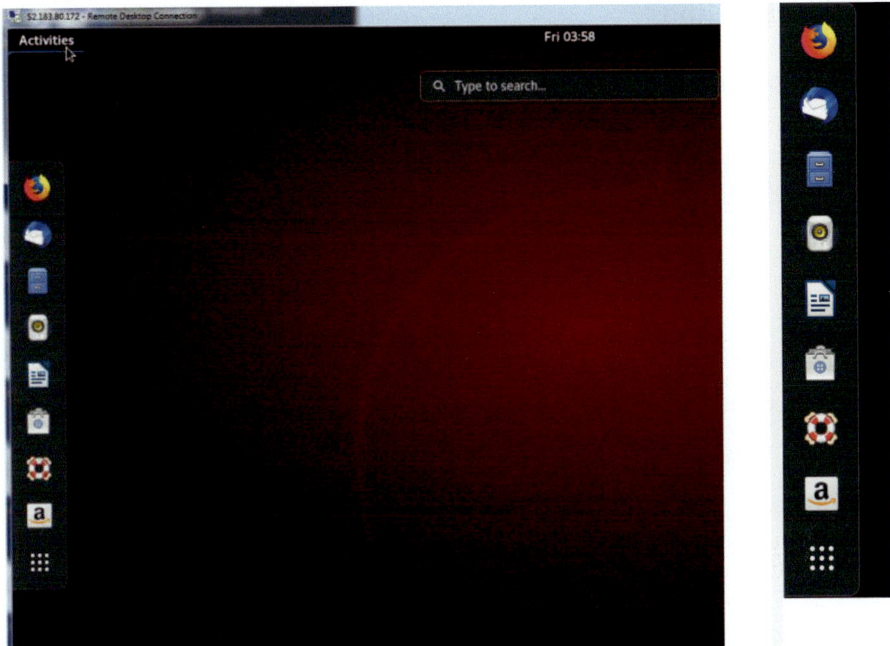

Fig. 6.21 Ubuntu desktop detail

However, there is no idea of the password. For completeness for those with the need for root access, this password has to be set up.

This is the Ubuntu workstation with internet connection and ready for work.

After all these contents have been covered, a break is called for.

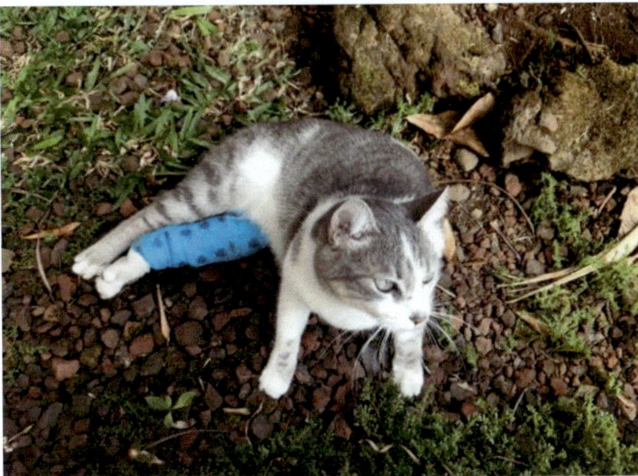

Fig. 6.22 Time for other activities

That's all folks

Further Readings

Quickstart. Create a Linux virtual machines in the Azure portal. Retrieved from https://docs.micro-
    soft.com/en-us/azure/virtual-machines/linux/quick-create-portal

Notes

© Springer Nature Switzerland AG 2020

B. A. Stradi-Granados, *Cloud Computing for Engineering Applications*,

https://doi.org/10.1007/978-3-030-40445-1

Solidworks

COMSOL Multiphysics©

Matlab

Operating System Windows

Operating System Linux

Index

© Springer Nature Switzerland AG 2020
B. A. Stradi-Granados, *Cloud Computing for Engineering Applications*,
https://doi.org/10.1007/978-3-030-40445-1

MIX
Papier aus verantwortungsvollen Quellen
Paper from responsible sources
FSC® C105338